Applied Stochastic Modelling

Second Edition

CHAPMAN & HALL/CRC
Texts in Statistical Science Series

Series Editors
Bradley P. Carlin, *University of Minnesota, USA*
Julian J. Faraway, *University of Bath, UK*
Martin Tanner, *Northwestern University, USA*
Jim Zidek, *University of British Columbia, Canada*

Analysis of Failure and Survival Data
P. J. Smith

**The Analysis of Time Series—
An Introduction, Sixth Edition**
C. Chatfield

**Applied Bayesian Forecasting and Time Series
Analysis**
A. Pole, M. West and J. Harrison

**Applied Nonparametric Statistical Methods, Fourth
Edition**
P. Sprent and N.C. Smeeton

**Applied Statistics — Handbook of GENSTAT
Analysis**
E.J. Snell and H. Simpson

Applied Statistics — Principles and Examples
D.R. Cox and E.J. Snell

Applied Stochastic Modelling, Second Edition
B.J.T. Morgan

Bayesian Data Analysis, Second Edition
A. Gelman, J.B. Carlin, H.S. Stern
and D.B. Rubin

**Bayesian Methods for Data Analysis,
Third Edition**
B.P. Carlin and T.A. Louis

Beyond ANOVA — Basics of Applied Statistics
R.G. Miller, Jr.

**Computer-Aided Multivariate Analysis,
Fourth Edition**
A.A. Afifi and V.A. Clark

A Course in Categorical Data Analysis
T. Leonard

A Course in Large Sample Theory
T.S. Ferguson

Data Driven Statistical Methods
P. Sprent

Decision Analysis — A Bayesian Approach
J.Q. Smith

**Elementary Applications of Probability Theory,
Second Edition**
H.C. Tuckwell

Elements of Simulation
B.J.T. Morgan

**Epidemiology — Study Design and
Data Analysis, Second Edition**
M. Woodward

Essential Statistics, Fourth Edition
D.A.G. Rees

**Extending the Linear Model with R: Generalized
Linear, Mixed Effects and Nonparametric Regression
Models**
J.J. Faraway

A First Course in Linear Model Theory
N. Ravishanker and D.K. Dey

**Generalized Additive Models:
An Introduction with R**
S. Wood

**Interpreting Data — A First Course
in Statistics**
A.J.B. Anderson

**An Introduction to Generalized
Linear Models, Third Edition**
A.J. Dobson and A.G. Barnett

Introduction to Multivariate Analysis
C. Chatfield and A.J. Collins

**Introduction to Optimization Methods and Their
Applications in Statistics**
B.S. Everitt

Introduction to Probability with R
K. Baclawski

**Introduction to Randomized Controlled Clinical
Trials, Second Edition**
J.N.S. Matthews

**Introduction to Statistical Methods for
Clinical Trials**
T.D. Cook and D.L. DeMets

Large Sample Methods in Statistics
P.K. Sen and J. da Motta Singer

Linear Models with R
J.J. Faraway

**Markov Chain Monte Carlo —
Stochastic Simulation for Bayesian Inference,
Second Edition**
D. Gamerman and H.F. Lopes

Mathematical Statistics
K. Knight

Texts in Statistical Science

Applied Stochastic Modelling

Second Edition

Byron J. T. Morgan

University of Kent
UK

CRC Press
Taylor & Francis Group
Boca Raton   London   New York

CRC Press is an imprint of the
Taylor & Francis Group an **informa** business

A CHAPMAN & HALL BOOK

MATLAB® is a trademark of The MathWorks, Inc. and is used with permission. The MathWorks does not warrant the accuracy of the text or exercises in this book. This book's use or discussion of MATLAB® software or related products does not constitute endorsement or sponsorship by The MathWorks of a particular pedagogical approach or particular use of the MATLAB® software.

Chapman & Hall/CRC
Taylor & Francis Group
6000 Broken Sound Parkway NW, Suite 300
Boca Raton, FL 33487-2742

© 2009 by Taylor & Francis Group, LLC
CRC Press is an imprint of Taylor & Francis Group, an Informa business

No claim to original U.S. Government works

ISBN 13: 978-1-58488-666-2 (pbk)

**Visit the Taylor & Francis Web site at
http://www.taylorandfrancis.com**

**and the CRC Press Web site at
http://www.crcpress.com**

Contents

CONTENTS

Preface to the Second Edition

The structure of the second edition of this book is very similar to that of the first edition; however, there have been numerous changes throughout. In particular, a large number of new exercises have been added, there is a new appendix on computational methods, and the discussion of Bayesian methods has been extended. The bibliography has been updated, throughout figures have been improved, and, where necessary, errors have been corrected. I am grateful for the many positive comments that the first version of the book received, and to those who have written to point out mistakes, and ways in which the book might be improved. I thank especially Ted Catchpole, Rachel Fewster, Ruth King, Rachel McCrea, David Miller, Karen Palmer and Martin Ridout. I thank MATLAB® for providing the latest version of the package, and I am also grateful for the help and patience of Rob Calver and colleagues at CRC Chapman & Hall. The book continues to be the set text for final year under-graduate and post-graduate courses in respectively Applied Stochastic Modelling and Data Analysis, and Computational Statistics in the University of Kent. The book successfully revises and integrates the probability and statistics methods of earlier lecture courses. At the same time it brings students into contact with modern computational methods; it provides students with practical experience of scientific computing at use in applied statistics, in the context of a range of interesting real-life applications. The book Web Site contains the data sets from the book, the MATLAB computer programs, as well as corresponding versions in R. There is a solutions manual for the exercises and the computer practical sheets that are used in Kent. The book has been the basis for a course on *Applied Stochastic Modelling* held at Pfizer Central Research in Sandwich, Kent, during 2007, and also for a Continuing Professional Development course with the same name, to be held at the Royal Statistical Society, London, in 2008. The slides for these courses are also available on the book Web site.

Canterbury

Preface

This book is being completed at the end of a millenium, the last 30 years of which have seen many exciting major developments in the theory and practice of statistics. The book presents many of the most important of these advances. It has its origins in a series of talks on aspects of modern statistics for fitting stochastic models to data, commissioned by statisticians at Pfizer Central Research in Sandwich in Kent. These talks gave rise to a 30-hour lecture course given to third year undergraduates, statistics MSc students and first-year statistics PhD students at the University of Kent, and this book has grown from the notes for that lecture course. These students have found that even the most recently developed statistical methods may be readily understood and successfully applied in practice. As well as covering modern techniques, the material of the book integrates and revises standard probability and statistical theory. Much modern statistical work is implemented using a computer. Thus it is necessary in a book of this nature to include computer instructions of some kind. The integrated computer language MATLAB has been selected for this purpose, and over 50 short MATLAB programs are included throughout the book. Their distribution and purpose are described in the index of MATLAB programs. They may all be accessed from the book site on the World Wide Web. They are designed to be illustrative, rather than completely efficient. Often doubts concerning theory are dissipated when one can see computer code for the theory. Students of the book material have certainly found the MATLAB programs to be a useful aid to learning. It has been uplifting to observe the satisfaction that students have gained from running the MATLAB programs of the book. Often the students had no previous knowledge of MATLAB and also little prior experience of scientific computing. The material of Appendix B, which summarises important features of MATLAB, and tutorial assistance were all that was needed by these students. However it should be stressed that while the computer programs are included as an aid to learning, the book may be read and used without reference to the programs. S-plus versions of the programs are available on the book site on the World Wide Web. There are also some references to the use of symbolic algebra packages such as MAPLE, as these provide powerful tools for stochastic modelling.

The target student audience for the book is final-year undergraduate and MSc students of mathematics and statistics. The book is also intended as a single convenient source of reference for research scientists and post-graduate students, using modern statistical methods which are currently described in depth in a range of single-topic textbooks. Prior knowledge is assumed at the level of a typical second-year university course on probability and statistics. Appendix A summarises a number of important formulae and results from probability and statistics. A small fraction of the book sections and exercises contain more advanced material, and these are starred. Kernel density estimation is a central aspect of modern statistics, and therefore the basic ideas are summarised in Appendix C. While a limited number of exercises have solutions included in the book, a more extensive set of solutions is to be found on the World Wide Web book site.

Statistical methods are all-pervasive, contributing significantly to subjects as diverse as geology, sociology, biology and economics. The construction, fitting and evaluation of statistical and stochastic models are not only vitally important in areas such as these, but they are also great fun. It is hoped that some of the enjoyment and fascination of the subject will be gained by readers of this book.

The book is motivated by real data and problems. The examples and exercises are often chosen from my own experience and, as can be seen from the index of data sets, many have arisen from biology. The areas covered are sometimes atypical, and sometimes classical, such as survival analysis, quantal assay and capture-recapture. Several of the examples recur at various points throughout the book. The data are available on the book site on the World Wide Web.

Acknowledgments

I owe a great debt of gratitude to the many scientists and statisticians with whom I have collaborated, and with whom I continue to collaborate. I am especially grateful to my MATLAB mentor, Ted Catchpole, who with Paul Terrill, contributed some of the MATLAB programs. The original proposal for a set of lectures at Pfizer Central Research was agreed with Trevor Lewis, Group Executive Director, Biometrics, at Pfizer. The Pfizer audience was typically challenging, with Ken Martin and John Parrott in particular regularly asking probing questions. Steve Brooks, Steve Buckland, Martin Ridout, and Paul Terrill all commented helpfully on draft chapters. At Arnold, Nicki Dennis, and Kirsty Stroud were patient and encouraging. Three years of Kent students provided a sympathetic test bed and launching pad. The many drafts of the book were typed enthusiastically by Lilian Bond, Julie Snook, and Mavis Swain.

Canterbury

'For the things we have to learn before we can do them, we learn by doing them'

Aristotle

CHAPTER 1

Introduction and Examples

1.1 Introduction

Reported falls in human sperm counts in many developed countries have serious implications for the future of mankind. In *fecundability* studies, data are collected on waiting times to conception in human beings, as well as on variables such as age and body mass index, which is a measure of obesity. For instance, the paper by Jensen et al. (1998) concluded that the probability of conception in a menstrual cycle was lowered if only five alcoholic drinks were taken by the woman each week. Data from studies such as this require appropriate statistical analysis, which quite often results from describing the data by means of models tailored specifically to the particular question of interest.

In this book we shall consider a wide range of data sets. In each case our objective is to find a succinct description of the data, which may then be used either as a summary or to provide the basis for comparison with other examples. We shall do this by proposing simple models, and then fitting them to the data. The models we shall consider are based on the axioms of probability theory, as described, for example, by Grimmett and Stirzaker (1992). A number of useful results and formulae are to be found in Appendix A. Our models may be described in various ways. We may describe them as *probability* models, when the emphasis is on the probabilistic formulation, or we may describe the models as *statistical*, when there is a greater emphasis on fitting the models to data. Many of the models we describe in this book develop over time or space, and they are then called *stochastic*. What is involved in statistical modelling is readily appreciated from considering a fecundability example.

Example 1.1: Fecundability

Table 1.1 describes the number of fertility cycles to conception required by fertile human couples setting out to conceive. The data were collected retrospectively, which means that information was only obtained from women who had conceived, and the women involved have been classified according to whether they smoked or not. Couples requiring more than 12 cycles are grouped together in a single category.

The couples in this study are essentially waiting for an event, and the simplest probability model for waiting times when they are integers, as here, is the geometric model. Let X denote the number of cycles to conception, and

Table 1.1 *Cycles to conception, classified by whether the female of the couple smoked or not. The data, taken from Weinberg and Gladen (1986), form a subset of data presented by Baird and Wilcox (1985). Excluded were women whose most recent method of contraception was the pill, as prior pill usage is believed to reduce fecundability temporarily. The definition of "smoking" is given in the source papers.*

Cycle	Women smokers	Women non-smokers
1	29	198
2	16	107
3	17	55
4	4	38
5	3	18
6	9	22
7	4	7
8	5	9
9	1	5
10	1	3
11	1	6
12	3	6
>12	7	12
Total	100	486

let p be the probability of conception per cycle. If we assume p is constant, then

$$pr(X = k) = (1 - p)^{k-1}p, \quad \text{for } k \geq 1.$$

Noting that $pr(X = 1) = p$ allows us to fit this model to each column of data very simply, to give the following estimates of p, which we denote by $\tilde{p}$:

	$\tilde{p}$
Smokers	0.29
Non-smokers	0.41

On the basis of these results we then have the immediate conclusion that each cycle non-smokers are 41% more likely to conceive than smokers (since $41 \approx (41 - 29)/29$). The usefulness of this model is therefore apparent in providing a framework for simply comparing smokers and non-smokers. However, there are evident shortcomings to the model. For example, can we legitimately suppose that all couples behave the same way? Of course we cannot, and simply checking the fit of this model to the data will reveal its inadequacies. We shall consider later how we can make the model more elaborate to improve its description of the data.   □

Standard terminology is that p in the above example is a *parameter*, and in fitting the model to the data we are producing the *estimator*, $\tilde{p}$. For this example, the parameter summarises the fecundability of a particular population. By itself, $\tilde{p}$ tells only part of the story, because it does not reflect the amount of data summarised by $\tilde{p}$. Naturally, because of the difference in sample sizes, the value of $\tilde{p}$ for non-smokers is more precise than that for smokers, and we can indicate the difference by constructing appropriate *confidence intervals*, or by associating with each estimator an estimate of its *standard error*, often just referred to as its *error*. We shall discuss ways of indicating precision, and present several examples, later in the book.

This book is devoted to the modelling process outlined in Example 1.1, but using a wide range of more sophisticated tools. We shall conclude this chapter with several more examples, which will be considered again later in the book.

1.2 Examples of data sets

Several of the models that we encounter in this book are general, with wide application, for instance multinomial models, models of survival and logistic regression models. Others have to be specially designed for particular problems and data sets, and we shall now consider a range of examples which require individually tailored models. However, the same basic statistical principles and approach are relevant in all cases. As we shall see, models for one application may also be of use in other areas which at first sight may appear to be quite different.

Example 1.2: Microbial infections

The data of Table 1.2 are taken from Sartwell (1950) and provide information on incubation periods of individuals who contracted streptococcal sore throat from drinking contaminated milk. Knowledge of when the contamination occurred allowed the incubation periods to be calculated.

Of interest here was fitting a stochastic model for the underlying behaviour of the infecting agents. Models of this kind are also used in the modelling of AIDS progression, in models of biological control and in models for the distribution of prions over cells. (Prions are the protein particles thought to result in mad cow disease.) □

Example 1.3: Polyspermy

Polyspermy occurs when two or more sperm enter an egg. It is relatively common in certain insects, fish, birds and reptiles. Fertilization of sea-urchin eggs has been of particular interest to biologists since the nineteenth century, and polyspermy in sea urchins *(Echinus esculentus)* has been examined by Rothschild and Swann (1950) and Presley and Baker (1970). An illustration of the experimental data that result is given in Table 1.3.

Thus for example, in the second of the four experiments, 84 eggs were exposed to sperm. After 15 seconds fertilisation was stopped by the addition of a spermicide, and it was found that 42 eggs had not been fertilised, 36 eggs

Table 1.2 *Sore throat incubation periods, given in units of 12 hrs.*

Incubation period	Number of individuals
0–1	0
1–2	1
2–3	7
3–4	11
4–5	11
5–6	7
6–7	5
7–8	4
8–9	2
9–10	2
10–11	0
11–12	1
>12	0

Table 1.3 *Polyspermy: distributions of sperm over eggs of sea urchins (data provided by Professor P.F. Baker).*

Experiment	No. of eggs in experiment	Length of experiment (secs)	Number of sperm in the egg					
			0	1	2	3	4	5
1	100	5	89	11	0	0	0	0
2	84	15	42	36	6	0	0	0
3	80	40	28	44	7	1	0	0
4	100	180	2	81	15	1	1	0

had been fertilised once and 6 eggs had been fertilised twice. The eggs in each experiment were different, and all four experiments took place under uniform conditions of temperature and sperm density.

Of interest here was a stochastic model incorporating the rates at which sperm enter the eggs. By means of the modelling it was possible to investigate how these rates vary over time.   □

Example 1.4: Sprayed flour beetles

The data of Table 1.4 document the progress of flour beetles *Tribolium castaneum* sprayed with a well-known plant-based insecticide (pyrethrins B), and record the cumulative number of beetles that have died since the start of the experiment. The insecticide was sprayed at the given rates of application over small experimental areas in which the groups of beetles were confined but allowed to move freely. The beetles were fed during the course of the experiment to try to eliminate the effect of natural mortality.

Table 1.4 *Death of sprayed flour beetles (from Hewlett, 1974).*

Dose (mg/cm^2)

		0.20		0.32		0.50		0.80	
	Sex	M	F	M	F	M	F	M	F
Day	1	3	0	7	1	5	0	4	2
	2	14	2	17	6	13	4	14	9
	3	24	6	28	17	24	10	22	24
	4	31	14	44	27	39	16	36	33
	5	35	23	47	32	43	19	44	36
	6	38	26	49	33	45	20	46	40
	7	40	26	50	33	46	21	47	41
	8	41	26	50	34	47	25	47	42
	9	41	26	50	34	47	25	47	42
	10	41	26	50	34	47	25	48	43
	11	41	26	50	34	47	25	48	43
	12	42	26	50	34	47	26	48	43
	13	43	26	50	34	47	27	48	43
	Group size	144	152	69	81	54	44	50	47

Statistical models for what are called the "end-point" mortalities on day 13, when the experiment ended, are very well known, and have been studied extensively. In this example we want to elaborate those models, in order to account for the effect of the insecticide over time. □

Example 1.5: Longevity of herons

The grey heron *Ardea cinerea* has been a protected species in Britain since 1954. In Table 1.5 we present the results of a ring-recovery experiment, spanning the years 1955–1971, taken from a larger example in North and Morgan (1979). Birds were ringed as nestlings, and enter the table if they die and their death was reported to the British Trust for Ornithology, whose address appears on the rings. Thus for example, of the cohort of birds ringed in 1956, 14 were found dead in their first year of life, 5 were found dead in their second year of life, and so on. We can see from the data the relatively high mortality of first-year birds. This is often the case for wild animals.

These data provide a means of estimating the probabilities of annual survival of the herons. In some cases, records of the numbers of birds ringed in each cohort are also available. Wild animals can be regarded as monitors of the environment. When the environment changes, due to the destruction of

Table 1.5 *Recoveries of dead grey herons (data provided by the British Trust for Ornithology).*

Year of ringing	Year after ringing in which recovery was made																
	1st	2nd	3rd	4th	5th	6th	7th	8th	9th	10th	11th	12th	13th	14th	15th	16th	17th
1955	31	5	0	0	3	1	0	0	1	1	0	0	0	0	0	0	0
1956	14	5	5	2	1	1	1	0	0	0	0	0	0	0	0	0	
1957	27	10	1	3	3	1	0	0	0	0	0	0	0	0	1		
1958	13	2	2	1	1	0	1	0	0	0	0	0	0	0			
1959	35	22	7	6	1	2	1	2	0	0	0	0	0				
1960	5	6	5	0	2	0	0	0	0	1	0	0					
1961	22	5	2	1	1	0	0	0	0	0	0						
1962	7	0	0	0	0	0	0	0	0								
1963	3	2	1	2	0	0	0	0	0								
1964	7	3	2	0	0	0	0	0									
1965	12	5	1	0	0	1	0										
1966	7	9	4	0	1	0											
1967	31	9	5	4	1												
1968	35	11	2	0													
1969	58	16	6														
1970	40	17															
1971	30																

hedgerows or climate change such as global warming, this may be reflected in corresponding changes in animal survival. Stochastic models may be used to investigate possible relationships between survival probabilities and measures of weather and/or characteristics of individual animals.   □

Example 1.6: Hatching of quail eggs

Mammals expand and aerate their lungs when they take their first breath after birth. However with birds the situation is more complex and far less spontaneous, as they begin breathing air before hatching. The study of the development of avian embryos within eggs has a long history – see, for example, Horner (1853). Detailed investigation of embryonic breathing in birds is described by Vince (1979). Somewhat remarkably, the development of embryos within eggs of certain birds can be affected by external clicking noises, of the kind made by neighbouring eggs as they prepare for hatching. The end-result, for ground-nesting game birds, is an approximate *synchronisation* of the hatching of the eggs in a clutch. Two sets of data are given in Tables 1.6 and 1.7.

The effects observed in Tables 1.6 and 1.7 are striking, and may be communicated simply through the estimated parameters of a suitable stochastic model of how the eggs mature through time.   □

Example 1.7: Disease in a forest

The data of Table 1.8 were presented by Pielou (1963a) and describe the incidence of diseased trees in a forest. The data have been obtained by quadrat sampling a map of a 2.88 acre Douglas fir plantation on Vancouver Island, and the disease was *Armillaria* root rot. In all there were 2,110 trees, of which 453 were infected. In total, 800 quadrats were sampled, of 1.2 inch radius, equivalent to 6.3 ft. on the ground (quadrats containing no trees were not included).

Table 1.6 *Stimulating quail eggs: duration (in hours) of lung ventilation (when the lungs are being used for the first time) in individual Japanese quail (Coturnix coturnix japonica) fetuses (from Vince and Cheng, 1970). Measurements were taken at 30-minute time intervals, and not continuously. Also, for eggs in contact it was only possible to take measurements on the single egg being stimulated, or not, by the other two eggs.*

		Mean
Isolated eggs	24, 37, 33, 20.75, 15.5	26.05
One egg in contact with two others given the same amount of incubation	23.75, 9.25, 20.5, 17, 18, 15.5, 25.75, 18, 9.5, 23	18.03
One egg in contact with two others given 24 hours more incubation	5.5, 13, 7, 2.5, 0.5	5.7
One egg in contact with two others given 24 hours less incubation	23.5, 13, 19, 14, 37.5	21.4

Table 1.7 *Stimulating quail eggs: data from Vince (1968), describing the effect of artificial stimulation on the hatching time of Japanese quail eggs isolated from clutch siblings. Shown are the hatching times of single stimulated eggs, relative to the mean hatching time of the clutch.*

Artificial stimulation rate (clicks/sec.)	Hatching time of stimulated eggs, in hours	Artificial stimulation rate (clicks/sec.)	Hatching time of stimulated eggs, in hours
1	7.50	10	−6.25
1	3.13	10	−11.25
1.25	0.00	10	−33.75
1.5	−23.75	15	−33.75
2	−31.25	15	−35.00
2	−33.13	20	−18.75
3.5	−28.75	20	−25.00
3.5	−31.25	30	−11.25
3.5	−36.25	50	−20.00
3.5	−40.00	60	−14.38
6	−23.75	60	−16.25
6	−38.75	80	−10.63
6	−45.00	80	13.13
		100	8.75
		100	15.00

Table 1.8 *Incidence of root rot in a forest. Here n denotes the number of trees in a quadrat, of which d were diseased.*

n	d	No. of quadrats	n	d	No. of quadrats
1	0	146	5	0	28
	1	32		1	16
				2	9
				3	5
2	0	151		≥ 4	4
	1	61			
	2	16			
			6	0	9
3	0	122		1	5
	1	57		2	7
	2	18		3	3
	3	5		≥ 4	2
4	0	47	7	0	2
	1	25			
	2	20	8	3	1
	3	4		5	1
	4	3			
			9	1	1

It was thought that a simple statistical model would be useful here for summarising the data through the estimated parameters of the model. □

Example 1.8: Shakespeare's vocabulary

Shakespeare wrote 31,534 different words in total. An interesting question posed by Efron and Thisted (1976) is: how many words did he know, but not use? The known word frequencies are shown in Table 1.9, apart from 846 words which appear more than 100 times. Thus, for example, 14,376 words appear just once, 4343 appear twice, etc.

Statistical models and methods provide an answer. Similar models may also be used to estimate the number of species of a particular type of insect, such as a butterfly for example, occupying a region, when a trapping exercise yields only a subset of those species. □

Example 1.9: The ant-guard hypothesis

Seed predators and herbivores can operate as strong selective agents in the evolution of defence mechanisms by plants. This has given rise to an 'ant-guard' hypothesis, which tries to explain the role of extrafloral nectaries on plants. Distributed on species in over 80 plant families, these nectaries occur

Table 1.9 *Shakespeare's word frequencies.*

x	1	2	3	4	5	6	7	8	9	10	Row Total
0+	14376	4343	2292	1463	1043	837	638	519	430	364	26305
10+	305	259	242	223	187	181	179	130	127	128	1961
20+	104	105	99	112	93	74	83	76	72	63	881
30+	73	47	56	59	53	45	34	49	45	52	513
40+	49	41	30	35	37	21	41	30	28	19	331
50+	25	19	28	27	31	19	19	22	23	14	227
60+	30	19	21	18	15	10	15	14	11	16	169
70+	13	12	10	16	18	11	8	15	12	7	122
80+	13	12	11	8	10	11	7	12	9	8	101
90+	4	7	6	7	10	10	15	7	7	5	78

on parts not directly associated with pollination. The idea behind the ant-guard hypothesis is that extrafloral nectar production attracts pugnacious 'bodyguards' (usually ants) which by their presence deter the activities of herbivorous insects and seed predators. The data of Table 1.10 are a small subset taken from a study reported in Catchpole et al. (1987) in which the plants studied were *Helichrysum bracteatum*. They analysed the data to investigate the ant-guard hypothesis. Three sites were chosen in clearings in the Tallaganda Forest, 40 km. southeast of Canberra in Australia, and at each site ten pairs of plants were studied. Plants within each pair were of similar size initially, and less than 1 metre apart. Within each pair ants were excluded from one plant, while the other was untouched, and served as a control. Over the reproductive season, plants were censused once a week. For each plant a record was made of the number of flowerheads (capitula), the number of capitula with ants and the total number of insects. Different species of ants (predominantly *Iridomyrmex* spp.) and other insects were observed, but in Table 1.10 they are pooled within each general category.

The full data set is extensive, and this is a clear instance of the potential value of a simple probability model which might condense a large body of information so that it is far easier to appreciate the salient features of the data. □

Example 1.10: Ant-lions

Ant-lions (Morisita, 1971) prefer to burrow in fine rather than coarse sand. However they also prefer to avoid other ant-lions. The data of Table 1.11 describe the numbers that settle in fine sand when groups of size n of ant-lions are placed in an environment containing fine and coarse sand.

A statistical model for these data is proposed in Exercise 2.21. The model

Table 1.10 *Ant-insect interactions:'Index' here denotes pairs of plants on site 1 (indices 1–10), site 2 (indices 11–20) and site 3 (indices 21–30). Pairs of plants are excluded when neither plant has insects, and when one of the plants has no capitula. A denotes ant access and E denotes ant exclusion.*

Week	Index	No. of capitula		No. with	No. of insects	
		A	E	ants	A	E
1	7	1	1	0	0	1
1	16	2	1	1	0	1
1	18	1	1	1	0	1
1	23	1	3	0	0	1
1	26	11	1	3	2	0
2	2	4	4	1	0	1
2	3	4	2	0	3	0
2	5	1	1	0	0	1
2	9	1	1	0	2	0
2	10	1	1	0	0	1
2	13	1	2	1	0	1
2	15	1	2	1	1	2
2	16	2	1	2	0	1
2	17	1	2	1	0	4
2	21	3	3	2	1	4
2	22	1	3	1	0	4
2	23	1	3	0	0	2
2	24	10	13	6	3	15
2	25	1	1	0	0	2
2	26	14	1	1	1	2
2	28	1	1	0	0	9
2	29	1	1	1	0	3
3	1	1	1	1	1	1
3	2	3	3	3	0	5
3	3	4	2	4	0	4
3	4	1	1	1	0	1
3	5	1	1	0	3	2
3	6	1	1	1	1	1
3	7	1	1	1	0	2
3	9	1	1	1	0	1
3	10	1	1	1	1	5
3	11	3	1	1	0	3
3	12	1	1	1	0	2
3	13	1	4	1	2	5
3	14	5	3	5	0	3
3	15	3	3	1	0	2

Table 1.11 *Distribution of ant-lions in sand.*

Numbers of ant-lions introduced (n)	Numbers settled in fine sand (x)							
	0	1	2	3	4	5	6	7
3	0	7	24	1				
4	0	3	17	10				
5	0	0	10	15	4			
6	0	2	4	5	9	2	0	
7	0	0	1	2	4	3	0	0

can provide a useful summary of the data, and shares a common basis with the model proposed later in Exercise 3.3. □

1.3 Discussion

Much has been written on mathematical models, both on the basic modelling philosophy, and in providing case studies demonstrating how useful the models can be. See for example the books by Clements (1989), Giordano and Weir (1985), Mesterton-Gibbons (1989), Thompson (1989), and Thompson and Tapia (1990).

Many models are deterministic, and exclude all references to random effects. However, as can be appreciated from the examples of this chapter, there is no shortage of situations for which only statistical models are appropriate. Many probability models are exploratory, explaining the consequences of certain rules. For an example, see Morgan (1993), which considers the predictions of different stochastic models for the way that individuals form groups. *Simulation* is the process of using suitable random numbers to mimic the rules of a probability model, and in general simulation models are exploratory. They provide a way of studying quite complicated processes (Morgan, 1984, chapter 8). Exploratory models can be constructed and studied for particular ranges and values of their parameters, without any reference to data. In this book we are more interested in using models in the context of data. Data can make modelling challenging and sometimes frustrating but, ultimately, often far more interesting and relevant.

There now exists a wide range of books describing stochastic models for particular areas of application. See, for example, Lawless (2002), Cressie (1991), Diggle (1983), Lebreton and North (1993), Morgan and Thomson (2002), and Morgan (1992). Our objective is to provide a general approach to statistical modelling which is not tied to any one particular area, or particular computer package such as GLIM (see for example Aitkin et al., 1989) which is designed for a particular family of models. However, certain applications, for instance to bioassay and ecology, will recur throughout the book. We aim to emphasise the most important general principles for fitting probability models to data.

1.4 Exercises

1.1   Locate and read the paper by Pielou (1963b), which uses the geometric distribution to model the runs of diseased trees in forest transects.

1.2   Try to construct a probability model for the ant-lion distribution data of Table 1.10.

1.3   Data which might otherwise be expected to result from a Poisson distribution are sometimes found to contain too many zeroes. Suggest one possible mechanism for this 'zero inflation,' and consider how you might describe the data by means of a probability model. For discussion, see the paper by van den Broek (1995), and Example 5.1.

1.4   A simple model for the data of Table 1.5 would be one in which each year a heron has a probability, ϕ say, of surviving that year. Consider whether you think this would be an adequate description of the data. What else is needed in order to be able to write down probabilities of birds being reported as dead at different ages? Records of birds reported dead in Britain have generally been declining in recent years. Why do you think that is?

1.5   Locate and read at least one of the papers by Brooks et al. (1991), Goldman (1993), Jørgensen et al. (1991), Royston (1982), and Ridout (1999), all of which provide examples of complex stochastic models and how they are fitted to data.

1.6   Each April, little blue penguins (*Eudyptula minor*) return to Somes Island in Wellington Harbour, New Zealand, to choose a mate, build a nest and rear chicks. Pledger and Bullen (1998) proposed a test for fidelity in the choice of mates. Read the paper and critically assess the suggested test.

1.7   The scientific *Journal of Agricultural, Biological and Environmental Statistics (JABES)* received about 100 submitted papers a year for the period, 2003–2006. These submissions are distributed over months as shown in Table 1.12.

By means of simple tabulations and plots, consider whether there are any patterns in the data, and predict the missing numbers of submissions for the last 3 months of 2006.

1.8   *The Independent* newspaper published in Britain on 23 March 1999 discussed the attempts of couples trying to give birth to 'millenium babies' as follows: 'The race to conceive a millenium baby is certain to end in tears, ... only one in three will conceive in the next month.' Discuss in the context of Section 1.1. How would these attempts distort the demands made of the National Health Service?

1.9   Discuss possible biological reasons for the synchronisation of hatching observed in the data of Table 1.6.

Table 1.12 *Papers submitted to JABES.*

Month	2003	2004	2005	2006
Jan	7	9	6	12
Feb	12	7	8	5
Mar	10	9	8	12
Apr	9	8	15	9
May	7	15	9	10
June	6	6	10	10
July	9	5	12	5
Aug	4	5	8	12
Sept	12	10	4	14
Oct	7	10	6	–
Nov	8	7	5	–
Dec	8	10	11	–
Total	99	101	102	–

1.10 In a study of the behaviour of bees, plants previously sprayed with pesticides (which are damaging to bees) were also sprayed with chemical repellents, denoted by A, B and C. These were arranged in eight dishes and counts taken of the numbers of bees arriving at the dishes at intervals of 1 minute. The results are shown in Table 1.13, and are taken from Ridout et al. (2006). Analyse these data.

1.11 The data of Table 1.14 continue the data presented in Table 1.10. Compare the two sets of data, and consider whether there are any striking temporal effects.

Table 1.13 *The effects of repellent chemicals on foraging bees.*

Time	Counts for dishes and treatments							
(min)	1(A)	2(C)	3(B)	4(C)	5(A)	6(C)	7(B)	8(C)
1	0	5	1	1	1	0	2	6
2	1	6	3	16	1	17	1	12
3	2	6	4	8	0	11	1	5
4	2	3	1	2	2	6	3	7
5	3	12	8	15	0	7	3	12
6	3	10	3	7	0	5	1	10
7	3	10	3	1	0	2	4	14
8	1	5	8	14	1	10	4	11
9	1	11	6	11	2	5	3	8
10	1	9	6	2	1	12	2	18
11	4	9	5	10	0	9	3	12
12	3	6	6	5	2	15	7	11
13	0	10	10	9	2	12	7	11
14	3	6	2	13	2	12	7	11
15	3	11	3	13	2	12	7	11
16	4	8	9	11	4	15	4	9
17	2	11	11	8	4	8	10	14
18	8	11	9	21	2	10	9	12
19	3	8	11	12	5	12	9	10
20	6	12	7	18	3	16	8	14
21	1	4	7	9	5	14	11	15
22	1	8	12	16	3	18	12	12
23	3	14	12	15	6	13	14	13
24	3	12	15	19	3	14	16	13
25	3	14	16	11	5	8	9	15
26	6	8	15	11	9	14	15	11
27	7	14	17	10	6	12	18	14
28	7	10	11	10	8	13	14	8
Total	84	253	221	298	82	300	204	317

Table 1.14 *Further data on ant-insect interactions.*

Week	Index	No. of capitula		No. with ants	No. of insects	
		A	E	ants	A	E
3	16	3	3	3	0	1
3	17	4	3	4	0	7
3	18	2	2	1	0	1
3	19	3	3	2	0	2
3	20	5	1	2	1	2
3	21	7	3	5	0	3
3	22	3	3	2	1	3
3	23	1	4	1	0	4
3	24	20	17	4	7	20
3	25	1	2	1	0	3
3	26	19	1	5	0	1
3	27	1	3	1	0	2
3	28	1	1	1	0	3
3	29	1	2	0	5	4
3	30	3	3	2	0	3
4	2	3	3	3	0	3
4	3	3	2	1	4	3
4	4	2	1	1	1	1
4	5	1	1	0	1	2
4	6	1	1	0	2	2
4	8	1	1	0	0	2
4	9	1	1	0	0	1
4	11	4	1	3	0	3
4	13	4	5	3	1	2
4	14	4	6	3	0	1
4	15	4	3	2	1	0
4	16	3	3	1	1	1
4	17	4	4	3	0	7
4	18	4	3	0	0	1
4	19	3	3	2	0	1
4	20	4	1	1	0	1
4	21	10	4	0	3	0
4	22	3	4	0	1	0
4	24	30	24	0	4	4
4	26	13	3	6	1	2
4	29	3	3	0	2	0

Table 1.15 *Times to cell-division for yeast cells (hours).*

Mother times									
3.00	2.30	3.50	3.85	3.75	2.50	2.35	2.75	2.10	1.85
5.70	2.20	2.05	1.65	1.75	1.75	5.00	4.95	1.70	2.80
2.20	1.45	1.60	2.45	1.65	1.20	1.45	1.25	1.35	2.10
1.75	1.75	1.85	2.10	1.85	2.65	1.70	1.05	1.20	1.05
1.75	1.40	1.10	1.15	1.10	1.60	2.00	2.25	2.00	1.60
1.80	1.60	1.70	1.60	1.35	1.70	1.95	1.05	1.75	1.60
2.05	1.45	1.70	1.60	1.70	1.67				

Daughter times									
4.20	3.95	3.65	4.95	3.65	2.55	3.75	4.15	4.10	4.60
7.40	6.25	2.70	3.85	5.35	3.70	3.85	1.85	2.70	4.40
4.20	3.55	4.00	3.05	2.50	3.15	3.05			

1.12   The data in Table 1.15 describe cell division times for mother and daughter cells of the budding yeast, *Saccharomyces cerevisiae*, taken from Ridout et al. (2006). Consider whether there is a difference between mothers and daughters.

CHAPTER 2

Basic Model-Fitting

2.1 Introduction

The method of fitting the model to data in Example 1.1 was *ad hoc*, and may well not have good properties. In addition, most model-fitting situations are more complicated and will not result in such simple estimators. What we need are general tools, which can be applied in most situations, and which produce estimators with desirable properties. The method of *maximum likelihood* is just such a technique, and in this chapter we shall use the method of maximum likelihood to fit probability models to the data of Examples 1.1 and 1.3. For a full discussion of the concept of likelihood, see Cox and Hinkley (1974, p.11), Edwards (1972), and Pawitan (2001). The simplest case is if one has a random sample of values $\{x_i\}$ from some distribution, with pdf, if the values are continuous, or probability function, if the values are discrete, given by $f(x, \boldsymbol{\theta})$, where $\boldsymbol{\theta}$ are the model parameters; the likelihood is readily obtained by forming $\prod_i f(x_i, \boldsymbol{\theta})$. The likelihood is central to both classical and Bayesian methods of inference, and from a classical perspective we regard the likelihood as a function of the parameters $\boldsymbol{\theta}$, with maximum-likelihood estimates being the values $\hat{\boldsymbol{\theta}}$ that maximise the likelihood for the given random sample of data. In the next section we shall present one example of forming a likelihood, which combines probability functions with cdfs. The chapter ends with a discussion of general modelling issues.

2.2 Maximum-likelihood estimation for a geometric model

We shall start by considering further the fecundability study of Example 1.1. We can write down the *likelihood, L*, for the data in Table 1.1 as:

$$L \propto \left\{ \prod_{k=1}^{12} pr^s(X = k)^{n_{sk}} \right\} Pr^s(X > 12)^{n_{s+}}$$

$$\times \left\{ \prod_{k=1}^{12} pr^a(X = k)^{n_{ak}} \right\} Pr^a(X > 12)^{n_{a+}}$$

where s is used to denote smokers, and a is used to denote non-smokers, $n_{sk}(n_{ak})$ is the number of smokers (non-smokers) who take k cycles, and $n_{s+}(n_{a+})$ is the number of smokers (non-smokers) who take at least 13 cycles. We are here assuming that the times to conception of different couples are independent and that the smokers' data are independent of the non-smokers' data. Both of these assumptions seem perfectly reasonable.

In this example the likelihood is the product of two multinomial probabilities, from which we have excluded the multinomial coefficients; see Appendix A.1. They do not contain model parameters and so need not be considered further. Had they been included, then the proportionality sign for L above would have been replaced by an equality sign. In addition, the likelihood is more complicated than the simple likelihoods referred to in the last section, as it describes two types of data, viz., the times that are known exactly, and the times that are greater than 12 cycles. In the latter case the information in the data enters the likelihood through the cdf. Let us now take just the smokers as an illustration: letting $p = pr^s(X = 1)$ and dropping the s superscript, we then have the likelihood written as a function of the single parameter p :

$$L(p) \propto \prod_{k=1}^{r} \{p(1-p)^{k-1}\}^{n_k}(1-p)^{rn_+},$$

since

$$Pr(X > r) = p \sum_{j=r+1}^{\infty} (1-p)^{j-1} = p(1-p)^r/p$$
$$= (1-p)^r,$$

where r now denotes the largest individually recorded number of cycles. In our case, we have $r = 12$. It is simpler (and equivalent, as $\log(x)$ is a monotonic function of x) to maximise $\ell(p)$, where $\ell(p) = \log_e\{L(p)\}$. For simplicity we let $n_+ = n_{r+1}$, so that ignoring the multinomial coefficient, which does not involve p, we can write

$$\ell(p) = \{\log(1-p)\} \left\{\sum_{j=1}^{r+1}(j-1)n_j\right\} + (\log p)\left(\sum_{j=1}^{r}n_j\right).$$

The *maximum-likelihood* estimate of p is the solution, $\hat{p}$, of the equation, $\frac{d\ell}{dp} = 0$. Differentiating with respect to p, we get:

$$\frac{d\ell}{dp} = -\frac{\sum_{1}^{r+1}(j-1)n_j}{(1-p)} + \frac{\sum_{1}^{r}n_j}{p}.$$

Setting $\frac{d\ell}{dp} = 0$ then results in the following equation for $\hat{p}$:

$$(1-\hat{p})\left(\sum_{1}^{r}n_j\right) = \hat{p}\left\{\sum_{1}^{r+1}(j-1)n_j\right\}, \qquad \text{which simplifies to give}$$

$$\sum_{1}^{r}n_j = \hat{p}\left(\sum_{1}^{r+1}jn_j\right) - n_{r+1}\hat{p}$$

$$\therefore \quad \hat{p} = \left(\frac{\sum_1^r n_j}{\sum_1^r j n_j + r n_{r+1}} \right). \tag{2.1}$$

We note that $\frac{d^2 \ell}{dp^2} < 0$, so that we have indeed maximised the likelihood.

It is interesting to see from expression (2.1) how we incorporate the n_{r+1} right-censored values into the formula for $\hat{p}$. If $n_{r+1} = 0$ then $1/\hat{p}$ is the sample mean, $(\sum_1^r j n_j)/(\sum_1^r n_j)$. When $n_{r+1} > 0$, in $1/\hat{p}$ we have added to the numerator a quantity $(r n_{r+1})$ that is less than the total (unknown) conception times for the n_{r+1} cases, but the denominator is not increased at all.

The n_{r+1} values are those conception times which are more than 12 cycles in length. They are said to be *right-censored*, an expression used frequently in the general analysis of survival data.

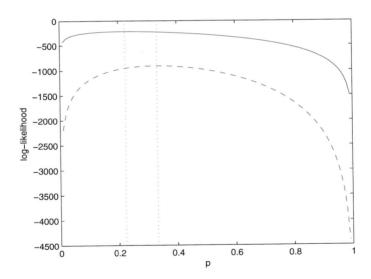

Figure 2.1 *Graphs of the log-likelihood, $\ell(p)$, when the geometric model is fitted to the smokers (solid line) and the non-smokers (dashed line) data of Table 1.1. Note that here we have excluded additive logarithms of combinatorial coefficients. They do not include p, and simply translate the $l(p)$ curves in the direction perpendicular to the p-axis. Also shown (dotted lines) are the locations of the maximum-likelihood estimates of p, one for each data set.*

Figure 2.1 provides graphs of the log-likelihood function $\ell(p)$ for the smokers and the non-smokers separately. We can see from Figure 2.1 that the log-likelihood is less flat for the larger set of data. This makes sense and suggests either using sections of the graph to give confidence regions for p and/or using a measure of curvature of $\ell(p)$ at the value $\hat{p}$. Convenient measures of curvature are $-\left(\frac{d^2\ell}{dp^2}\right)$ and $-\mathbb{E}\left[\frac{d^2\ell}{dp^2}\right]$, where we write $\mathbb{E}[\ ]$ to denote expectation. We discuss this topic fully in Chapter 4, but we may note here that had the data set been uniformly larger, say by a multiple of κ, then that would not have changed the maximum-likelihood estimates, but it would have increased the curvature of the graphs in Figure 2.1.

The values of the maximum-likelihood estimates, $\hat{p}$, for the data in Table 1.1 are: $\hat{p}$(smokers) $= 0.2241$ and $\hat{p}$ (non-smokers) $= 0.3317$. Although they have the same ordering as the values of $\tilde{p} = (n_1/n.)$, obtained in Chapter 1, where $n. = \sum_1^{r+1} n_j$, they are clearly different from those values. For a variety of reasons to be discussed later, it is important to use an established method of parameter estimation such as maximum likelihood.

Using $\hat{p}$, we can write down the numbers we would expect under the fitted model (E_i) for comparison with those which we observe (O_i), and these are shown in Table 2.1. The appropriate distribution to describe how the data are allocated to the categories corresponding to the different numbers of cycles is the multinomial distribution, and so in this case these expected numbers are easily calculated, by multiplying the total numbers observed in each case by the corresponding estimated geometric probability; for instance, $22.41 = 0.2241 \times 100$, etc. Because of the geometric decline of geometric probabilities, we can see that the expected numbers also decline geometrically. It is interesting now to calculate a formal measure of how well the observed and expected numbers match up. A standard way of doing this is by using the Pearson goodness-of-fit test. The Pearson goodness-of-fit statistic is given by:

$$X^2 = \sum_{i=1}^{k} \frac{(O_i - E_i)^2}{E_i} ,$$

when there are k categories involved (in our case $k = 13$). The benchmark against which to judge X^2, to see if it is too large, which would imply an inadequacy in the model, is provided by an appropriate chi-square distribution. The justification for using this benchmark is an 'asymptotic' one, which means that for the test to be valid, cell values need to be 'large.' The particular chi-square distribution we use is determined by the degrees of freedom for the distribution, and in general, the formula for the degrees of freedom (df) is:

$$df = k - 1 - t ,$$

where t is the number of parameters appropriately estimated (for example by maximum likelihood), and the parameters have been estimated from using the multinomial form of the problem (see the comments below).

Table 2.1 *Evaluating the geometric model fitted by maximum likelihood, using the standard Pearson X^2 test.*

Smokers

Cycle (j)	Observed (O_j)	Expected (E_j)	Components of X^2 $\frac{(O_j - E_j)^2}{E_j}$
1	29	22.4	1.938
2	16	17.4	0.111
3	17	13.5	0.913
4	4	10.5	3.996
5	3	8.1	3.230
6	9	6.3	1.155
7	4	4.9	0.162
8	5	3.8	0.383
9	1	2.9	1.283
10	1	2.3	0.722
11	1	1.8	0.336
12	3	1.4	1.920
> 12	7	4.8	1.053
Total	100		17.204

Non-smokers

Cycle	Observed	Expected	Components of X^2
1	198	161.2	8.398
2	107	107.7	0.005
3	55	72.0	4.013
4	38	48.1	2.127
5	18	32.2	6.232
6	22	21.5	0.012
7	7	14.4	3.774
8	9	9.6	0.037
9	5	6.4	0.312
10	3	4.3	0.386
11	6	2.9	3.431
12	6	1.9	8.718
> 12	12	4.8	11.009
Total	486		48.454

We note here that several of the expected values are 'small' (say < 5), and it is a common practice to pool over categories, to obtain larger expected values. This approach is somewhat arbitrary, and should be avoided if possible (Everitt 1977, p.40). Pooling also poses problems with selection of degrees of freedom — see here Bishop, Fienberg, and Holland (1977, p.523) and de Groot (1986, p.530). Other work (see Fienberg, 1981, p.172) suggests that the chi-square approximation to the distribution of X^2 may be acceptable even with expected frequencies as small as unity. In our current example we do not pool. We have estimated the probability p from the multinomial structure of the data, and so the degrees of freedom are: $(13 - 2) = 11$. Standard chi-square percentage points for these degrees of freedom are: $\chi^2_{11:10} = 17.28$, and $\chi^2_{11:5} = 19.68$, where we use $\chi_{df:\nu}$ to indicate the value (percentage point) that a chi-square random variable with df degrees of freedom exceeds with probability $\nu/100$. Here we see that we cannot formally reject the geometric model at the 5% level of significance for the smokers, but we can for the larger sample of non-smokers. The lack of significance for the smaller sample may well be due to smaller power resulting from the smaller sample in that case. The geometric model is certainly too simple for the data, as we have already acknowledged in Chapter 1. We can improve the fit by making the model more complex, and that is done in the next section.

2.3 Maximum-likelihood for the beta-geometric model

Clearly individual couples can be expected to exhibit variation in the probability of conception per menstrual cycle, and it is natural to give this probability a distribution over the population considered. The obvious choice for this distribution, which we call the *mixing* distribution and which must have the $(0,1)$ interval for its range, is the beta distribution $\text{Be}(\alpha, \beta)$, with probability density function (pdf),

$$f(p) = \frac{p^{\alpha-1}(1-p)^{\beta-1}}{B(\alpha,\beta)}, \quad 0 < p \le 1, \ \alpha, \beta > 0$$

where $B(\alpha, \beta)$ is the beta function:

$$B(\alpha,\beta) = \Gamma(\alpha)\Gamma(\beta)/\Gamma(\alpha+\beta).$$

Here the gamma function $\Gamma(\kappa)$ is given by $\Gamma(\kappa) = \int_0^\infty e^{-x}x^{\kappa-1}dx$ and satisfies the relationship, $\Gamma(\kappa+1) = \kappa\Gamma(\kappa)$, which we find useful below in simplifying the expression for the distribution of a beta-geometric random variable X. Note that for this beta distribution, the mean and variance are, respectively,

$$\frac{\alpha}{\alpha+\beta} \quad \text{and} \quad \frac{\alpha\beta}{(\alpha+\beta)^2(\alpha+\beta+1)}.$$

We can see that in changing from the geometric model we have moved from a model with a single parameter, p, to one with two parameters, α and β. A convenient alternative parameterisation is in terms of the two parameters

$$\mu = \frac{\alpha}{\alpha+\beta}$$
$$\text{and} \quad \theta = \frac{1}{\alpha+\beta} .$$

Thus we have replaced the single probability, p, of conception per cycle by the mean probability of conception, μ. We may regard the parameter θ as a "shape" parameter; when this parameter is zero then the beta distribution can be seen to have zero variance, and there is then no heterogeneity: all couples have the same probability of conception per cycle and the geometric model results. Integrating over the mixing beta distribution gives the *beta-geometric* distribution:

$$pr(X = k) = \mu \, \frac{\displaystyle\prod_{i=1}^{k-1}\{1 - \mu + (i-1)\theta\}}{\displaystyle\prod_{i=1}^{k}\{1 + (i-1)\theta\}}, \quad \text{for } k \geq 1,$$

where we take $\displaystyle\prod_{i=1}^{0} = 1$. We see this as follows:

We shall write the probability function of X for a particular value of p as the conditional probability: $pr(X = k|p)$. We now seek the unconditional probability, $pr(X)$, obtained as:

$$
\begin{aligned}
pr(X = k) &= \int_0^1 pr(X = k|p) f(p) dp = \int_0^1 (1-p)^{k-1} \, \frac{p p^{\alpha-1}(1-p)^{\beta-1}}{B(\alpha,\,\beta)} dp \\[2mm]
&= \int_0^1 \frac{p^\alpha (1-p)^{\beta+k-2}}{B(\alpha,\,\beta)} dp \\[2mm]
&= \frac{B(\alpha+1,\,\beta+k-1)}{B(\alpha,\,\beta)},
\end{aligned}
$$

making use of the fact that the integral of a beta pdf over its range is unity.

$$
\begin{aligned}
\text{Thus } pr(X = k) &= \frac{\Gamma(\alpha+1)\Gamma(\beta+k-1)\Gamma(\alpha+\beta)}{\Gamma(\alpha+\beta+k)\Gamma(\alpha)\Gamma(\beta)} \\[2mm]
&= \frac{\alpha(\beta+k-2)(\beta+k-3)\ldots\beta}{(\alpha+\beta+k-1)(\alpha+\beta+k-2)\ldots(\alpha+\beta)},
\end{aligned}
$$

making use of the recursion for gamma functions stated above. This expression for $pr(X = k)$ readily simplifies to the above expression for $pr(X = k)$ in terms of the alternative parameterisation of $(\mu \ \theta)$ – see Exercise 2.18. When $\theta = 0$ we can see that the model reduces to the geometric form. The kind of integration involved here is identical to the integration encountered in Bayesian analysis, to be described in Chapter 7. Like the geometric distribution, the

beta-geometric distribution describes a discrete random variable. For the case
$\alpha > 1$, it has mean and variance given respectively by:

$$\mathbb{E}[X] = \left(\frac{1-\theta}{\mu-\theta}\right) , \quad \text{Var}(X) = \frac{\mu(1-\mu)(1-\theta)}{(\mu-\theta)^2(\mu-2\theta)} .$$

If this model is appropriate then for the case $\alpha > 1$ the expected remaining
waiting time for couples who have had j failed cycles can be shown to increase
linearly with j:

$$\mathbb{E}[(X-j)|X>j] = \left(\frac{1-\theta}{\mu-\theta}\right) + \frac{j\theta}{(\mu-\theta)} .$$

See Exercise 2.10.

This formula shows how we can use models to make predictions. In particular, the formula simply expresses the fact that the longer couples have been
waiting to conceive, then the longer they can expect to wait until conception.
Unlike the geometric model, the beta-geometric model has *memory*. If we set
$\theta = 0$ in the above equation then the beta-geometric model reduces to the ge-
ometric model, and $\mathbb{E}[(X-j)|X>j]$ is a constant, which no longer involves
j.

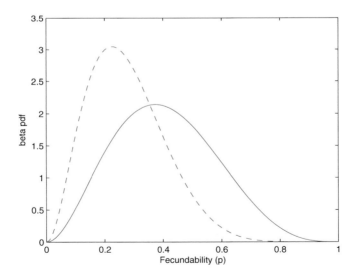

Figure 2.2 *(Weinberg and Gladen, 1986). Beta pdfs for smokers (dashed line) and
non-smokers (solid line), following maximum-likelihood estimation.*

A distribution which shares essentially the same derivation as the beta-
geometric distribution is the beta-binomial distribution (Exercise 2.7), which
results when the probability of success in the binomial distribution is given a

Table 2.2 *The maximum-likelihood fit of the beta-geometric model to the data of Table 1.1. Pearson goodness-of-fit statistics are now : for smokers: $X^2 = 15.45$ (df = 10); for nonsmokers: $X^2 = 12.57$ (df = 10). Parameter estimates are given later in Table 4.1.*

Cycle	Smokers		Non-smokers	
	Observed	Expected	Observed	Expected
1	29	27.5	198	198.3
2	16	18.3	107	103.3
3	17	12.6	55	59.1
4	4	9.0	38	36.3
5	3	6.6	18	23.5
6	9	4.9	22	15.9
7	4	3.8	7	11.1
8	5	2.9	9	8.0
9	1	2.3	5	5.9
10	1	1.8	3	4.5
11	1	1.5	6	3.5
12	3	1.3	6	2.7
>12	7	7.6	12	13.9
Total	100		486	

beta-distribution. Williams (1975) used the beta-binomial model to describe litter mortality data which were *overdispersed* relative to the binomial distribution — that is, the sample variance was appreciably larger than would have been the case if the data had been described by a binomial distribution. The overdispersion is due in this case to the different mortality rates experienced in different litters.

A key difference between fitting the beta-geometric model to data by maximum likelihood, and fitting the simpler geometric model, is that for the beta-geometric model, the likelihood cannot be maximised explicitly, and a numerical algorithm is needed. This is frequently the case when models are fitted to data by maximum likelihood, and in Chapter 3 we shall discuss the various options available. Ridout and Morgan (1991) consider extensions of the beta-geometric model, in which individuals misreport their durations, for example reporting 6 instead of 5 cycles, and we shall investigate these extensions in Section 4.2. Figure 2.2 displays the beta pdfs constructed from the estimated parameter values following maximum-likelihood fitting of the beta-geometric distribution to the data of Table 1.1. The goodness of fit is now much improved, especially for the non-smokers, as can be seen from the expected values of Table 2.2. The 5% point for a χ^2_{10} distribution is $\chi^2_{10:5} = 18.31$. We see from Figure 2.2 that the beta pdf for the smokers has less spread, as well as a smaller mean, than that for the non-smokers.

2.4 Modelling polyspermy

2.4.1 A Poisson process model

The fecundability data were *overdispersed* relative to the geometric distribution. We now consider a stochastic model for the data of Table 1.3, resulting from an investigation of polyspermy. We shall see that in this case the data are *underdispersed*, that is, the variability in the data is less than would be expected under a particular simple model.

The simplest probability model for data of this kind is the *Poisson process* see, e.g., Clarke and Cooke (2004, p.517), and Appendix A. In the polyspermy context, under the Poisson process, fertilisations take place at random over time at a constant rate λ. The prediction of the Poisson process is that at any time t, the probability $p_k(t)$ that an egg has been fertilised k times in the interval $(0, t)$ is given by the *Poisson distribution*, $P_o(\lambda)$:

$$p_k(t) = \frac{e^{-\lambda t}(\lambda t)^k}{k!} \quad (k = 0, 1, 2, ...) . \tag{2.2}$$

Two questions now arise:

(i) If we assume that fertilisation can be described by a Poisson process, what value of λ, the fertilisation rate, should we take to describe this particular set of data?

(ii) Once that λ is chosen, is the description of the data provided by the model a reasonable one?

We answer (i) by again using the method of maximum likelihood: under the model, the likelihood is

$$L \propto \prod_{i=1}^{4} \prod_{k=0}^{\infty} p_k(t_i)^{n_k(t_i)}$$

in which $t_1 = 5$, $t_2 = 15$, $t_3 = 40$ and $t_4 = 180$, $n_0(5) = 89$, $n_1(5) = 11$, $n_0(15) = 42$, $n_1(15) = 36$, $n_2(15) = 6$, and so on. Once again, we ignore constant combinatorial terms which are not functions of the model parameter λ. In writing down this likelihood we have assumed that the fertilisations of different eggs, between and within experiments, are all independent, which does not seem unreasonable. L is a function of the single parameter λ, and the method of maximum likelihood estimates λ by the value $\hat{\lambda}$ that maximises the likelihood L. As before, we take logarithms to get

$$\begin{aligned}
\ell(\lambda) \;=\; \log L \;&=\; \sum_{i=1}^{4}\sum_{k=0}^{\infty} n_k(t_i) \log p_k(t_i) \\
&=\; C - \lambda \sum_{i=1}^{4}\left(t_i \sum_{k=0}^{\infty} n_k(t_i) \right) + (\log \lambda)\sum_{i=1}^{4}\sum_{k=0}^{\infty} k n_k(t_i) ,
\end{aligned}$$

in which the constant term C does not involve λ. In search of stationary values of ℓ we set $d\ell/d\lambda = 0$, to obtain the maximum-likelihood estimate of the fertilisation rate λ,

$$\hat{\lambda} = \frac{\sum_{i=1}^{4}\sum_{k=0}^{\infty}kn_k(t_i)}{\sum_{i=1}^{4}t_i\sum_{k=0}^{\infty}n_k(t_i)} \ .$$

By obtaining $\frac{d^2\ell}{d\lambda^2}$, it is simple to verify that $\hat{\lambda}$ does in fact maximise ℓ (Exercise 2.9), and for the data of Table 1.3, we obtain the point estimate, $\hat{\lambda} = 0.0104$.

For an experiment at time t, involving $N(t) = \sum_{j=0}^{\infty}n_j(t)$ eggs altogether, the expected number of eggs with k fertilisations is given by

$$N(t)\,\frac{e^{-\hat{\lambda}t}(\hat{\lambda}t)^k}{k!} \quad (k = 0, 1, 2, \ldots)\ .$$

This gives, for the Poisson process model, the expected numbers shown in Table 2.3(a). The expected numbers are rounded to the nearest integer, which explains the small round-off error seen to be present when we form expected totals.

In answer to question (ii) above regarding goodness of fit, from comparison of expected numbers with the data of Table 1.3, the model very clearly does not provide an adequate description of the data. On the one hand λ should be larger than 0.0104 in order to reduce the expected number of eggs that have still to be fertilised. However on the other hand that would, particularly for the last experiment, result in expected values of $n_k(t_i)$, for $k > 1$, which are far too large.

A possible explanation of this poor fit is that once an egg has been fertilised for the first time then subsequent fertilisations take place at some *reduced* rate, μ, say, relative to λ This is, in fact, what the biologists believe to take place; the form of the $p_k(t)$ under this new model was given in Morgan (1975), and an alternative derivation is presented below. It is this adaptation of the simple Poisson process which results in underdispersion relative to the simpler model.

2.4.2 Poisson process with two rates

The Poisson process has two important properties, stated in Appendix A. The first, that the number of events (which are fertilisations in our case) in a fixed time interval has a particular Poisson distribution, was given in Equation (2.2), and used above. The second property, which we shall use shortly, is that the time intervals between events are independent, continuous random variables with an exponential pdf

$$f(\theta) = \beta e^{-\beta\theta}, \quad \theta > 0\ , \tag{2.3}$$

Table 2.3 *Goodness of fit of models to the polyspermy data of Table 1.3. Expected values for: (a) single Poisson process; (b) Poisson process with two rates; (c) Poisson process with two rates and a fertilisation membrane at time* $\tau = 15$ *seconds after the first fertilisation.*

Experiment	No. of eggs in experiment	Length of experiment (in seconds)	Number of sperm in the egg						
			0	1	2	3	4	5	6
1	100	5	95	5	0	0	0	0	0
2	84	15	72	11	0	0	0	0	0
3	80	40	53	22	5	1	0	0	0
4	100	180	15	29	27	17	8	3	1

(a)

Experiment	No. of eggs in experiment	Length of experiment (in seconds)	Expected number of sperm in the eggs				
			0	1	2	3	4
1	100	5	87	13	0	0	0
2	84	15	56	28	0	0	0
3	80	40	26	51	3	0	0
4	100	180	1	72	23	4	0

(b)

Experiment	No. of eggs in experiment	Length of experiment (in seconds)	Expected number of sperm in the egg				
			0	1	2	3	4
1	100	5	87	13	0	0	0
2	84	15	54	27	3	0	0
3	80	40	25	46	8	1	0
4	100	180	1	81	17	2	0

(c)

for a Poisson process of rate $\beta > 0$. The Poisson process is the continuous-time analogue of a system of independent Bernoulli trials that gives rise to the geometric distribution, so that the exponential distribution here describes a waiting time, as does the geometric distribution in discrete time.

Returning now to the model with two fertilization rates, clearly $p_0(t) = e^{-\lambda t}$, precisely as before. However if the first fertilisation takes place at some time θ, $0 \leq \theta \leq t$, then, conditional upon this event,

$$p_n(t) \quad = \quad Pr\{n-1 \text{ events occur in the remaining time interval, } (t-\theta)\}$$

$$= \quad \frac{e^{-\mu(t-\theta)}\{\mu(t-\theta)\}^{n-1}}{(n-1)!} \qquad (n \geq 1),$$

where the fertilisation rate is now μ. But of course θ is itself a random variable, with pdf given by Equation (2.3), when $\beta = \lambda$. To obtain the unconditional form for $p_n(t)$ we therefore integrate the above expression with respect to this density, to obtain the solution

$$p_n(t) = \int_0^t \frac{e^{-\mu(t-\theta)}\{\mu(t-\theta)\}^{n-1}}{(n-1)!} \; \lambda e^{-\lambda\theta} d\theta$$

$$= \frac{\lambda\mu^{n-1}e^{-\mu t}}{(n-1)!} \int_0^t e^{(\mu-\lambda)\theta}(t-\theta)^{n-1} d\theta \qquad (n \geq 1). \qquad (2.4)$$

The problem now is to choose λ and μ to fit this model to the data. Once again the method of maximum likelihood may be used. Explicit expressions for $\hat{\lambda}$ and $\hat{\mu}$ are not available, and numerical optimisation gives the point estimates, $\hat{\lambda} = 0.0287$, $\hat{\mu} = 0.0022$.

As could be anticipated from the way that we have adapted the model, $\hat{\lambda}$ has now increased, in fact to three times its previous value, and $\hat{\mu}$ is much smaller (less than one-tenth of $\hat{\lambda}$). The (rounded) expected values for the Poisson process with two rates are shown in Table 2.3(b).

This model provides a much better description of the data, though there still remain obvious discrepancies. In Morgan (1975) these discrepancies are reduced by allowing λ and μ to be functions of time. Further investigation of the underlying biology reveals, however, that while the biologists anticipate a reduction in fertilisation rate, from λ to μ, they also expect the secondary fertilisations (by which we mean those at rate μ) to last only for some *limited* time, τ, say, after the time of the first fertilisation. A 'fertilisation membrane,' detectable under the microscope, is then formed which bars all further fertilisations.

2.4.3 *Including a fertilisation membrane*

After the first fertilisation has taken place, the time-window for subsequent fertilisations, τ, is likely to be a random variable. However we shall simply assume here that it takes some fixed value, which is constant for all eggs. (There are biological difficulties in determining the time of the first fertilisation for any egg, and, consequently, in finding a precise value for τ.) Under these conditions, we have

$$p_0(t) = e^{-\lambda t},$$

as before.

If $t \leq \tau$ then the values for $\{p_n(t)\}$ remain as in Equation (2.4), since the membrane has not had time to operate. But if $t \geq \tau$ then one of two events may occur:

(i) Either the first fertilisation takes place before time $(t - \tau)$, which from expression (2.3) has the probability

$$\int_0^{t-\tau} \lambda e^{-\lambda x} dx = 1 - e^{-\lambda(t-\tau)},$$

and then $p_n(t)$ is just the probability of $(n-1)$ fertilisations in the time τ, i.e.,

$$p_n(t) = e^{-\mu\tau}(\mu\tau)^{n-1}/(n-1)! \quad \text{for } n = 1, 2, \ldots, \text{ by Equation (2.2)}.$$

(ii) Alternatively, the first fertilisation takes place after time $(t - \tau)$, at some time θ, say, leaving just a time $(t - \theta)$ for further fertilisations, exactly as in the previous model.

Taking account of the time of the first fertilisation, we find:

$$p_n(t) = (1 - e^{-\lambda(t-\tau)})\frac{e^{-\mu\tau}(\mu\tau)^{n-1}}{(n-1)!} + \frac{\lambda\mu^{n-1}}{(n-1)!}e^{-\mu t}\int_{t-\tau}^t e^{(\mu-\lambda)\theta}(t-\theta)^{n-1}d\theta,$$

for $n = 1, 2, \ldots,$ and for $0 < \tau \leq t$.

A useful check that this expression is correct is that if we set $\tau = t$, we return to the earlier model.

This model was fitted to the data for selected values of τ. Interestingly, the values taken for τ had little effect on $\hat{\lambda}$, and also little effect on the expected cell values; their effect on $\hat{\mu}$ was appreciable, however, as the following table indicates.

τ (seconds)	$\hat{\lambda}$	$\hat{\mu}$
15	0.026	0.0140
40	0.029	0.0064

For $\tau = 15$ seconds, the (rounded) expected values for this final model with a fertilisation membrane are given in the Table 2.3(c). The fit to the data is clearly further improved, though some deficiencies remain, mainly in the data of the second experiment.

Further investigation of these and other models is described by Morgan (1982), both for additional experiments of the type considered here, and for other experiments in which nicotine, which is thought to increase μ, is present. For the standard experiments, biologists believed that the ratio of $\lambda : \mu$ was in the region of 10 : 1, as we found earlier for the model with two rates but no fertilisation membrane. However, the results above suggest that this ratio might well be only 4:1 or even 2:1. The effect of a fertilisation membrane is to confine the secondary fertilisation to a limited time, with the obvious result that μ must be larger than if no such membrane were present. This work shows the power of a statistical model in both predicting and *quantifying* the consequences of a simple biological mechanism.

2.5 Which model?

Moving from an atypical data set to a simple summary, by fitting a suitable stochastic model, involves several stages. While we may acquire an extensive toolbox for the later stages, and the objective of later chapters is to describe such material, this is of little use if we do not know how to start building a model.

In this chapter we have considered two stochastic models. The first was of waiting times, in discrete time. The obvious first model in such a case is geometric. We found that the geometric model needed elaboration, to describe overdispersion, and we shall extend the model further still in Chapter 4. The second model of this chapter was for counts in time. The obvious starting point was the Poisson process. This, too, was found to be too simple, in this case due to underdispersion, and a better model resulted from trying to model the real-life situation more closely. Further extensions of this model will also be considered in Chapter 4. For models to be realistic, it is often necessary for them to describe heterogeneity resulting from differences between individuals. One way of doing this was encountered in Section 2.3; see also Exercise 2.13. Another is to introduce covariates which describe individual differences. For further discussion, see Morgan and Ridout (2008a,b).

In the two examples of this chapter, we initially tried simple models that are appropriate for the situations considered. General guidelines can be obtained from studies of stochastic processes (see, e.g., Cox and Miller, 1965). The models were then made more complex in order to provide a better description of the data. Consideration of how a model matches the data can lead us to produce modified models which are then, in turn, fitted to the data by maximum likelihood, a process called *model criticism*. We need to take care that we do not *overfit* the data, by including too many parameters, and thereby, to some extent, diminish the value of the model, which is in simplification. Our aim in modelling is to provide a concise and useful summary of the data. Ideally the modelling process proceeds in consultation with scientists who would provide guidance and advice.

One good example of this is the area of modelling data such as those of Table 1.5, describing the annual mortality of British grey herons. Throughout this book we shall present a number of illustrative models for the survival of wild animals. These models are only of any real scientific value when they are constructed in close collaboration with ecologists. For instance, young animals typically have a lower annual survival probability than older animals, and this feature should be incorporated into appropriate ecological models. Models involving age-dependent annual survival probabilities of wild animals are considered in the examples in Chapter 4.

The data sets presented in this book have either been analysed in discussion with the scientists who produced the data, or they are 'historical,' in the sense that they have been taken from published sources without reference to the full experimental context. Historical data sets often provide useful benchmarks, but greater satisfaction is to be derived from working with current data, and with scientists who can readily advise on whether or not certain models are appropriate — see for example Exercise 2.15 — and respond to efficiently designed proposals for the collection of data.

2.6 What is a model for?

Within the confines of this book we shall be fitting models to single, sometimes small sets of data. In an ideal world, statisticians work closely with scientists who can sometimes supply additional data if necessary, and in some cases also a range of replicated data sets. A good test of a model is whether it is useful for new data, and not just the data which gave rise to it in the first place. To give three examples from this chapter and from Chapter 1: the models developed for the fecundability data of Table 1.1 were also found to be useful for the independent data of Exercise 2.2 and to some extent for the somewhat different data of Exercise 2.1; the polyspermy data of Table 1.3 are just one example of many obtained by Professor P.F. Baker, and further examples are given in Exercise 2.15; the data of Exercise 2.14, provided by Dr. D. Taylor, are replicated for several years — see for example Ridout et al. (1999). Section 6 of Chatfield (1995) is entitled, 'Collecting more data,' and provides a comprehensive discussion of this issue.

2.7 *Mechanistic models

In Exercise 3.7 a particular stochastic process is suggested for the microbial infection data of Example 1.2. In this case, a model for the small-scale mechanism of infection can be fitted to large-scale date in order to try to estimate the parameters of the small-scale process. As discussed by Morgan and Watts (1980), alternative plausible mechanisms can also be proposed. In this case mechanistic modelling is a somewhat academic exercise unless good scientific arguments can be made in favour of any particular mechanism. The same point is made by Link (2003) in the context of estimating the size of wild animal populations. He shows that different models that fit the data equally well can make different predictions. The process of infection is further complicated if the host can die and release large numbers of infecting organisms which can then infect further hosts. A particular form of biological control, which seeks to kill pests using natural pathogens such as viruses and bacteria, rather than by using pesticides (as in Example 1.4), operates in this way. For further discussion and examples of fitting mechanistic models of biological control, see Terrill et al. (1998).

Later, in Example 4.8, a model making use of ideas from the area of survival analysis is used to describe the data of Example 1.4. This model is more *empirical*, rather than mechanistic, but, as we shall see in Chapter 4, it provides a very convenient framework for statistical inference. In some cases the choice between empirical and mechanistic models is simple to make. For instance, a bird survival model for data like those in Table 1.5 is bound to include parameters which are probabilities of annual survival, so that we would regard the modelling as mechanistic. In other cases, such as the flour-beetle data of Example 1.4, both types of model may be used. Laurence and Morgan (1989) discuss the relative merits of mechanistic and empirical modelling for the flour beetle data. They show that while it is possible to specify an attractive model of the underlying mechanism, the model makes assumptions which cannot be tested using available data, resulting effectively in over-parameterisation: the model contains more parameters than can be precisely estimated from the data. This is a general problem which is discussed in detail in Section 5.2. The following example provides a further illustration of the failure of a mechanistic model.

Example 2.1: The patch-gap model (Jarrett and Morgan, 1984)

Pielou (1963a) proposed an attractive, *patch-gap* model for describing the incidence of diseased trees in a forest. It is assumed that the forest consists of patches, within which each tree has probability p of having the disease, independently of all other trees, and gaps, in which no trees have the disease. It is supposed that a proportion ϕ of the forest is in patches, and that circular quadrats laid down at random on the forest are sufficiently small that they do not hit a patch boundary.

The likelihood can then be written as:

$$\text{L} \propto \prod_{i=1}^{\infty} \left\{ (1 - \phi + \phi q^i)^{n_{0i}} \phi^{N_i - n_{0i}} \prod_{j=1}^{i} (q^{i-j} p^j)^{n_{ji}} \right\} \tag{2.5},$$

in which $q = 1 - p$, N_i is the number of quadrats containing i trees and n_{ji} is the number containing i trees, j of which are diseased. For the data of Table 1.7, the likelihood is a simple function of the two parameters, p and ϕ, and maximum-likelihood estimates, together with estimates of error, are $\hat{\phi} = 0.613$ (0.0356) and $\hat{p} = 0.329$ (0.0196). The model fits the data well, with a Pearson X^2 value of $X^2 = 7.55$, which is satisfactory when referred to χ^2_{13} tables – the 5% point is $\chi^2_{13;5} = 22.36$. The case $\phi = 1$ corresponds to a homogeneous forest, in which case $\hat{p} = 0.2041$. However we then have $X^2 = 59.55$, indicative of significant clumping of the disease when referred to χ^2_{14} tables – the 0.05% point is $\chi^2_{14;0.05} = 38.11$.

What is wrong with this model is that the assumption of non-overlap of quadrats with patch boundaries is likely to be violated for the physical area surveyed to produce the data of Table 1.7. Jarrett and Morgan (1984) conclude that

> Although a model seems quite appropriate at first sight, produces an excellent fit to the data and results in precise parameter estimators, it could be based on an incorrect assumption.

□

The illustration of Example 2.1 is extreme, but it demonstrates the care that needs to be exercised in constructing models. Necessarily, all models make assumptions. The challenge lies in judging which ones can be safely made so that the models remain useful.

2.8 Discussion

In this chapter we have discussed basic aspects of modelling, and introduced the general tool of the method of maximum likelihood for parameter estimation. The idea of estimating parameters by maximising a likelihood was proposed by Fisher (1925); see also Aldrich (1997) and Stigler (2007) for historical background. The construction of likelihood functions is usually straightforward, but it still requires a certain amount of practice. Students who encounter likelihood for the first time frequently find it hard to construct a likelihood for a general sample, rather than a particular set of data. Thus the concepts of likelihood and of maximum likelihood require thought and study. Relevant exercises are 2.3, 2.15, 2.21, and 2.23. The likelihood combines the model with the data; any attempted likelihood construction that does not contain the data is wrong! The validity of the approach when one has a sample comprising both censored and non-censored data is explained in Lawless (2002), which is important for the analysis of survival data in general. At an elementary level, the justification of the method of maximum likelihood flows from observing, as in Figure 2.1, how well the method can work in practice. We defer until later further discussion of properties of the method, and of how we construct estimates of standard error, and of confidence intervals.

The method of maximum likelihood frequently requires non-linear function optimisation. We have seen this demonstrated in the two main examples of this chapter, once we introduced elements of complexity and reality into the models. Model-fitting then becomes dependent on methods for function optimisation, and for this reason we devote the next chapter to this important topic. It is at this point that we shall need to consider an appropriate computer language to use, since in practice function optimisation can only take place with the aid of a computer.

2.9 Exercises

2.1 In a study of recently qualified drivers, the following data were obtained on the number of tests taken before success, classified by whether the drivers were taught by a driving school or not.

Number of tests	Other	Driving school
1	99	71
2	39	39
3	17	14
4	13	10
5	6	7
6	6	3
7	5	0
8	1	0
> 8	14	3
	200	147

A possible model for these data is a geometric distribution with a constant probability p of success, that is,

$$pr(X = x) = (1 - p)^{x-1}p \qquad x = 1, 2, \ldots,$$

where X is the number of tests taken. Verify that

$$Pr(X > r) = (1 - p)^r \qquad r = 0, 1, 2, \ldots .$$

1. Obtain an estimate of p for each group based on N_1, the number of observations for which $X = 1$. What is the distribution of N_1? Use this information to approximate standard errors for your estimates. What can you say about the success rates for the two groups of drivers?

2. Calculate the maximum-likelihood estimates for the two groups. Consider how you would obtain standard errors for these estimates. Using the maximum-likelihood estimate, perform a Pearson X^2 goodness-of-fit test for the driving school group, and comment on the result. What other information might it have been useful to collect in the survey?

2.2 The data below are taken from Harlap and Baras (1984). Here we have cycles to conception, following the cessation of contraception use, which was either use of the contraceptive pill or some other method. Discuss the data, and consider suitable stochastic models. Write down a likelihood.

Cycle	Pill users	Non-pill users
1	383	1,674
2	267	790
3	209	480
4	86	206
5	49	108
6	122	263
7	23	54
8	30	56
9	14	21
10	11	33
11	2	8
12	43	130
>12	35	191
Total	1,274	4,014

2.3 For the flour beetle data of Table 1.4, we may focus on the endpoint mortality data, corresponding to the observations on day 13. A *logistic* or *logit* analysis of these data is based on the model,

$$\Pr(\text{death by day 13 on dose } d_i) = [1 + \exp\{-(\alpha_{1s} + \alpha_{2s}\log d_i)\}]^{-1},$$

where $s = 1, 2$ is the index used to indicate sex. Draw appropriate graphs to investigate whether this may be a suitable way to describe the endpoint mortality data. Comment on the differences observed between the responses of the two sexes. Write down the likelihood.

2.4 Many species of insect parasite lay their eggs in live hosts, and try to do this to avoid *superparasitism*, that is, to try to avoid parasitising an already parasitised host. Consider how you would model the following data, from Simmonds (1956), which relate to the parasitic behaviour of *Spalangia drosophilae*:

Number of eggs laid	Number of hosts observed
0	19
1	169
2	50
3	6
4	1
5	1
>5	0
Total	246

Write down a likelihood.

2.5 For any pair of plants at any census point, Catchpole et al. (1987) model the ant-guard data of Table 1.10 by means of the trinomial distribution

$$\mathrm{pr}(n_1, n_2, n_3) = \left(\begin{array}{ccc} n_1 + n_2 + n_3 \\ n_1, \quad n_2, \quad n_3 \end{array} \right) \frac{(m_1\alpha)^{n_1}(m_2\beta)^{n_2}(m_3\gamma)^{n_3}}{(m_1\alpha + m_2\beta + m_3\gamma)^{n_1+n_2+n_3}},$$

for parameters α, β, γ, subject to the constraint, $\alpha + \beta + \gamma = 1$, where m_1 is the number of capitula on the ant access plant with ants, and n_1 insects in all; m_2 is the number of capitula on the ant-access plant without ants, and n_2 insects in all; m_3 is the number of capitula on the ant-exclusion plant with n_3 insects in all. Discuss the roles played by the parameters, α, β and γ, and whether they might be expected to be constant, or to vary. If they might vary, discuss which one(s) and how this might occur. Write down an expression for the likelihood.

2.6 The data below provide the numbers of responses a set of human subjects made when hearing the consonant phonemes, p, t and k (Clarke, 1957). Write down a probability model for the data, and try to fit the model to the data using maximum likelihood.

		Response		
		p	t	k
	p	440	179	177
Stimulus	t	290	259	251
	k	188	344	268

2.7 Derive the probability function of the beta-binomial distribution. An urn contains r red and b black balls. Balls are removed at random, one at a time, and each is replaced, together with an extra ball of the same colour as the ball drawn. Show that the number of red balls drawn after n drawings in all has a beta-binomial distribution.

2.8 Verify that the value of $\hat{p}$ in Equation (2.1) maximises the relevant likelihood. Note that the problem here is equivalent to finding the maximum-likelihood estimate of the binomial probability p when r successes occur out of n independent Bernouilli trials. Why is this?

2.9 In this question we consider maximum-likelihood estimation for the polyspermy example under the single Poisson process model. Suppose that we write the log-likelihood as:

$$l(\lambda) = C - \lambda A + B(\log \lambda)$$

for positive constants A, B and C. Obtain the value of λ that results in $\frac{dl}{d\lambda} = 0$, and show that this corresponds to a maximum.

2.10* Verify the results of Section 2.3, that under the beta-geometric model, the expected remaining waiting time for couples who have had j failed cycles is given by:

$$\mathbb{E}[(X - j)|X > j] = \left(\frac{1 - \theta}{\mu - \theta}\right) + \frac{j\theta}{(\mu - \theta)} \ .$$

Hint: First of all show that under the beta-geometric model, conditional upon the event $X > j$, the distribution of the parameter p is $Be\left(\frac{\mu}{\theta}, \frac{1 - \mu}{\theta} + j\right)$.

2.11   Verify the expressions for the mean and variance of the beta-geometric distribution, given in Section 2.3, using the following expressions for the conditional mean and variance of random variables:

$$\begin{aligned} \mathbb{E}[X] &= \mathbb{E}_p\left[\mathbb{E}[X|p]\right] \ , \text{ and} \\ \text{Var}(X) &= \mathbb{E}_p[\text{Var}\,(X|p)] + \text{Var}_p\left[\mathbb{E}[X|p]\right] \ , \end{aligned}$$

when the subscript p denotes the random variable that is used in forming the expectation and variance.

2.12   Suppose that the random variable $X \sim N(0, \theta^{-1})$, and that $\theta \sim \Gamma(\frac{1}{2}, \frac{2}{\gamma})$, where γ is a positive integer. Show that the marginal distribution of X is the t-distribution, t_γ.

2.13   Suppose that the random variable $X \sim P_o(\lambda)$ and that $\lambda \sim \Gamma(\alpha, \beta)$, where α is a positive integer. Show that the marginal distribution of X has the negative-binomial form:

$$pr(X = k) = \binom{k + \alpha - 1}{\alpha - 1} \frac{1}{(1 + \beta)^\alpha} \frac{\beta^k}{(1 + \beta)^k} \ , \quad k \geq 0 \ .$$

2.14   (Ridout et al., 1999.) Figure 2.3 shows the topological branching structure of a typical strawberry inflorescence. Consider how you would form a probability model for the data below. The data describe the numbers of inflorescences having a given number of flowers of rank r classified by the number of flowers at the previous rank (m), known as 'parents,' for one year of study.

Rank (r)	Number of parents (m)	Number of flowers of rank r								
		0	1	2	3	4	5	6	7	8
3	2	0	0	37	104	222	12	3	1	
4	2	3	6	22	9	3	1			
4	3	0	6	22	35	34	7	2		
4	4	0	3	13	43	92	38	25	7	2

2.15   (Morgan, 1982.) Two further examples of data resulting from studies of polyspermy are shown below. In (a) the eggs were sea urchins *Echinus esculentus*, with a sperm density of 6×10^7 sperm/ml at $15°C$. In (b) the

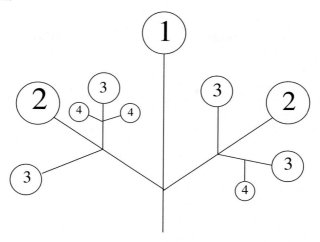

Figure 2.3 *The branching structure of a typical strawberry inflorescence. The numbers denote the ranks of the flowers. Figure reproduced with permission from the Royal Statistical Society.*

only difference is that a dilute solution (approximately 1mM) of nicotine was present.

(a)

Length of experiment	Number of fertilisations			
(seconds)	0	1	2	3
5	97	3		
8	83	16	1	
15	55	40	3	2
40	32	58	9	1
90	5	73	18	4

(b)

Length of experiment	Number of fertilisations							
(seconds)	0	1	2	3	4	5	6	7
5	96	4						
12	82	17	1					
20	33	56	9	2				
40	17	68	12	2	1			
80	9	56	25	7	2	1		
180	1	59	27	9	2	1		1

(i) Fit a single Poisson process model to these data by the method of maximum likelihood, and examine the goodness of fit of the model using a Pearson X^2 test.

(ii) An alternative mechanism for the process of fertilisation is size-limited with

$$ p_n(t) = \binom{m}{n} \exp(-\lambda tn)\{1 - \exp(-\lambda t)\}^{m-n} , \quad \text{for} \quad 0 \le n \le m , $$

where $p_n(t)$ is the probability that at time t after the start of the experiment an egg has n sperm. In this case, there is an upper bound, m, on the number of fertilisations that can take place. Consider what underlying mechanism results in such a model. How might you build a fertilisation membrane into the model? Write down the likelihood corresponding to the data from an experiment of one duration only, and obtain the maximum-likelihood estimator of the single parameter λ in this case. Examine the goodness of fit of this model, and make comparisons with the model of (a).

Note that Professor Baker, who provided the data of this question, favoured the model of (i), on biological grounds.

2.16   Explain why the multinomial distribution is appropriate in Section 2.2.

2.17   For a random sample, $(u_1, \ldots, u_n)$ from a $U(0, \theta)$ distribution, find the maximum-likelihood estimate of θ.

2.18   In Section 2.3, two alternative expressions are given for the probability function of the beta-geometric distribution. Verify that they are equivalent.

2.19   The number of male rabbits in 100 litters, each containing 3 rabbits, are described in the table below:

Number of males in litter	0	1	2	3
Number of litters	25	39	27	9

Consider how you would model these data, and construct a likelihood.

2.20   A study of the mortality of grey herons involves the release of a cohort of birds marked soon after hatching in a single year. The probabilities of birds being reported dead in each of the following four years are *partially* shown below in temporal order, for a model with a constant reporting probability λ, a probability ϕ_1 that any bird survives its first year of life and a probability ϕ_a that any bird survives any later year of life

$$ (1 - \phi_1)\lambda \qquad \phi_1(1 - \phi_a) * \qquad \phi_1\phi_a(1 - *) * \qquad \phi_1\phi_a * (1 - \phi_a)\lambda. $$

(i) Complete the last three expressions, by replacing the $*$'s by the appropriate probabilities. Show that the probability of a bird not being found during these four years is given by $1 - \lambda(1 - \phi_1\phi_a^3)$.

(ii) The data below relate to a cohort of grey herons marked in Britain in 1957, and form a small subset of the data in Table 1.5.

Year of reported death

1957	1958	1959	1960
27	10	1	3

Derive the conditional multinomial distribution for these data, which may be used to estimate model parameters, and comment on the number of parameters that may be estimated.

(iii) A more complex model has a conditional probability ϕ_2 that a bird survives its second year of life. Explain why this model is saturated – that is, the model fits the data exactly, and there are no degrees of freedom to test for goodness-of-fit.

2.21   When introduced into an area containing both fine and coarse sand, ant-lions are thought to prefer to dig burrows in fine sand, but also to avoid other ant-lions. The data below describe the results of 62 experiments in which either 3 or 4 ant-lions are introduced into an area with fine and coarse sand, and form a subset of the data in Table 1.10. Explain how you would describe the data by means of a binomial model with probability p of burrowing in fine sand.

Number of ant-lions introduced	Total number of experiments	Number of ant-lions burrowing in fine sand				
		0	1	2	3	4
3	32	0	7	24	1	
4	30	0	3	17	10	0

Obtain the maximum-likelihood estimate $\widehat{p}$, and use it to form expected values corresponding to the observed values in the above table.

Without formally computing a goodness-of-fit statistic, discuss the goodness of fit of the binomial model to the data by visually comparing observed and expected cell numbers in the above table.

2.22   (Continuation of Exercise 2.4.) A species of wasp lays its eggs on larvae. Suppose that any larva has X encounters with wasps where X has the Poisson distribution,

$$pr(X = k) = \frac{e^{-\lambda}\lambda^k}{k!}, \quad \text{for} \quad k = 0, 1, 2, \ldots.$$

Suppose that an egg is always laid at the first encounter, but that at all subsequent encounters, single eggs are laid independently with probability $\delta < 1$ at each encounter. If p_r denotes the probability that a larva receives r eggs, $(r = 1, 2, \ldots)$ then

$$p_r = \delta^{r-1}(\delta - 1)^{-r}e^{-\lambda\delta} \sum_{i=r}^{\infty} \frac{\lambda^i(\delta - 1)^i}{i!}.$$

Verify this for $r = 1$ and $r = 2$.

2.23   For the period 1920–1979, lengths of 'very warm spells' (periods of three or more consecutive days with maximum temperature more than $4°C$ above the long-term mean) have been recorded at Edgbaston, Birmingham. The results are as follows:

Length of spell (days):	3	4	5	6	7	8	9	10	11	12	> 12
Number of spells:	149	78	49	20	17	7	4	2	4	3	1

The total number of warm spells is 334, and the single spell of > 12 days actually lasted 17 days.

(i)   Explain the assumptions you have to make in order to model warm spell data using a geometric random variable X, with probability function,

$$pr(X = j) = p(1 - p)^j, \quad j = 0, 1, 2, \ldots, \quad 0 < p < 1.$$

(ii)   Define Y as the random variable that results when X is truncated, so that only values of $X \geq 3$ are recorded. Show that Y has probability function,

$$pr(Y = j) = p(1 - p)^{j-3}, \quad j = 3, 4, \ldots.$$

(iii)   By fitting the probability function of Y to the warm spell data, obtain the maximum-likelihood estimate of p.

2.24   The data below describe the mortality of adult flour beetles (*Tribolium confusum*) after 5 hours' exposure to gaseous carbon disulphide (CS_2); from Bliss (1935).

Dose(mg/l)	49.06	52.99	56.91	60.84	64.76	68.69	72.61	76.54
No. of beetles	59	60	62	56	63	59	62	60
No. killed	6	13	18	28	52	53	61	59

Plot estimated logits vs dose levels, and use this plot to discuss whether you think the logit model is appropriate for the data. Note that $\text{logit}(p) = \log\{p/(1 - p)\}$.

The ED_{50}, μ, for an experiment such as that above is defined as that dose corresponding to a 50% response. Provide a rough estimate of μ from the data.

2.25   The data below describe the onset of menopause for women classified according to whether or not they smoke (Healy, 1988, p.85.)

Age(years)	Non-smokers		Smokers	
	Total	Number menopausal	Total	Number menopausal
45 – 46	67	1	37	1
47 – 48	44	5	29	5
49 – 50	66	15	36	13
51 – 52	73	33	43	26
53 –	52	37	28	25

A statistical analysis involving logistic modelling concludes that *parallelism* is acceptable, and that the age-shift for the same expected proportion of menopausal women is 1.23 years. Explain what you think is meant by parallelism. Provide in outline only a suitable graphical analysis of these data which would allow you to draw these conclusions.

2.26 On each of k occasions, a closed population of wild animals, of fixed but unknown size N, is sampled at random and any animals in the sample that have not been marked previously are marked and returned to the population. The objective is to estimate N.

The general form for the likelihood is

$$L(N, \boldsymbol{\eta}) \propto \binom{N}{D} \prod_{j=0}^{k} p_j^{f_j}, \qquad (2.1)$$

where f_j denotes the number of distinct animals that have been captured j times and D is the number of distinct animals caught, given by $D = \sum_{j=1}^{k} f_j$, so that $f_0 = N - D$. The probability p_j denotes the probability that an animal is caught j times out of the k occasions, and the vector $\boldsymbol{\eta}$ denotes the set of model parameters, excluding N.

1. If all animals share the same recapture probability p at each occasion, events at sampling occasions are independent and animals are caught independently of one another, write down an expression for p_j.
2. An alternative model allows the recapture probability to vary between animals. In this case the probability of an animal being caught j times has the expression given below.

$$p_j \propto \frac{\prod_{r=0}^{j-1}(\mu + r\theta) \prod_{r=0}^{k-j-1}(1 - \mu + r\theta)}{\prod_{r=0}^{k-1}(1 + r\theta)}.$$

Explain how you think this expression is derived, and provide an interpretation of the two model parameters.
3. The expression for the likelihood can be factorised as follows:

$$L(N, \boldsymbol{\eta}) \propto \left\{ \binom{N}{D} p_0^{N-D}(1 - p_0)^D \right\} \left\{ \prod_{j=1}^{k} \left(\frac{p_j}{1 - p_0} \right)^{f_j} \right\}.$$

By maximising the second part of the likelihood alone, in terms of the $\{p_j\}$, it is then possible to substitute the resulting estimate of p_0 in the first part of the likelihood and obtain the maximum-likelihood estimate of N by maximising the first part of the likelihood solely as a function of N. The table below compares the estimates of N obtained this way (denoted *Approximate*) with the exact maximum-likelihood estimates, for a range of surveys of different animals. Discuss the performance of the approximate method, and why it is not exact.

Animals	Approximate	Exact
Voles, *Microtus pennsylvanicus*	103.2	102.7
Chipmunks, *Tamias striatus*	74.1	73.5
Hares, *Lepus americanus*	75.4	74.7
Skinks, *Oligosoma lineoocellatum*	171.3	170.7
Rabbits, *Sylvilagus floridamus*	97.2	96.3

For more discussion of this area of modelling, see Morgan and Ridout (2008a,b) and Fewster and Jupp (2008).

2.27   Suggest a suitable model for the cell division time data of Table 1.15.

CHAPTER 3

Function Optimisation

3.1 Introduction

It is clear from Chapter 2 that much model fitting involves numerical function optimisation methods, which proceed by a succession of iterative steps. This is true for the method of maximum likelihood, but also necessary for other methods that we shall encounter later, such as the method of least squares, and methods based on empirical transforms. For simplicity, in this chapter we shall consider maximising likelihoods, but the ideas are of general application.

If a model only has a small number of parameters then contour plots of the log-likelihood surface can be very revealing. Figure 3.1 shows the log-likelihood contours for the beta-geometric model of Chapter 2, fitted to the fecundability data of Table 1.1. This amplifies features already present in Figure 2.1 and provides a reassuring check of the point estimates. We can also readily obtain confidence regions from such plots, as will be described in Chapter 4.

Graphical methods are also valuable, through profile log-likelihoods for example (a topic which is also covered in Chapter 4), when the number of parameters is large, and may, for instance, reveal useful information when parameters are estimated at boundaries of the parameter space.

For many years the needs of the statistician wishing to maximise a likelihood were in principle satisfied by the numerical analyst, and complex multi-parameter likelihoods were handed over to black-box *deterministic* optimisation routines. We describe these routines as deterministic because they proceed in a predetermined manner, are totally predictable and exclude any random features. Many examples of these are provided by the NAG (2006) (Numerical Algorithms Group) library of FORTRAN routines. Using deterministic optimisation routines is still the standard approach, though frequently routines which involve iteration fail to converge. Much time can be spent by statisticians struggling with imperfectly understood numerical procedures. This is the source of much frustration, and can lead to promising directions for research being abandoned. Even when an optimum is obtained, there is no guarantee that it is a global optimum. To provide reassurance on this point when a deterministic search method is used, the method should be run from a variety of different starting points, a practice which can be quite time-consuming.

Stochastic search methods now also exist, which provide a far more flexible approach to maximising multi-dimensional surfaces. We shall, in this chapter, describe features of both deterministic and stochastic search methods. Their understanding is crucial to successful applied stochastic modelling. In the next section we shall introduce MATLAB and meet simple but useful numerical procedures which may be used in deterministic optimisation.

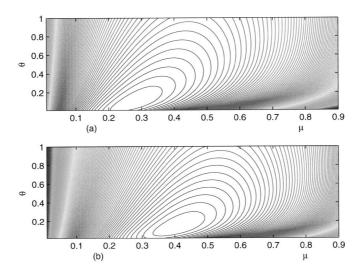

Figure 3.1 *Log-likelihood contours when the beta-geometric model is fitted to the data for (a) smokers and (b) non-smokers from Table 1.1.*

3.2 Matlab; graphs and finite differences

The graphs of Figures 2.1 and 3.1 are easily produced from programs written in MATLAB. Some of the basic features of MATLAB are described in Appendix B. It provides an integrated programming environment that is well suited to the needs of statisticians who are fitting stochastic models by maximum likelihood. It is not necessary to understand fully the MATLAB programs that occur in this book; they are provided as an aid to understanding the theory. The programs needed to produce Figures 2.1 and 3.1(a) are given in Figure 3.2. It is especially attractive to see how simply the likelihoods are formed in Figure 3.2(a); similar examples are to be found later in the book. In outline here, in Figure 3.2(a), the variable p is a vector, which provides the values at which the function will be evaluated, prior to plotting. A summation such as $\sum(i-1)n_i$ from Section 2.2 is then formed simply by means of the vector multiplication: s2 = int*data1. In Figure 3.2(b), the contour command plots the log-likelihood over the gird of values that has been established by the meshgrid command. Here we operate with matrices, rather than vectors, of parameter values, and this requires the elementwise operations described in Appendix A. The end result is succinct.

(a)
```
% Program to produce Figure 2.1
%------------------------------------------------------------------
data1=[29 16 17 4 3 9 4 5 1 1 1 3 7]; s1=sum(data1(1:12));
data2=[198 107 55 38 18 22 7 9 5 3 6 6 12]; s3=sum(data2(1:12));
int=0:12;                 % produces the vector, int=[0,1,...,12];
p=0.01:0.01:.99;          % produces the vector,
s2=int*data1';            % p=[0.01, 0.02, ..., 0.99];
s4=int*data2';            % sums are formed without using loops
                          % now we form the two log-likelihoods
loglik1=s2*log(1-p)+log(p)*s1;
loglik2=s4*log(1-p)+log(p)*s3;
plot(p,loglik1, '-',p,loglik2,'--')
xlabel p; ylabel loglikelihood;
```

(b)
```
% Program to produce Figure 3.1(a)
% Note that here we use the MATLAB elementwise
% operations, '.*' and './'
%------------------------------------------------------------
data=[29 16 17 4 3 9 4 5 1 1 1 3 7];
[mu,theta]=meshgrid(0.01:0.01:.9, 0.01:0.01:1.0);
loglik=log(mu)*data(1);        % mu and theta are now matrices
p=mu;                          % of the same dimensions : the
s=p;                           % rows of m are copies of the
for i=2:12                     % vector (0.1, 0.11,...,0.9), and
  p=p.*(1-(mu+theta)./(1+(i-1)*theta));
                               % the columns of theta are copies
  s=s+p;                       % of the vector (0, 0.1,...,1)'.
  loglik=loglik+log(p)*data(i); % recursive specification of the
end                            % beta-geometric probabilities
p=ones(size(mu))-s;            % from Section 2.3.
loglik=loglik+log(p)*data(13); % completes log-likelihood
contour(0.01:0.01:.9, 0.01:0.01:1.0,loglik,20);
xlabel \mu; ylabel \theta;
```

Figure 3.2 MATLAB *programs to produce Figures 2.1 and 3.1(a).*

Some of the deterministic search methods of this chapter make use of first- and second-order partial derivatives of a log-likelihood function, with respect to its parameters. It is therefore very useful to have numerical procedures to approximate these derivatives. We give simple MATLAB routines in Figure 3.3, kindly provided by Ted Catchpole. What the routines do is use simple differences, based on an interval of length 2δ, to approximate the appropriate differential. Thus for example, for a function $f(\theta)$, we form the approximation,

$$\frac{df}{d\theta} \approx \frac{f(\theta + \delta) - f(\theta - \delta)}{2\delta} \ .$$

The same approach may be applied to approximating derivatives of higher order, though we only require first-order and second-order partial derivatives. The matrix of second-order derivatives with respect to the parameters is called the *Hessian* matrix, and this plays an important rôle in the estimation of standard errors, as we shall see in the next chapter. The order of error in these simple routines, resulting from using central differences and an interval of length δ, is δ^2. Better routines can be constructed, for example, with error of order δ^4 — see Exercise 3.1. Additionally, if parameters differ appreciably in size, an improvement would be to make step sizes proportional to the values of individual parameters. Care must always be exercised when using these numerical methods, because of the possibility of round-off error (see, e.g., Lindfield and Penny, 1995, §4.2), but they are useful general-purpose tools. Ridout (2009) investigates the complex-step method, which avoids round-off error in numerical approximation to the first-derivative; see Lai and Crassidis (2008). More sophisticated numerical derivatives which are available in certain computer packages have, on occasion, been found to perform less well than the simple routines of Figure 3.3. An alternative approach is to use a computer algebra package such as Maple to obtain explicit formulae for derivatives — see Currie (1995).

If these programs are applied to a function which uses random numbers, it is important that the function sets the seed of the random number generator, so that each time the function is called it uses the same stream of random numbers — see Chapter 6.

A user of finite-difference methods to estimate derivatives needs to specify the value δ, used in the approximation. It is not possible to give a universal specification of δ which would work well in all instances. In practice we would normally try two or three alternatives; usually small changes in δ make little or no difference to the approximation. We shall encounter several varied instances later in the book where essentially we have to decide on the value of a free parameter such as δ. This occurs in kernel density estimation, simulated annealing, Gibbs sampling, etc. Quite often the end-product of an analysis can be sensitive to the value selected for a free parameter, and the statistician needs to exercise careful judgement and sometimes be prepared to produce not just one final result but a number of alternatives.

3.3 Deterministic search methods

These methods are designed to minimise functions, moving iteratively from a pre-determined starting point in the parameter space, with the objective of finding a local, ideally global, minimum. In order to obtain maximum-likelihood estimates one can apply minimising routines to minus the log-likelihood. The simplest of these methods are well described in a range of texts — see, for example, Everitt (1987), Adby and Dempster (1974), Gill et al. (1981), Bunday (1984), Ross (1990), Maindonald (1984), and Nash and

(a)
```
function g = grad(funfcn,x)
%
%GRAD Calculates a central difference approximation to the first-order
%  differential of 'funfcn' with respect to 'x'.
%  The difference width used is twice 'delta'.
%  Code due to Ted Catchpole.
%_____
delta = 10^(-6);
t = length(x);
g = zeros(t,1);
Dx = delta*eye(t);              % eye :identity matrix
for i = 1:t
   g(i) = (feval(funfcn,x+Dx(i,:)) - feval(funfcn,x-Dx(i,:)))/(2*delta);
end
% For information on feval, see Hunt et al (2006, p98).


(b)
function h = hessian(funfcn,x)
%
%HESSIAN Calculates a central difference approximation to the
%  hessian matrix of 'funfcn', with respect to 'x'
%  The width of the difference used is twice 'delta'.
%  Code due to Ted Catchpole.
%_____
delta = 10^(-6);
t = length(x);
h = zeros(t);
Dx = delta*eye(t);
for i = 1:t
  for j = 1:t
  h(i,j)=(feval(funfcn,x+Dx(i,:)+Dx(j,:))...
        - feval(funfcn,x+Dx(i,:)-Dx(j,:))...
        - feval(funfcn,x-Dx(i,:)+Dx(j,:))...
        + feval(funfcn,x-Dx(i,:)-Dx(j,:))...
         )/(4*delta^2);
  end
end
```

Figure 3.3 *General* MATLAB *routines to estimate (a) the gradient vector and (b) the Hessian of a function (funfcn). Note that these functions are defined in general for a vector x. Thus while they will approximate first-order and second-order derivatives for a scalar x, they are designed to produce all partial first-order and second-order derivatives when x is a vector. The value of the interval length, here based on $\delta = 10^{-6}$, may be changed by the user. It is explained in Appendix B how these and similar functions are used in practice. Appendix B also explains the use of the 'feval' command in* MATLAB.

Walker-Smith (1987). We shall start by considering a standard method in one dimension.

3.3.1 Newton-Raphson in one dimension

Suppose a log-likelihood $\ell(\theta)$ is a function of a scalar parameter θ. To obtain the maximum-likelihood estimate $\hat{\theta}$, we try to solve the equation, $\frac{d\ell}{d\theta} = 0$.

More generally, suppose we need to solve the equation,

$$F(\theta) = 0 , \quad \text{for some function } F(\theta) .$$

From a first-order Taylor series expansion,

$$F(\theta) \approx F(x) + (\theta - x)F'(x),$$

and so setting $F(\theta) = 0$ provides:

$$\theta \approx x - F(x)/F'(x).$$

Thus if we regard x as a guess at a solution to the equation $F(\theta) = 0$, then the guess is updated by θ as shown.

This is the basis of the Newton-Raphson iterative method for obtaining a solution to the equation $F(\theta) = 0$, viz.,

$$\theta^{(r+1)} = \theta^{(r)} - F(\theta^{(r)})/F'(\theta^{(r)}), \quad r \geq 0,$$

where $\theta^{(r)}$ is the value for θ at the rth iterate, and iterations start from $\theta^{(0)}$. There is no guarantee that the iterations converge in general, and good choice of $\theta^{(0)}$ can be important. A short MATLAB function for the Newton-Raphson method is given by Lindfield and Penny (1995, p.95). For finding a stationary value of a log-likelihood, ℓ, we have $F(\theta) = \dfrac{d\ell}{d\theta}$, and so the iteration is:

$$\theta^{(r+1)} = \theta^{(r)} - \frac{d\ell}{d\theta}(\theta^{(r)}) \bigg/ \frac{d^2\ell}{d\theta^2}(\theta^{(r)}) . \tag{3.1}$$

As well as resulting in the Newton-Raphson method, first-order Taylor series expansions underpin much of classical statistical inference, as we shall see in Chapter 4.

3.3.2 Line searches and the method of steepest ascent

Usually we are interested in maximising a function of several variables, but this can involve a sequence of *line searches*, each one requiring optimisation in a particular direction, which is equivalent to optimising with respect to a single variable. It is useful now to establish a two-parameter likelihood, which we shall use several times in this chapter for illustration; in this case there is not usually an explicit expression for the maximum-likelihood estimate.

Example 3.1: The two-parameter Cauchy likelihood

Consider a random sample, $x_1, \ldots, x_n$, from the Cauchy pdf:

$$f(x) = \frac{\beta}{\pi\{\beta^2 + (x - \alpha)^2\}} , \qquad -\infty < x < \infty.$$

The parameter β is called the scale parameter, and clearly we must have $\beta > 0$. The Cauchy pdf is in fact also t_1, that is a t-distribution with one degree of freedom. The log-likelihood is given by

$$\ell(\alpha, \beta; \mathbf{x}) = n \log \beta - \sum_{i=1}^{n} \log\{\beta^2 + (x_i - \alpha)^2\} - n \log \pi ,$$

and is illustrated, for the particular sample (1.09, −0.23, 0.79, 2.31, −0.81), in Figure 3.4. This is one of two Cauchy samples that we shall use for illustration. The Cauchy distribution is an example of a stable law; stable laws are used for describing data with heavy tails, as sometimes occurs with financial data — see for example Adler et al. (1998). Stable laws are sometimes fitted to data by methods other than maximum likelihood; see for example Besbeas and Morgan (2001, 2004, 2008), and also the work of Section 5.7.4. □

Note that we now specify the likelihood as explicitly resulting from the sample $\mathbf{x}$, and we shall usually adopt this practice in the future. We shall provide a number of approaches for locating the surface maximum in Example 3.1, at $\alpha = 0.728$, $\beta = 0.718$. The first of these is the method of steepest ascent.

It is a common experience, when we climb a hill, that some paths are steeper than others, and the shortest are the steepest. In mathematical terms this means that functions increase fastest along the line of the gradient vector. Consider a surface $z = f(x, y)$: for a small step from (x, y) to $(x + \delta x, y + \delta y)$, we can write the change in the surface value as

$$\delta z = \frac{\partial f}{\partial x}\delta x + \frac{\partial f}{\partial y}\delta y = \left(\frac{\partial f}{\partial x}, \frac{\partial f}{\partial y}\right)\begin{pmatrix}\delta x \\ \delta y\end{pmatrix}$$

$$= \mathbf{g}'\begin{pmatrix}\delta x \\ \delta y\end{pmatrix}, \quad \text{where } \mathbf{g}' \text{ is the gradient vector,}$$

$$\mathbf{g}' = \left(\frac{\partial f}{\partial x}, \frac{\partial f}{\partial y}\right) .$$

Thus $\delta z = ||\mathbf{g}|| (\delta x^2 + \delta y^2)^{\frac{1}{2}} \cos \theta$, where θ is the angle between the two vectors $\mathbf{g}'$ and $(\delta x, \delta y)$ — see Carroll and Green (1997, p.96) — and $||\mathbf{g}||$ is the length of the vector $\mathbf{g}$. But the slope of the line connecting points z and $(z + \delta z)$ on the surface is

$$\frac{\delta z}{(\delta x^2 + \delta y^2)^{\frac{1}{2}}} = ||\mathbf{g}|| \cos \theta .$$

We see therefore that the slope is maximised when $\theta = 0$, that is when $(\delta x, \delta y)$ is parallel to $\mathbf{g}'$. The optimisation method of *steepest ascent* uses this principle.

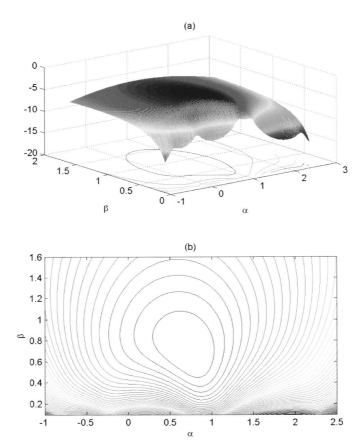

Figure 3.4 *(a) Isometric projection and (b) contour plot for the log-likelihood* $\ell(\alpha, \beta; \mathbf{x})$ *of Example 3.1.*

Suppose now that the function $F(\boldsymbol{\theta})$ has a vector argument $\boldsymbol{\theta}$ of dimension d. The steepest ascent method produces the iteration:

$$\boldsymbol{\theta}^{(r+1)} = \boldsymbol{\theta}^{(r)} + \lambda_r \mathbf{g}(\boldsymbol{\theta}^{(r)}), \tag{3.2}$$

where $\mathbf{g}$ is the gradient vector:

$$\mathbf{g}' = \left(\frac{\partial F}{\partial \theta_1}, \ \frac{\partial F}{\partial \theta_2}, \ldots, \ \frac{\partial F}{\partial \theta_d} \right) \ .$$

The scalar value λ_r is chosen to maximise F along the line of search, $\boldsymbol{\theta}^{(r)} + \lambda_r g(\boldsymbol{\theta}^{(r)})$. We can obtain each λ_r by using any one-dimensional optimisation

(a)
```
% Program for steepest ascent of Cauchy log-likelihood
%--------------------------------------------------------
gr=1;
global data start gr
data=[1.09    -0.23    0.79    2.31 -0.81];
start=[.5 .5];
while norm(gr)>.0001
  gr=grad('cauchy', start);
  lambda=fminbnd('linesearch', -0.5,0.5);
  % Thus lambda maximises the function along the line of search
  start=start+lambda*gr';
  % This is equivalent to equation (3.2)
end
```

(b)
```
function w=linesearch(x)
%
% LINESEARCH calculates the Cauchy log-likelihood along a particular
% line in the parameter space.
% This is a stage in the method of steepest ascent applied to this
% surface.
%------------------------------------------------------------------
global data start gr
w=cauchy(start+x*gr');
```

Figure 3.5 MATLAB *programs for the method of steepest ascent applied to the log-likelihood function of Example 3.1. The Cauchy likelihood is given in* MATLAB *in Figure 3.8(b). Note that rather than use a Newton-Raphson routine for the one-dimensional line search, it is easier to use the* MATLAB *fminbnd function based on golden section search and parabolic interpolation (see Forsythe et al., 1976, who provide a FORTRAN program).*

method, such as the one-dimensional Newton-Raphson method of Equation (3.1). How this may be done in MATLAB is shown in Figure 3.5, and the resulting path is given in Figure 3.6. Clearly here we make use of the numerical approximation for first-order derivatives of Figure 3.3. Successive directions in the steepest ascent method are orthogonal. A *conjugate gradient* improvement is described by Lindfield and Penny (1995, Section 8.3), who also provide MATLAB code. Line searches, as above, are a common feature of deterministic optimisation methods. For clear descriptions see Everitt (1987, Section 2.2).

3.3.3 Newton-Raphson in higher dimensions

The p-dimensional version of the Newton-Raphson method for maximising log-likelihoods gives us the iteration:

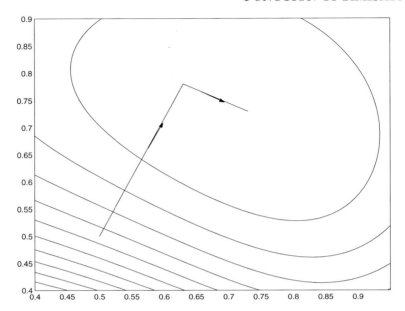

Figure 3.6 *An illustration of the progress of the steepest ascent method applied to the log-likelihood function of Example 3.1. Note that as the derivatives are evaluated numerically, the path lines only approximate lines of steepest ascent, and are not orthogonal.*

$$\boldsymbol{\theta}^{(r+1)} = \boldsymbol{\theta}^{(r)} - \mathbf{A}^{-1}(\boldsymbol{\theta}^{(r)})\mathbf{g}(\boldsymbol{\theta}^{(r)},) \quad r \geq 0 \,, \tag{3.3}$$

starting from $\boldsymbol{\theta}^{(0)}$, where now $\mathbf{g}$ is the gradient vector,

$$\mathbf{g}' = \left(\frac{\partial \ell}{\partial \theta_1}, \cdots, \frac{\partial \ell}{\partial \theta_p} \right),$$

and $\mathbf{A}$ is the Hessian matrix, with (i, j)th element,

$$a_{ij} = \frac{\partial^2 \ell}{\partial \theta_i \partial \theta_j} \quad , \quad 1 \leq i, \ j \leq p \,.$$

An obvious difference between the Newton-Raphson method (Equations (3.1) and (3.3)) and the method of steepest ascent (Equation (3.2)) is the fact that the latter method only uses first-order derivatives, while the former method uses the surface curvature information in $\mathbf{A}$.

Equation (3.3), in providing $\boldsymbol{\theta}^{(r+1)}$, gives us both a direction and a magnitude of the step-length in that direction. The approach can be made more sophisticated by carrying out a line-search in the direction given by (3.3) in order to modify the step length. Additionally, matrix inversion can be time-consuming, and can be avoided, when functions are to be minimised, through a sequence of numerical positive definite symmetric approximations to $\mathbf{A}^{-1}$, which ultimately converge to $\mathbf{A}^{-1}$ as the iteration proceeds. For *maximising* functions, negative definite approximations are used. Such methods are called quasi-Newton methods — see Everitt (1987, p.24).

A modification is to replace $\mathbf{A}$ by its expectation, $\mathbb{E}[\mathbf{A}]$, resulting in what is called the *method of scoring*, which is widely used in statistics. In some cases (e.g., generalised linear models with a canonical link function — see Morgan, 1992, p.373, and Chapter 8) the Newton-Raphson and method-of-scoring approaches are identical. Note also that

$$\mathbb{E}\left[\frac{-\partial^2 \ell}{\partial \theta_i \partial \theta_j}\right] = \mathbb{E}[-a_{ij}] = \mathbb{E}\left[\frac{\partial \ell}{\partial \theta_j}\frac{\partial \ell}{\partial \theta_k}\right]$$

— see for example Dobson (1990, Appendix A).

Example 3.1 continued

We shall now illustrate both of the Newton-Raphson and method-of-scoring procedures for the Cauchy likelihood of Example 3.1.

It is readily verified that the first-order derivatives are of the form:

$$\frac{\partial \ell}{\partial \alpha} = 2\sum_{i=1}^{n} \frac{(x_i - \alpha)}{\{\beta^2 + (x_i - \alpha)^2\}} ,$$

$$\frac{\partial \ell}{\partial \beta} = \sum_{i=1}^{n} \frac{\{(x_i - \alpha)^2 - \beta^2\}}{\beta\{\beta^2 + (x_i - \alpha)^2\}} ,$$

leading to the Hessian:

$$\mathbf{A} = \begin{bmatrix} 2\displaystyle\sum_{i=1}^{n} \frac{\{(x_i - \alpha)^2 - \beta^2\}}{\{\beta^2 + (x_i - \alpha)^2\}^2}, & -4\beta\displaystyle\sum_{i=1}^{n} \frac{(x_i - \alpha)}{\{\beta^2 + (x_i - \alpha)^2\}^2} \\ -4\beta\displaystyle\sum_{i=1}^{n} \frac{(x_i - \alpha)}{\{\beta^2 + (x_i - \alpha)^2\}^2}, & -\frac{n}{\beta^2} + 2\displaystyle\sum_{i=1}^{n} \frac{\{\beta^2 - (x_i - \alpha)^2\}}{\{\beta^2 + (x_i - \alpha)^2\}^2} \end{bmatrix}$$

In this example, the method of scoring results in a substantial simplification, as we obtain:

$$\mathbb{E}[\mathbf{A}] = -\frac{n}{2\beta^2}\begin{bmatrix} 1 & 0 \\ 0 & 1 \end{bmatrix} .$$

Computer algebra packages, such as Maple, may be used to obtain the above expressions for derivatives, and to carry out the integration needed to form $\mathbb{E}[\mathbf{A}]$ — see Exercise 3.2.

Detailed programming of the Newton-Raphson method in MATLAB is shown in Figures 3.7 and 3.8. Alternatively, if one uses the numerical approximations of Figure 3.3, the translation of the iteration of Equation (3.3) into MATLAB is straightforward, as shown in Figure 3.8. This figure provides examples of both script and function programs in MATLAB. Function programs were encountered earlier in Figure 3.5, and script programs in Figure 3.2. The difference between the two types of program is explained in Appendix A. □

```
% Program for the Newton-Raphson method for the
% two-parameter Cauchy likelihood of Example 3.1,
% giving full detail of the gradient and Hessian
% construction.
%_____
x0=[.7,.3];                      % starting value for the iteration
x0=x0';
data=[1.09,-0.23,0.79,2.31,-0.81];
one=ones(size(data));
g=[1,1];                         % starting value for the 'while'
while norm(g)>0.00001            % command
   a=x0(1);b=x0(2);
   b2=b^2;
   a=a*one;
   b2=b2*one;
   x1=data-a;
   den=b2+x1 .^2;
   den2=den .^2;
   g(1)=2*sum(x1 ./den);         % the gradient vector components
   g(2)=sum((x1 .^2-b2) ./(b*den));
   h(1,1)=2*sum((x1 .^2-b2) ./den2);
   h(1,2)=-4*b*sum(x1 ./den2);
   h(2,1)=h(1,2);
   h(2,2)=-5/b^2-h(1,1);         % completes specification of the
   xn=x0-inv(h)*g';              % Hessian matrix
   x0=xn;
end
disp(x0)
```

Figure 3.7 *A* MATLAB *program to provide the detailed iterations of the Newton-Raphson method for finding the maximum-likelihood estimates of the parameters of the two-parameter Cauchy distribution described in Example 3.1.*

The Newton-Raphson method can be obtained as the function *fminunc* in MATLAB, through the MATLAB numerical optimisation toolkit. While *fminunc* allows the user to specify the forms of first- and second-order derivatives, the default is that they are evaluated numerically — see Grace (1994).

A comparison of the Newton-Raphson and method of scoring procedures is provided by Morgan (1978), some of whose conclusions are repeated in Table 3.1.

All iterative methods require a stopping rule, and a crude one has been used in producing the results of Table 3.1 — see for example Figures 3.7 and 3.8. Something of this kind is usually adequate for statistical modelling work, since we go on, as will be described in Chapter 4, to consider how well models fit the data, and also any one model will usually be considered as just one out of a range of alternatives. An iterative method which has converged too soon should therefore be readily spotted.

(a)

```
% Program to perform the Newton-Raphson iteration of Equation (3.3),
% using the gradient and Hessian numerical approximations of Figure 3.3
% 'data' are global, so that they can be accessed by the 'cauchy'
% function.
% Calls the functions GRAD, HESSIAN and CAUCHY.
%------------------------------------------------------------------------
global data
data=[-4.2 -2.85 -2.3 -1.02 0.7 0.98 2.72 3.5];
x0=[.7 .5];               % starting value for Newton Raphson
g=[1,1];                  % value selected to start the 'while'
while norm(g)>0.00001     % loop
  g=grad('cauchy',x0);
  h=hessian('cauchy',x0);
  xn=x0'-inv(h)*g;
  x0=xn';
end
disp(x0)                  % the maximum-likelihood estimate
disp(g)                   % the gradient at the maximum
disp(eig(h))              % the eigen values of the Hessian at
                          % maximum-likelihood estimate
```

(b)

```
function y=cauchy(x)
%CAUCHY calculates minus the Cauchy log-likelihood.
% 'alpha', 'beta' are the standard Cauchy parameters;
% 'data' are global, and set in the calling program.
%----------------------------------------------------
global data
alpha=x(1); beta=x(2);
alphv=alpha*ones(size(data));
datashift=(data-alphv).^2;
beta2v=(beta^2)*ones(size(data));
arg=log(beta2v+datashift);
loglik=length(data)*log(beta)-sum(arg);
y=-loglik;
```

Figure 3.8 *By using gradient and Hessian approximations, the Newton-Raphson it-erations of Equation (3.3) can be seen to translate directly into* MATLAB *code. The program of (a) uses the function of (b) for evaluating minus the Cauchy log-likelihood. A different sample is now used, from that resulting in Figure 3.4.*

Table 3.1 *(Morgan, 1978). Numbers of iterations until $\frac{\partial \ell}{\partial \alpha} = \frac{\partial \ell}{\partial \beta} = 0$ to three decimal places. The approximate time units per iteration are, for Newton-Raphson, 1.34 and for the method of scoring, 0.88, though these figures are dependent on the particular computing environment used. Convergence may be obtained at $\alpha = \hat{\alpha}$, $\beta = -\hat{\beta}$, when $(\hat{\alpha}, \hat{\beta})$ is the maximum likelihood estimate of (α, β), and this occurred for the (vii) starting point for the Newton-Raphson Method. This is because the equation $\partial \ell / \partial \beta = 0$ is a function of β^2.*

	Starting-point		Newton-Raphson (NR)		Scoring
	$\alpha^{(0)}$	$\beta^{(0)}$			
(i)	0.7	0.3	6	NR better	10
(ii)	0.0	1.0	4	(overall time)	10
(iii)	0.5	0.7	6		10
(iv)	0.1	0.9	5		10
(v)	0.2	0.8	8	NR worse	10
(vi)	0.3	0.3	13	(overall time)	11
(vii)	1.0	1.0	14		11
(viii)	0.6	0.4	Diverges		7

More sophisticated stopping rules may be employed — see Kennedy and Gentle (1980, p.437) for discussion. Convergence to a point which is not a maximum can be checked by seeing whether the Hessian of the log-likelihood is negative-definite at the optimum (as it should be). This is readily done by considering the eigenvalues of this Hessian (by use of the MATLAB eig command, which is illustrated later in the program of Figure 4.1), since a negative-definite matrix has eigenvalues which are all negative. If the iteration converges to a maximum, we do not usually know if it is the global maximum. Two-dimensional surfaces can, of course, be examined through computer plots or isometric projections, but it is more difficult to inspect surfaces in higher dimensions. We discuss the use of profile log-likelihoods for such cases in Chapter 4. The standard approach when using iterative methods to optimise a multi-dimensional surface is to carry out the iterations from a range of alternative starting points. For example, we need to do this when fitting mixture models (McLachlan, 1987), which are notorious for producing bumpy likelihood surfaces. An automatic way to produce sets of starting points is described in Brooks and Morgan (1994), following a simulated annealing search we shall describe in Section 3.4.

3.3.4 Simplex search methods

The use of numerical approximations to derivatives, or of computer algebra packages to form expressions for derivatives, means that deterministic optimisation methods making use of first- or second-order derivatives need not involve the statistician in more labour than simply specifying the form of the function to be optimised. As we shall see later, at times just specifying the function can be quite complicated, and require novel approaches. Simplex search methods also do not require anything other than specification of the function to be optimised, and are therefore described as *direct search* methods.

A simplex is a set of $n+1$ points in n-dimensional space. For example, in two dimensions the corners of a triangle form a simplex, while in three dimensions it would be the apexes of a tetrahedron, and so forth. Simplex methods started with the paper by Spendley et al. (1962), and were extended by Nelder and Mead (1965). In the basic method, the function is evaluated at the corners of a regular simplex, which is one which has sides of equal length, and then a new simplex is formed, by dropping from the simplex the corner at which the function value is the largest (or smallest), if the objective is to minimise (maximise). The replacement vertex in the simplex is formed by reflecting the rejected vertex through the centroid of all of the original vertices except for the rejected one. The Nelder-Mead extension of the method is to allow expansion and contraction of the simplex — see Nelder and Mead (1965). Suppose that our objective is to minimise a function. If the function value at a reflected point is lower than at any other point in the previous simplex, then the simplex is expanded in the same new direction. If that also produces a lowest yet value, then that new point enters the simplex. However, if the expansion does not result in a lowest yet value then the method returns to the reflected point, as expansion has not improved matters. Contraction occurs when reflection does not produce an improved position, with contraction towards the centroid. See for example the algorithm in Bunday (1984).

As an illustration of what is involved, we shall just consider reflection. Suppose we start with the simplex $(\boldsymbol{\theta}_1, \boldsymbol{\theta}_2, \ldots, \boldsymbol{\theta}_{n+1})$ in n-dimensional space, and we want to maximise some function $\psi(\boldsymbol{\theta})$. We form $\psi(\boldsymbol{\theta}_i)$, for $i = 1, 2, \ldots, n+1$. If the lowest value of the function $\psi(\boldsymbol{\theta})$ is at $\boldsymbol{\theta}_\ell$ then reflection drops $\boldsymbol{\theta}_\ell$ from the simplex, replacing it by the new point,

$$\boldsymbol{\theta}_m = (1 + \alpha)\boldsymbol{\theta}_c - \alpha\boldsymbol{\theta}_\ell,$$

where the centroid $\boldsymbol{\theta}_c$ is given as:

$$\boldsymbol{\theta}_c = \frac{1}{n} \sum_{\substack{i=1 \\ i \neq \ell}}^{n+1} \boldsymbol{\theta}_i \, ,$$

and α is a suitable reflection coefficient.

Termination occurs when the sample standard deviation of the function values at the simplex vertices is regarded as suitably small, or the simplex size is acceptably small. For more detail, see Everitt (1987, pp.16–20). Bunday (1984) provides a flow diagram for the method and translates this into a BASIC program. He recommends restricting use of the method to problems with no more than six parameters. However it may be used successfully for far

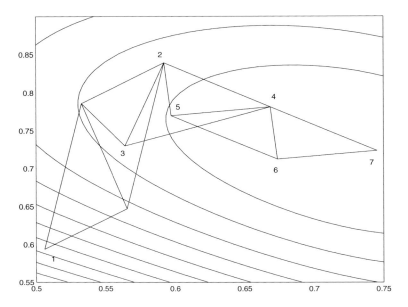

Figure 3.9 *An illustration of the progress of a basic simplex iteration, applied to the log-likelihood of Example 3.1. The numbers refer to the order in which the simplex vertices are selected, as the iteration proceeds.*

higher-dimensional problems. MATLAB carries out a simplex search by means of the *fminsearch* function, which is remarkably simple to use.

3.4 Stochastic search methods

3.4.1 The case of multiple optima

This section follows closely the material in Brooks and Morgan (1995). We start by considering how deterministic optimisation methods perform when a surface has many local optima.

Example 3.2: The Bohachevsky surface

Consider the surface of Figure 3.10. This surface is the function,

$$f(x,y) = x^2 + 2y^2 - 0.3\cos(3\pi x) - 0{\cdot}4\cos(4\pi y) + 0{\cdot}7, \quad \text{for } -1 \leq x, y \leq 1,$$
$$(3.4)$$

suggested by Bohachevsky et al. (1986). The global surface minimum is at (0,0), but there are clearly a large number of local minima, maxima and saddlepoints. This surface is therefore a good one for testing methods of function optimisation.

We randomly selected 10,000 starting points, uniformly distributed over the square of Figure 3.10, and recorded both those that led to the correct global

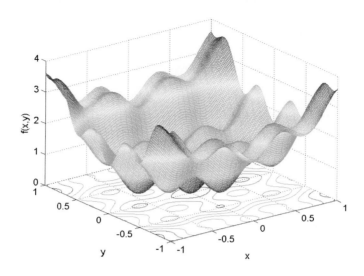

Figure 3.10 *Isometric projection of the surface given by Equation (3.4).*

minimum, and those that did not, when the simplex algorithm was used. Figure 3.11 indicates those starting points which led to the global minimum. The similar plot of those starting points for which the algorithm failed to converge to the global optimum confirms that the entire parameter space was uniformly covered, and that any features within these graphs are features of the problem at hand, and not of the random number generator which was used to generate the starting points. We can see that the points are essentially symmetrical about the axes, due to the symmetry of the surface.

A standard source for FORTRAN coded numerical optimisation routines is the NAG (2001) library, the manual for which recommends two deterministic search algorithms in particular. The first is a Quasi-Newton (QN) algorithm, E04JAF, and the second is a Sequential Quadratic Programming (SQP) algorithm, E04UCF, described by Gill et al. (1981, p.237). It is interesting also to see how these algorithms perform when applied to the surface of Figure 3.10. It was found that fewer than 20% of the starting points lead to the QN algorithm converging to the global minimum, and these are mainly within a central rectangle defined by the four peaks surrounding the global minimum, roughly as in Figure 3.11. Beyond this rectangle, the algorithm searches for other valleys, though in some particularly steep places local valleys can be overshot. A similar picture results when the SQP algorithm is used, and of the two, the SQP algorithm has the better performance (Brooks and Morgan, 1995). □

It is not surprising that deterministic methods can converge to local optima when they exist. However, complex models with many parameters may well give rise to likelihood surfaces which would possess such local optima. One *ad hoc* way to proceed is to run deterministic search methods from several alternative starting points, as we have already described. An alternative approach is to introduce a stochastic element into the search, allowing the method to jump at random to different parts of the surface. One way of accomplishing this is to use simulated annealing, which proceeds by analogy with thermodynamics, and the way in which metals (and some liquids) cool and crystallise.

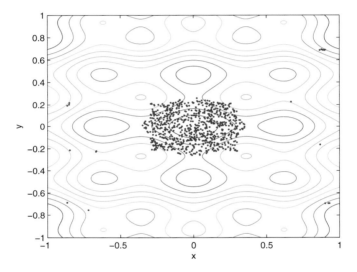

Figure 3.11 *Starting points for which the simplex algorithm converged to the true minimum of the Bohachevsky surface of Figure 3.10, with contours superimposed.*

3.4.2 The general simulated annealing algorithm

What follows is specified in terms of 'temperatures' because of the analogy with thermodynamics. In terms of function optimisation, a 'temperature' is simply a constant which governs the probability of the system moving to an unfavourable point in the search for the global optimum.

We may implement a general simulated annealing algorithm for minimising a function by the following steps:

(i) Beginning at an initial temperature, T_0, we pick an initial set of parameter values with function value, E_{old}.

(ii) Next, we randomly select another point in the parameter space, within a neighbourhood of the original, and calculate the corresponding function value, E_{new}.

(iii) We then compare the two points in terms of their function value, using what is called the *Metropolis Criterion* as follows. If $\triangle = E_{\text{new}} - E_{\text{old}}$, the iteration moves to the new point if and only if a uniform random variable U from the $U(0, 1)$ distribution satisfies the inequality:

$$U \leq \exp(-\triangle/T)$$

or equivalently

$$E_{\text{new}} \leq E_{\text{old}} - T \log U,$$

where T denotes the current temperature.

As we shall see in Section 6.2, the random variable $-T \log U$ has an exponential distribution with mean T. Thus $E_{\text{new}} - E_{\text{old}}$ is compared with such an exponential random variable. Note that the iteration always moves to the new point if its corresponding function value is lower than that of the old point, and that at any temperature there is a chance for the iteration to move 'upwards.' This chance decreases with temperature.

(iv) Whether the system has moved or not, we repeat steps 2–3. At each stage we compare the function value of a new point with the function value of the old point until the sequence of accepted points is judged, by some criterion, to have settled down and reached a state of equilibrium. We shall consider one possible way of doing this in the next section.

(v) Once an equilibrium state has been achieved for a given temperature, the temperature is lowered to a new temperature as defined by what is called the *annealing schedule*. The effect of this, as we saw above, is to reduce the chance of moving upwards. The process then begins again from step 2, taking as the initial state the point following the last iteration of the algorithm, until some stopping criterion is met, and the system is considered to have 'frozen.'

Note that since we continue steps 2–3 until a state of equilibrium is attained, the starting values in step 1 are arbitrary, and have no impact upon the solution.

The adoption of the Metropolis criterion in step 3 maintains a further link with thermodynamics. Alternative rules may be used here, but this one has been found to work in a satisfactory manner. We shall see in Chapter 7 how this procedure is similar to a method of Markov chain Monte Carlo iteration.

3.4.3 A simple implementation

Various choices have to be made before the general algorithm can operate in practice. The following approach was suggested by Brooks and Morgan (1995).

If we wish to minimise a function of p parameters, $f(\theta_1, \theta_2, \ldots, \theta_p)$, then this particular implementation of the algorithm works as follows: at step 2 of the general simulated annealing algorithm, the new point is chosen by first selecting one of the p variables at random, and then randomly selecting a new value for that parameter within bounds set for it by the problem at hand. Thus the new point takes the same parameter values as the old point except for one. This is one way to choose new points so that they are in a neighbourhood of the old one; other methods are discussed later. At step 4, the repetition of steps 2–3 occurs exactly N times, after, say, s successful moves from one point to another. N should, if possible, be selected so that it could be reasonably assumed that the system is in equilibrium by that time. At step 5, if $s > 0$ then we decrease the temperature by letting T become ρT, where $0 \leq \rho \leq 1$ defines what we call the *annealing schedule*, and then begin again. If, however, $s = 0$ then we can consider the system to have frozen, as no successful moves were made, and the algorithm stops. Note that far more complicated annealing schedules have been devised — see for example Osman (1993).

This implementation has been found to be both fast and reliable for a range of functions. The convergence time is problem-dependent, as the complexity of the problem directly affects the number of temperature reductions necessary for convergence. This number is usually found to be several hundred. It is also highly dependent upon the values of the parameters ρ, N and the starting temperature T_0. A large N gives an accurate solution, but at the expense of convergence time: doubling the value of N more than doubles the execution time. Increasing ρ increases the reliability of the algorithm in reaching the global optimum, and corresponds to a slower cooling of the system. The value of the initial temperature, T_0, is also important: T_0 must be sufficiently large for any point within the parameter space to have a reasonable chance of being visited, but if it is too large then too much time is spent in a 'molten' state. In practice it may be necessary to try the algorithm for several values of T_0 before deciding upon a suitable value. A range of alternative approaches are described in Brooks and Morgan (1995). Stochastic search procedures are reviewed in Fousakis and Draper (2002). The technique of trans-dimensional simulated annealing extends stochastic search to include searching over models as well as parameter values for particular models — see Brooks et al. (2003).

Example 3.1 concluded

When the scale parameter β is fixed, the maximum-likelihood estimate for α is that which minimises

$$-\ell(\alpha; \mathbf{x}) = \sum_{i=1}^{n} \log\{\beta^2 + (x_i - \alpha)^2\} .$$

We shall now fix the scale parameter $\beta = 0.1$, and consider the random sample, $(-4.20, -2.85, -2.30, -1.02, 0.70, 0.98, 2.72, 3.50)$. This provides a log-likelihood with more local optima than the previous sample — see Figure 3.12. For fixed β, $\ell(\alpha; \mathbf{x})$ is made up of the components, $\log[1 + \{(x_i - \alpha)/\beta\}^2]$, the ith of which, as a function of α, is centered on x_i. Different βs produce different degrees of smoothing, as shown in Figure 3.13. Copas (1975) provides relevant discussion. There are connections here with non-parametric density estimation

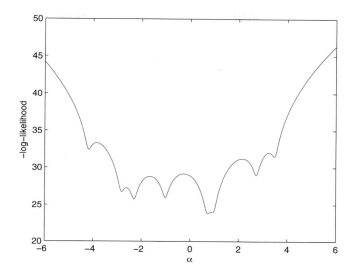

Figure 3.12 *A graph of minus the Cauchy log-likelihood for a sample of size 8, with* $\beta = 0.1$. *The global minimum is at* $\alpha = 0.737$.

which we shall find useful in later sections of the book (see Silverman, 1986, and Appendix C).

The simulated annealing algorithm of this section was then used to produce the minimum value of $-\ell(\alpha; \mathbf{x})$, beginning from 1000 randomly chosen starting points. The annealing parameters, T_0, N and ρ, were set to values giving an expected convergence time similar to those of the deterministic routines applied to this problem. For this example, experience suggests suitable values to be: $T_0 = 10.0$, $N = 300$ and $\rho = 0.95$. This produces a fairly slow cooling rate and with this value of N, which experience also suggests is quite low, we have a high chance of stopping the algorithm before the system has frozen. Thus we would not expect the solution to be accurate to many decimal places. We found that 100% of the solutions lay in the range [0·70, 0·94] and that only 1% were outside the interval [0·70, 0·80] corresponding to the tiny 'well' containing the true minimum. The accuracy of the solution is correct to only the first decimal place, but the performance is superior to the QN algorithm, which had a success rate of less than 27% in finding the global optimum, and to the SQP algorithm, which had a success rate of < 35%. (Both the QN and SQP algorithms were introduced in Example 3.2.)

Figure 3.14 shows all of the points which were both selected and accepted by the Metropolis criterion during a typical execution of the simulated annealing algorithm. Note the high concentration of accepted points around the bottom of the large well containing both the global minimum and the local minimum

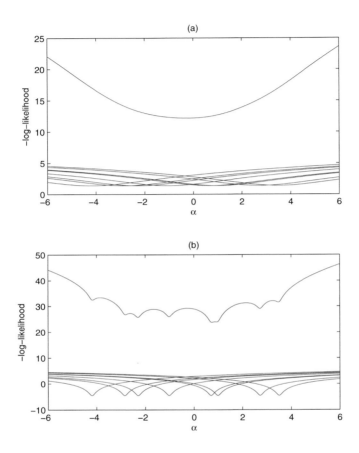

Figure 3.13 *The effect of β on the occurrence of multiple optima, when minus the Cauchy log-likelihood is graphed as a function of α for fixed β. We show minus the log-likelihood components, $log[1 + \{(x_i - \alpha)/\beta\}^2]$, as well as their sum, for (a) $\beta = 2$, (b) $\beta = 0.1$. The sample in this case is $(-4.20, -2.85, -2.30, -1.02, 0.70, 0.98, 2.72, 3.50)$.*

closest to it. Here the points indicating the values that were accepted are so close together that they reproduce well the form of the curve of the log-likelihood in this region. Note also that each of the local minima has been visited by the algorithm and thus each has been considered and subsequently rejected in favour of the global minimum. The outline of the function profile is quite distinct but, of course, the exploration of the parameter space has only been partial.   □

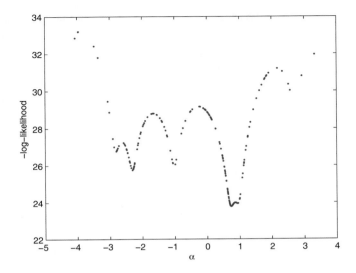

Figure 3.14 *Values of minus the Cauchy log-likelihood, $-\ell(\alpha; \mathbf{x})$, for points accepted by the simulated annealing algorithm with parameter settings of $T_0 = 6$, $N = 250$, $\rho = 0.92$.*

The performance of the simulated annealing algorithm when applied to the surface of Example 3.2 is illustrated in Figure 3.15.

The simulated annealing algorithm is very reliable in that all of the solutions are within the central well containing the global minimum. We can also see that the final solution is accurate only to the first decimal place. Accuracy can be improved by altering the algorithm parameters, but this would be at the expense of the time taken for convergence.

3.5 Accuracy and a hybrid approach

From the two examples considered in the last section, we can draw the following conclusions. The simulated annealing algorithm appears to be very reliable, in that it always converges to within a neighbourhood of the global minimum, and the size of this neighbourhood can be reduced by altering the algorithm parameters, but this can be expensive in terms of time.

In most cases we would like to find the minimising solution to several decimal places of accuracy. It is clear that the simulated annealing algorithm could produce results with such accuracy, but that the execution time would then be prohibitive. A deterministic algorithm is able to produce solutions to machine accuracy, that is, governed only by the limitations of the computer used, but can have considerable difficulty in finding the correct solution in the case of multiple optima, as we have already demonstrated.

Brooks and Morgan (1994) suggested a hybrid method involving two components, in which the simulated annealing component is stopped prematurely,
but instead of taking a single endpoint to start the second (deterministic)
component, we take each of the points accepted at the final temperature, together with the best point overall, as starting points. The second, deterministic
component is then run once from each starting point and the best solution
generated from these points is given as the solution. Here the choice of the
value of N is not so important. As we have seen, as long as N is sufficiently
large for the algorithm to have settled down to the equilibrium distribution
then we can reasonably expect at least one of these points to be within the
required 'well' and thus lead to the second component finding the global minimum. Brooks (1995) discusses this hybrid algorithm in greater detail, as well
as providing the FORTRAN code, and additional examples are provided by
Brooks and Morgan (1994, 1995).

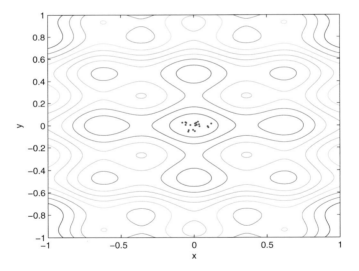

Figure 3.15 *Final solutions given by the simulated annealing algorithm for 20 random
starts, with parameter settings of $T_0 = 1$, $N = 500$, $\rho = 0.9$ for the Bohachevsky
surface.*

Kvasnicka and Pospichal (1997) integrate a random element into the simplex
method.

3.6 Discussion

Fitting probability models to data, whether by maximum likelihood or by
one of the several alternative methods we discuss elsewhere in this book,

often reduces to an exercise in multi-dimensional function optimisation. We have seen in this chapter how established deterministic methods, while well developed and efficient for well-behaved likelihoods with single optima, need to be given several runs from different starting values if multiple optima are suspected. Function optimisation by a deterministic rule is akin to flying an aircraft — the difficult parts are take off (selecting a good starting value) and landing (deciding when the algorithm has converged).

An alternative approach is to use, alone or in combination with a deterministic search, a form of stochastic search mechanism. Simulated annealing provides such a mechanism which has been shown to work well in practice. Judging whether a system has reached equilibrium is generally difficult, and there is additional discussion of this issue in Chapter 7, in the context of Markov chain Monte Carlo procedures. Whatever method has been used to maximise a likelihood, as we shall discuss in the next chapter, we shall be interested in the surface curvature at the maximum. For a method such as the Newton-Raphson method, the Hessian will have been used during the iterative search procedure. For other methods this will not be true and then additional computation will be needed — see Exercise 3.16.

No numerical methods can be expected to work well for every possible problem. The value of δ in the difference approximations of Figure 3.3 will need to be changed for some applications. Similarly, simple termination rules will prove inadequate in certain cases, and there are no general optimal settings of the simulated annealing parameters. In all cases a certain amount of care is needed, but experience is likely to produce the desired end-result.

An attraction of the MATLAB *fminsearch* function is its good performance, using default settings, for optimising straightforward non-linear functions with ease. A number of the exercises which follow provide opportunities to verify this. In some cases of model-fitting we need to impose inequality constraints on parameters, and this is discussed in Section 5.3.

3.7 Exercises

3.1 The numerical approximation to the gradient of a function $f(x)$ for a scalar quantity x, given in Figure 3.3, is $f'(x)$, given by:

$$2\delta f'(x) = f(x + \delta) - f(x - \delta).$$

Use MATLAB to compare the performance of this method with the alternative which results in:

$$12\delta f'(x) = f(x - 2\delta) - 8f(x - \delta) + 8f(x + \delta) - f(x + 2\delta).$$

3.2* Use a computer algebra package such as Maple to verify the forms for $\mathbf{A}$ and $E[\mathbf{A}]$ given in Section 3.3.2 for the Cauchy likelihood of Example 3.1.

3.3* (Brooks et al., 1997.) The data below describe the number of dead fetuses in litters of mice. Use the MATLAB routine *fminsearch* to fit the beta-binomial model, which has the probability function:

$$\mathrm{pr}(X = x) = \frac{\binom{n}{x} \prod_{r=0}^{x-1}(\mu + r\theta) \prod_{r=0}^{n-x-1}(1 - \mu + r\theta)}{\prod_{r=0}^{n-1}(1 + r\theta)},$$

$$= Q(x, n, \theta), \quad \text{say, for} \quad 0 \le x \le n,$$

where n denotes the number of fetuses, and X is the number that die. (See Exercise 2.7.) Also shown are the expected values resulting from the maximum-likelihood estimates, $\hat{\mu}$ and $\hat{\theta}$.

Litter	Number of dead fetuses (x)									
Size n	0	1	2	3	4	5	6	7	8	9
1	1									
	0.9	0.1								
2	1									
	0.8	0.2								
3										
4	3									
	2.1	0.7	0.2							
5	1	1								
	1.3	0.5	0.1							
6	1	2								
	1.9	0.8	0.3	0.1						
7	1	3								
	2.9	1.3	0.5	0.2						
8	2	1								
	1.6	0.8	0.4	0.1						
9	5	5	2	1						
	6.7	3.6	1.7	0.7	0.2	0.1				
10	17	6	2							
	12.2	7.0	3.4	1.5	0.6	0.2	0.1			
11	14	9	3	1	1	2				
	13.9	8.3	4.3	2.1	0.9	0.4	0.1			
12	19	17	7	2	2					
	20.6	12.9	7.1	3.6	1.7	0.7	0.3	0.1		
13	9	14	7	3			2	1		
	15.0	9.8	5.6	3.0	1.5	0.7	0.3	0.1		
14	10	5	5	1	1				1	
	9.2	6.2	3.7	2.0	1.1	0.5	0.2	0.1		
15	4	1	2				1			
	3.0	2.1	1.3	0.7	0.4	0.2	0.1			
16		2	1							
	1.1	0.8	0.5	0.3	0.2	0.1				
17	2									
	0.7	0.5	0.3	0.2	0.1	0.1				

Use the maximum-likelihood estimates, $(\hat{\mu}, \hat{\theta})$, to verify the expected values shown. Extend your program to try to fit the mixture model:

$$\Pr(X = x|n) = \gamma Q(x, n, \theta) + (1 - \gamma)\binom{n}{x}\nu^x(1 - \nu)^{n-x}, \quad \text{for} \quad 0 \le x \le n.$$

3.4 Consider the microbial infection data of Example 1.2. A possible model for the progress of the infection is that a single infecting bacterium living according to a stochastic linear birth-and-death process (see Cox and Miller, 1965, p.156) with respective birth and death rates λ and μ, with $\lambda > \mu$, produces the symptoms of a sore throat if and only if the resulting colony grows to size n. Then the time, T, to the onset of symptoms is governed by:

$$\text{pr}(T \le t) \approx (1 - \alpha)\beta^{n-1}(1 - \frac{\mu}{\lambda})^{-1}$$

(Morgan and Watts, 1980), where

$$\alpha = \mu\{1 - e^{(\mu-\lambda)t}\}\{\lambda - \mu e^{(\mu-\lambda)t}\}^{-1},$$
$$\beta = \lambda\{1 - e^{(\mu-\lambda)t}\}\{\lambda - \mu e^{(\mu-\lambda)t}\}^{-1}.$$

Verify that the likelihood is given by:

$$L(\lambda, \mu, n) = \prod_{i=1}^{13}\{\text{pr}(T \le i) - \text{pr}(T \le i - 1)\}^{r_i},$$

where r_i is the number of individuals in the ith class interval, and try to obtain the maximum-likelihood estimates of λ, μ and n.

3.5 Stimulus-response experiments in psychology result in data such as that illustrated below:

	No	Unsure	Yes
Noise	30	10	15
Signal	9	7	35

At each of a number of trials, a subject was either given a stimulus (for example a light on a noisy screen) or just noise alone, and responded either 'Yes,' 'No,' or 'Unsure.' A model for the data results in the following probabilities, where $F(\ )$ denotes a suitable cumulative distribution function.

	No	Unsure	Yes
Noise	$F(z_1)$	$F(z_2) - F(z_1)$	$1 - F(z_2)$
Signal	$F(bz_1 - a)$	$F(bz_2 - a) - F(bz_1 - a)$	$1 - F(bz_2 - a)$

Try to motivate this choice of model. By adopting a logistic form for $F(\ )$, fit the model to the data using the MATLAB *fminsearch* routine. Compare observed and expected values, and comment on the results.

3.6 Use maximum-likelihood to fit the following model:

$$pr(x) = \frac{\beta\Gamma(x + \alpha)\gamma^{x-1}}{x!\Gamma(1 + \alpha)}, \quad \text{for} \quad x \ge 1,$$

in which α, β and γ are the parameters to be estimated, to the word type frequencies of Table 1.8. Obtain the expected values, and comment on the quality of the fit.

3.7 Use the MATLAB *tic* and *toc* commands to compare the time taken by the program of Figure 3.2(a) with that of a more standard version in which the sums are formed via a loop.

3.8* In Section 3.3.4 we see how reflection operates for the simplex method. Write down the corresponding equations for expansion and contraction. Program the simplex method in MATLAB, using the flow diagram in Bunday (1984).

3.9 Check the MATLAB function below for the simulated annealing algorithm. Modify it for use with a two-parameter Cauchy log-likelihood. Consider how you might adapt it to carry out the hybrid optimisation procedure.

```
function x=annealing(funfcn,x0)
%
%ANNEALING Performs General Simulated Annealing,
%  to minimise a function of several variables, specified in 'funfcn'.
%  Code due to Paul Terrill.
%------------------------------------------------------------------------
%  x0 contains the initial parameter estimates
%  T is the current temperature
%  N is the number of iterations at each temperature
%  s is the number of times we accept a new point for a given
%  temperature
%  r defines the annealing schedule
%  p is the number of parameters
%------------------------------------------------------------------------
p=length(x0);
T=100; N=5500; r=0.90;
x=x0;
Eold=feval(funfcn,x0);
s=1; while s>0
  s=0;
  for i=1:N
    rand('state',sum(100*clock));% sets a different seed for
                               % the random number generator.
    u1=rand;                   % u1 is used to select a particular
                               % parameter.
    for i=1:p
      if u1 < i/p              % sets a new value for one
        x(i)=4*abs((rand-0.5)); % of the parameters according
        u1=2;                  % to the range of x(i). This needs
      end                      % careful consideration for each
    end                        % problem considered. Cf. 'fminbnd'.
    Enew=feval(funfcn,x);
    u2=rand;
    del=Enew-Eold;
```

```
    crit=exp(-del/T);
    if u2<=crit                  % checks criterion for accepting
       x0=x;                     % new point. If the point is accepted,
       Eold=feval(funfcn,x0);    % we set s=s+1.
       s=s+1;
    end
  end
  T=r*T;
  fprintf('The current value of the temperature is:');
  disp (T);x0
  disp(s)
 end
 x=x0;
```

3.10 In this question we look at the logit model for quantal assay data. At
each of a number of doses, x_i, groups of n_i individuals are given the dose,
and r_i respond, $1 \le i \le k$. In the logit model, the probability of response
at dose x_i is written as

$$P(x_i) = 1/(1 + e^{-(\alpha + \beta x_i)})$$

(i) Study the following MATLAB program for evaluating the negative log-
likelihood.

```
function y=logit(t)
%
%LOGIT Calculates the negative log-likelihood for a logistic
%   model and quantal response data
%   x is the 'dose'
%   n is the number of individuals at each 'dose'
%   r is the number of individuals that respond at each 'dose'
%_____
x=[49.06 52.99 56.91 60.84 64.76 68.69 72.61 76.54];
n=[59 60 62 56 63 59 62 60];
r=[6 13 18 28 52 53 61 60];
w=ones(size(x));
alpha=t(1); beta=t(2);
y=w ./(1+exp(-(alpha+beta*x)));
z=w-y;
loglik=r*(log(y))'+(n-r)*(log(z))';
y=-loglik;
```

(ii) Obtain the maximum likelihood estimate of the pair of parameters (α, β),
using the MATLAB *fminsearch* routine.

(iii) Display the data, by plotting r_i/n_i vs x_i, $1 \le i \le k$.

(iv) Retain the plot on the screen by using the command *hold*, and then plot
the fitted logistic curve.

3.11 Construct a MATLAB m-file to evaluate the beta-geometric negative log-likelihood. You can base this on the code in Figure 3.2(b), but note that as the function is to provide a single value of the negative log-likelihood for a given pair of scalar quantities, (m, t), you do not need the elementwise operations. Use the MATLAB *fminsearch* routine to obtain the maximum-likelihood parameter estimates.

3.12 Use the MATLAB *fminsearch* routine to obtain the maximum-likelihood estimates for the Cauchy likelihood, using the function in Figure 3.8(b).

3.13 Use the *grad* function of Figure 3.3 to check the gradient at the maximum-likelihood estimate in Exercises 3.10 and 3.12.

3.14 Set up m-files to produce the geometric model likelihoods as explained in Section 2.2 for the two sets of fecundability data. Obtain the maximum likelihood estimates of p numerically, using the MATLAB routine *fminbnd* as follows. Suppose one of the negative log-likelihoods is specified in the m-file: *geo.m*, then, in MATLAB, use the command:

$$\mathsf{p} = \mathsf{fminbnd} \; (\text{`geo'}, \; 0, \; 1)$$

Note the format for the m-file:

$$\mathsf{function} \; y = \mathsf{geo}(p)$$
$$\mathsf{data} = \ldots$$
$$\vdots$$
$$\mathsf{y} = \ldots$$

3.15 Suppose data arose from an exponential distribution, with pdf

$$f(x) = \lambda e^{-\lambda x}, \quad \text{for } x \geq 0 \, .$$

We have a random sample of values: $x_1, \ldots, x_m$, and further $(n-m)$ values (with $n > m$) which are right-censored at value τ. Thus the censored values are not known exactly — all we know is that they exceed τ in value. This is analogous to the censoring observed in Table 1.1, where the precise times to conception were not recorded for some individuals, only that they required greater than 12 cycles. Write down the likelihood for the data, and obtain the maximum-likelihood estimate of λ.

3.16* Study the MATLAB code below, provided by Paul Terrill, to calculate an approximate covariance matrix, following use of the MATLAB routine *fminsearch*.

```
% Code to be added to  'fminsearch.m' in order to also
% calculate an approximate variance-covariance matrix after minimising
% minus a log-likelihood
% Code due to Paul Terrill.
%-----------------------------------------------------------------
xestimate=x;                              % Store parameter estimates
vbar = (sum(v(:,1:n+1)')/(n+1))';         % Centroid
x(:)=vbar;
fvbar=eval(evalstr);                      % Function value at centroid
for i=1:n+1;
  exceed=abs(fv(i)-fvbar;                 % Check simplex 'large'enough
    while exceed<eps*10^(4);
    disp('Note: Simplex point expanded to allow ...
        for machine precision');
    v(:,i)=2*(v(:,i)-vbar)+vbar,          % Double the distance
    x(:)=v(:,i);
    fv(i)=eval(evalstr);
    exceed=abs(fv(i)-fbar);
  end
end
t=0; Q=v(:,ot)-(v(:,1))*onesn; Y=zeros(n+1,n+1); for i=1:n
  for j=1+i:n+1
    x(:)=(v(:,i)+v(:,j))/2;
    Y(i,j)=eval(evalstr);                 % calculate Pij
  end
end
Y+Y+flipud(rot90(Y))+diag(fv);            % Y is symmetric
                                          % See Hunt et al (2006, p.75)
B=zeros(n,n);
for i=1:n
  for j=1:n
    B(i,j}=4*(Y(i+1,j+1)+Y(1,1)-Y(1,i+1)-Y(1,j+1));
  end
end
vc=Q*inv(B)*Q';                           % variance-covariance matrix
x=xestimate;                              % parameter estimates
```

3.17 Verify the expression for the likelihood in Equation (2.5) and use the MATLAB routine *fminsearch* to obtain the maximum-likelihood estimates of ϕ and p.

3.18 Run the MATLAB program of Figure 3.7, and record the maximum-likelihood estimates of the two parameters. By changing the starting value, verify that the Newton-Raphson method may diverge.

3.19 Study the MATLAB program below, which produces Figure 3.4.

```
% Program to produce the isometric projection and contours
% of the Cauchy log-likelihood surface
%_____
data=[1.09,-0.23,0.79,2.31,-0.81]; n=length(data);
[a,b]=meshgrid(-1:0.01:2.5, 0.1:0.01:1.6);
loglik=n*log(b);
k=ones(size(a));
for i=1:n
  loglik=loglik-log(b .^2 +(data(i)*k-a) .^2);
end
subplot(2,1,1)
  [cs,h]=contour(a,b,loglik, 50);
subplot(2,1,2)
  meshc(a,b,loglik);
```

3.20 Run the program of Figure 3.2(b), but replacing *contour* by *mesh*. Obtain different views of the surface from rotating it, and present the results by using the *subplot* command.

3.21 (Examination question, University of Nottingham.) An electrical component has failure time given by the gamma, $\Gamma(2, \lambda)$, pdf:

$$f(t) = \lambda^2 t e^{-\lambda t} \quad \text{for} \quad t > 0, \quad f(t) = 0 \quad \text{for} \quad t \le 0,$$

where $\lambda > 0$.
The corresponding cdf is $F(t) = 1 - e^{-\lambda t}(1 + \lambda t)$ for $t > 0$,

$$= 0 \text{ for } t \le 0.$$

For m components the failure times are known exactly, and given by $\{t_i, \ 1 \le i \le m\}$. The failure times of n components are all right-truncated, with truncation times, $\{t_i, \ m + 1 \le i \le m + n\}$.

(i) Show that apart from an additive constant, the log-likelihood may be expressed as

$$\ell(\lambda) = 2m \log \lambda - \lambda \sum_{i=1}^{m+n} t_i + \sum_{i=m+1}^{m+n} \log(1 + \lambda t_i).$$

(ii) The Newton-Raphson iterative method is to be used to find the maximum likelihood estimate of λ. Let $\lambda^{(r)}$ be the value of λ at the rth iteration. Show that

$$\lambda^{(r+1)} = \lambda^{(r)} \left[1 + \frac{2m - \lambda^{(r)} \sum_{i=1}^{m+n} t_i + \lambda^{(r)} \sum_{i=m+1}^{m+n} z_i^{(r)}}{2m + (\lambda^{(r)})^2 \sum_{i=m+1}^{m+n} (z_i^{(r)})^2} \right],$$

where

$$z_i^{(r)} = \frac{t_i}{1 + \lambda^{(r)} t_i}.$$

3.22 Fit appropriate models to the data of Table 1.15.

CHAPTER 4

Basic Likelihood Tools

4.1 Introduction

4.1.1 Standard results

Essential components of the statistician's tool-kit involve procedures for dealing with likelihoods, and our objective in this chapter is to expand on the basic maximum-likelihood approach presented in Chapter 2. Classical statistical inference is basically made up of the three components: point estimation, interval estimation, where we construct confidence intervals or regions, and hypothesis testing. Point estimation by maximum-likelihood was introduced in Chapter 2. Now we consider all three aspects together. Standard references for this material are Cox and Hinkley (1974) and Silvey (1975), and in this chapter we shall outline the basic results. We shall not here provide the detail of the various regularity conditions which need to be satisfied in order for the results to hold. However, non-regular cases will be discussed in Section 5.8.

We shall use the following general notation. Suppose our model has parameters $\boldsymbol{\theta}$, with d elements. The likelihood is denoted by $L(\boldsymbol{\theta})$, and the log-likelihood is given by

$$\ell(\boldsymbol{\theta}) = \log L(\boldsymbol{\theta}) .$$

The vector of length p of *efficient scores*, which we call the *scores vector*, is then given by

$$\mathbf{U}(\boldsymbol{\theta}) = \left(\frac{\partial \ell(\boldsymbol{\theta})}{\partial \theta_1} \frac{\partial \ell(\boldsymbol{\theta})}{\partial \theta_2} \cdots \frac{\partial \ell(\boldsymbol{\theta})}{\partial \theta_d} \right)' .$$

This is the statistical terminology for the gradient vector of Chapter 3.

Note that if we write $\ell(\boldsymbol{\theta}) = \ell(\boldsymbol{\theta}; \mathbf{y})$, to indicate dependence on a vector of data $\mathbf{y}$, with, say, joint pdf $f(\mathbf{y}; \boldsymbol{\theta})$, then $\ell(\boldsymbol{\theta}; \mathbf{y}) = \log f(\mathbf{y}; \boldsymbol{\theta})$. Because of the involvement of the data $\mathbf{y}$, we can form the expectation of the scores vector. We obtain

$$\mathbb{E}\left[\frac{\partial \ell(\boldsymbol{\theta})}{\partial \theta_i}\right] = \mathbb{E}\left[\frac{\partial}{\partial \theta_i} \log f(\mathbf{y}; \boldsymbol{\theta})\right]$$

$$= \mathbb{E}\left[\frac{1}{f} \frac{\partial f}{\partial \theta_i}\right]$$

$$= \int \frac{\partial f}{\partial \theta_i} d\mathbf{y}$$

$$= \frac{\partial}{\partial \theta_i} \int f(\mathbf{y}; \boldsymbol{\theta}) d\mathbf{y} = 0,$$

the first equality in the final line following from the assumption that we can interchange the order of integration and differentiation, and the second from the fact that the integral of the joint pdf is unity.

If $\hat{\boldsymbol{\theta}}$ is the maximum likelihood estimate of $\boldsymbol{\theta}$, then, subject to certain regularity conditions,

$$\mathbf{U}(\hat{\boldsymbol{\theta}}) = \mathbf{0} \ ,$$

and asymptotically — that is for large samples — $\mathbf{U}(\boldsymbol{\theta})$ has the multivariate normal distribution,

$$\mathbf{U}(\boldsymbol{\theta}) \sim N_d(\mathbf{0}, \mathbf{J}(\boldsymbol{\theta})) \ , \tag{4.1}$$

where $\mathbf{J}(\boldsymbol{\theta})$ is *minus the expected Hessian matrix*, with (i, k)th element given by

$$j_{ik}(\boldsymbol{\theta}) = -\mathbb{E}\left[\frac{\partial^2 \ell(\boldsymbol{\theta})}{\partial \theta_i \partial \theta_k}\right] \ .$$

Evaluated at $\hat{\boldsymbol{\theta}}$, the maximum likelihood estimate, the matrix $\mathbf{J}(\hat{\boldsymbol{\theta}})$ is called the *expected Fisher information matrix*.

Also, asymptotically, a very useful and important result for maximum-likelihood estimators is that $\hat{\boldsymbol{\theta}}$ has the multivariate normal distribution,

$$\hat{\boldsymbol{\theta}} \sim N_d(\boldsymbol{\theta}, \mathbf{J}^{-1}(\boldsymbol{\theta})). \tag{4.2}$$

The variance-covariance matrix is now $\mathbf{J}^{-1}(\boldsymbol{\theta})$, whereas it was $\mathbf{J}(\boldsymbol{\theta})$ in expression(4.1). Therefore we often use $\mathbf{J}^{-1}(\hat{\boldsymbol{\theta}})$ to approximate the variance-covariance matrix of $\hat{\boldsymbol{\theta}}$. Also, in practice it is sometimes convenient to use minus the Hessian matrix $\left\{-\frac{\partial^2 \ell(\boldsymbol{\theta}))}{\partial \theta_i \partial \theta_k}\right\}$, evaluated at $\hat{\boldsymbol{\theta}}$, that is, the *observed Fisher information matrix*, rather than $\mathbf{J}(\hat{\boldsymbol{\theta}})$, in order to approximate the asymptotic distribution of $\hat{\boldsymbol{\theta}}$. As we have seen in Figure 3.3, we can also use a numerical approximation to approximate the Hessian matrix. As observed in Section 3.3.3, in some cases we obtain identical estimates of standard errors and correlations from using the Hessian and the expected Hessian, and we provide examples of that in Chapter 8. In Table 3.1 we made one comparison between the performance of the two information matrices, in the context of function optimisation. In a particular case, Efron and Hinkley (1978) found it better to use the Hessian, but Catchpole and Morgan (1996) found that the expected Hessian performed better when used for particular score tests, the topic of Section 4.6; see also Morgan et al. (2008).

Of course, in practice we only ever approximate the asymptotic situation, but it is standard to use the result of expression (4.2) to give an indication of confidence intervals or regions whenever maximum likelihood estimation is used. This is certainly a useful approach to adopt; however, as we shall see, the asymptotic approximation is sometimes not very good, and can be improved.

4.1.2 Derivation of distributional results

The full details of why maximum-likelihood estimators are usually asymptotically normally distributed, why the likelihood-ratio test statistic usually has a chi-square distribution asymptotically, etc., are provided by Cox and Hinkley (1974, Chapter 9) and Silvey (1975, pp.74, 112). However it is useful here to have a brief heuristic explanation, as the material is both very important and basically straightforward. The key is provided through the use of Taylor series expansions.

In the case of a scalar parameter θ, if we truncate a Taylor-series expansion for $\frac{d\ell}{d\theta}$, we obtain, subject to regularity conditions,

$$0 = \frac{d\ell}{d\theta}(\hat{\theta}) \approx \frac{d\ell}{d\theta}(\theta_0) + (\hat{\theta} - \theta_0)\frac{d^2\ell}{d\theta^2}(\theta_0),$$

for some fixed value of θ_0.
Hence

$$\hat{\theta} - \theta_0 \approx -\frac{\dfrac{d\ell}{d\theta}(\theta_0)}{\dfrac{d^2\ell}{d\theta^2}(\theta_0)},$$

which generalises when we have a vector $\boldsymbol{\theta}$ of parameters to the expression:

$$\hat{\boldsymbol{\theta}} - \boldsymbol{\theta}_0 \approx \mathbf{I}^{-1}(\boldsymbol{\theta}_0)\,\mathbf{U}(\boldsymbol{\theta}_0),$$

when $\mathbf{I}$ is *minus* the Hessian matrix, i.e., the observed Fisher information matrix. This corresponds to the Newton-Raphson iteration of Equation (3.3). We shall assume that $\mathbf{I}(\boldsymbol{\theta}_0)$ is non-singular. See Dobson (1990, p.50) for discussion of the singular case.

When we have a random sample of size n, the log-likelihood is the sum of n terms, the jth of which is the contribution of the jth member of the sample to the log-likelihood, $1 \leq j \leq n$. Thus in this case, the last result can then be written in the form,

$$\mathbf{I}(\boldsymbol{\theta}_0)(\hat{\boldsymbol{\theta}} - \boldsymbol{\theta}_0) \approx \sum_{j=1}^{n}\mathbf{U}_j(\boldsymbol{\theta}_0),$$

where $\mathbf{U}_j(\boldsymbol{\theta}_0)$ is the score vector from the jth element of the sample, and application of the central limit theorem to the summation explains why $\hat{\boldsymbol{\theta}}$ has an approximate normal distribution for large n.

Let us now consider the scalar case again. The null hypothesis that $\theta = \theta_0$, when tested by a likelihood-ratio test, produces the test statistic:

$$-2\log\lambda = 2\{\ell(\hat{\theta}) - \ell(\theta_0)\}\,,$$

where λ traditionally denotes the ratio of likelihoods. For discussion of likelihood-ratio tests, see Silvey (1975, p.109). If we now apply a truncated Taylor series expansion to $\ell(\theta_0)$, we obtain:

$$-2\log\lambda \approx 2\left\{\ell\left(\hat{\theta}\right) - \ell(\hat{\theta}) - (\hat{\theta}-\theta_0)\frac{d\ell}{d\theta}\left(\hat{\theta}\right) - \frac{1}{2}\left(\hat{\theta}-\theta_0\right)^2\frac{d^2\ell}{d\theta^2}\left(\hat{\theta}\right)\right\}.$$

In this case, we have taken three terms in the truncated Taylor series, for the good reason that $\dfrac{d\ell}{d\theta}(\hat{\theta}) = 0$, by the definition of $\hat{\theta}$, subject to regularity conditions, as stated above. Hence

$$-2\log\lambda \approx -(\hat{\theta}-\theta_0)^2\frac{d^2\ell}{d\theta^2}\left(\hat{\theta}\right),$$

which generalises when we have a vector $\boldsymbol{\theta}$ of parameters, to the expression:

$$-2\log\lambda \approx (\hat{\boldsymbol{\theta}} - \boldsymbol{\theta}_0)'\,\mathbf{I}(\hat{\boldsymbol{\theta}})\,(\hat{\boldsymbol{\theta}} - \boldsymbol{\theta}_0)$$

(recall that $\mathbf{I}$ is minus the Hessian matrix).

The well-known asymptotic chi-square result for the sampling distribution of $-2\log\lambda$ is then the consequence of the multivariate normal distribution of $\hat{\boldsymbol{\theta}}$, given in expression (4.2) — see Appendix A. We shall see later that the result of expression (4.2) also justifies the asymptotic distribution given later for the Wald test, in Section 4.6. Because we showed above that $(\hat{\boldsymbol{\theta}} - \boldsymbol{\theta}_0) \approx \mathbf{I}^{-1}(\boldsymbol{\theta}_0)\,\mathbf{U}(\boldsymbol{\theta}_0)$ and as $\mathbf{I}^{-1}$ is a symmetric matrix, we can also write

$$-2\log\lambda \approx \mathbf{U}(\boldsymbol{\theta}_0)'\,\mathbf{I}^{-1}(\boldsymbol{\theta}_0)\,\mathbf{U}(\boldsymbol{\theta}_0) \tag{4.3},$$

from essentially approximating $\mathbf{I}(\hat{\boldsymbol{\theta}})$ by $\mathbf{I}(\boldsymbol{\theta}_0)$. It is the expression of (4.3) which justifies the asymptotic result given later for the score test, also in Section 4.6.

4.2 Estimating standard errors and correlations

As commented in Section 2.2, error estimation based on expression (4.2) for the simple case of the geometric model with one parameter, p, just involves obtaining

$$\left(\frac{d^2\ell}{dp^2}\right)^{-1},\qquad\text{or}\qquad\left(\mathbb{E}\left[\frac{d^2\ell}{dp^2}\right]\right)^{-1},$$

and evaluating it at the maximum-likelihood estimate, $\hat{p}$.

Intuitively these are sensible measures to take, as they incorporate the flatness of the log-likelihood curve at the maximum. The flatter the curve is, then the smaller $d^2\ell/dp^2$ will be, resulting in a larger estimate of error.

In order to illustrate the general use of expression (4.2) for estimating standard errors and correlations in the case of several parameters, we return now to the beta-geometric model applied to the fecundability data of Table 1.1.

Example 4.1: Error-estimation for the beta-geometric model

From using expression (4.2) we obtain the results of Table 4.1. It is the point estimates of Table 4.1 that produce the fitted values in Table 2.2.

Table 4.1 *Maximum-likelihood estimates of (μ, θ) when the beta-geometric model of Section 2.2 is fitted to the data of Table 1.1. Also shown are the asymptotic estimates of standard error and of correlation, obtained from expression (4.2).*

(i) Smokers data

Parameter	Estimate	Errors/correlation	
		μ	θ
μ	0.275	0.036	
θ	0.091	0.739	0.060

(ii) Non-smokers data

Parameter	Estimate	Errors/correlation	
		μ	θ
μ	0.408	0.020	
θ	0.137	0.651	0.032

□

The convention of Table 4.1 is a convenient one for reporting the consequences of expression (4.2), viz: the estimates are presented in the initial column, and then the diagonal of the following triangular matrix presents appropriate asymptotic estimates of standard error, from taking the square roots of the diagonal entries of $\mathbf{J}^{-1}(\hat{\boldsymbol{\theta}})$. The remaining lower triangle of the matrix then gives appropriate asymptotic estimates of product-moment correlations, obtained in the usual way from the estimated variances and covariances in $\mathbf{J}^{-1}(\hat{\boldsymbol{\theta}})$. In Figure 4.1, we provide a short MATLAB program to produce approximate errors and correlations, and to present them in the format of Table 4.1.

The estimated errors for the non-smokers are roughly half those for the smokers, as we might expect from observing that there were 486 non-smokers and 100 smokers.   □

Estimates of correlation between estimators are often useful in suggesting a re-parameterisation of the model, a subject to which we return in Chapter 5, or even, if they are close to ±1, in indicating a ridge to the log-likelihood surface, which may in turn suggest an over-parameterised model. Rough 95%

```
function l=fmax(funfcn,x0)
%
%FMAX  minimises a negative log-likelihood, provided by 'funfcn'
%  and also produces, (i) an estimate of the gradient at the
%  optimum, to check for convergence; (ii) An estimate
%  of the Hessian at the maximum, and its eigen-values;
%  (iii) an estimate of standard errors and correlations,
%  using the observed Fisher information - see expression (4.2);
%  x0 is the starting-value for the simplex method used in
%  'fminsearch'
%  Calls GRAD, HESSIAN and FMINSEARCH.
%_____
x=fminsearch((funfcn),x0);     % x is the maximum likelihood
g=grad((funfcn),x)             % estimate
h=hessian((funfcn),x); d=eig(h); % eig finds eigen values
disp(d)
cov=inv(h);                    % this now approximates the
                               % dispersion (var/cov) matrix

k=sqrt(diag(cov));             % the approximation to the std.,
                               % errors
k1=diag(k);k2=inv(k1);
cor=k2*cov*k2;                 % this now approximates
                               % the correlation matrix
tril(cor)                      % tril extracts the lower triangular
                               % part of a matrix;
                               % tril(cor -1) removes the
cor=tril(cor,-1 )+k1;          % diagonal prior to replacement
                               % with std., errors.
out=[x' cor];                  % gives output in standard format
disp(out)
```

Figure 4.1 *A* MATLAB *program to obtain maximum-likelihood estimates and approximations to standard errors and correlations, using expression (4.2) but with the observed, rather than expected, information matrix, and numerical approximations to derivatives.*

confidence intervals for each of the parameters taken singly can be obtained by assuming that asymptotic normality holds, and taking ± 1.96 times the standard error for the parameter, though this ignores the effect of the correlation structure, a point to which we shall return later. When estimates of single parameters are reported it is usual to write, for example, $\hat{\mu} = 0.408(0.020)$, so that we present simultaneously the point estimate of the parameter, and in parentheses its estimated standard error. This was done in Example 2.1. The value of the above way of representing results can be appreciated when there are several parameters, as in the next example.

Example 4.2: A model for misreporting

We now return to the fecundability data of Exercise 2.2. This demonstrates substantial digit preference, with individuals tending to respond 3, 6 or 12 more frequently than other digits. Ridout and Morgan (1991) suggest the following probability model for this situation:

We suppose that having waited k fertility cycles until conception, individuals in fact *report* values of 3, 6, or 12 instead, with the following probabilities:

$$p_{k,3} = \begin{cases} \gamma^{|3-k|}, & \text{for} \quad k = 4,5, \\ 0, & \text{otherwise}, \end{cases}$$

$$p_{k,6} = \begin{cases} \alpha^{|k-6|}, & \text{for} \quad 4 \le k \le 11, \\ 0, & \text{otherwise}, \end{cases}$$

$$p_{k,12} = \begin{cases} \beta^{|12-k|}, & \text{for} \quad 7 \le k \le 17, \\ 0, & \text{otherwise}, \end{cases}$$

for suitable values of the parameters, $0 \le \alpha, \beta, \gamma \le 1$. This is just one possible, hypothesised mechanism for digit preference. It appears to work well in practice, but others are also possible. See Copas and Hilton (1990) for related work.

Suppose now that the reported numbers of cycles have probabilities $\{p_i\}$ as follows:

Cycles	1	2	$\cdots$	12	> 12
Observed number	n_1	n_2		n_{12}	n_+
Probability	p_1	p_2		p_{12}	$1 - \sum_{i=1}^{12} p_i$

The likelihood, which in this case is a function of the five parameters, $\mu, \theta, \alpha, \beta$ and γ, can be written:

$$L(\mu, \theta, \alpha, \beta, \gamma; \mathbf{n}) \propto \left(\prod_{i=1}^{12} p_i^{n_i} \right) \left(1 - \sum_{i=1}^{12} p_i \right)^{n_+},$$

and we complete the model specification as follows:

$$p_1 = \phi_1$$
$$p_2 = \phi_2$$
$$p_3 = \phi_3 + \phi_4 p_{4,3} + \phi_5 p_{5,3}$$
$$p_4 = \phi_4(1 - p_{4,3} - p_{4,6})$$
$$p_5 = \phi_5(1 - p_{5,3} - p_{5,6})$$
$$p_6 = \sum_{j=4}^{11} \phi_j p_{j,6}$$
$$p_j = \phi_j(1 - p_{j,6} - p_{j,12}), \quad 7 \leq j \leq 11,$$
$$p_{12} = \sum_{j=7}^{17} \phi_j p_{j,12},$$

where the $\{\phi_i\}$ correspond to the true beta-geometric model. For example, the value 3 results either from 3 being the correct number of cycles waited, with probability ϕ_3, or from 4 being correct and 3 being misreported, with probability $\phi_4 p_{4,3}$, or from 5 being correct and 3 being misreported, with probability $\phi_5 p_{5,3}$.

The results from fitting this model to the data of Exercise 2.2 are given in Table 4.2. We might *a priori* expect confusion to increase with the length of the time to conception, and indeed we do find that $\hat{\gamma} < \hat{\alpha} < \hat{\beta}$, for both of the examples considered here. There is a further discussion of this example in Exercise 4.1.   □

4.3 Looking at surfaces: profile log-likelihoods

Near their maxima, log likelihood surfaces can be well approximated by a quadratic surface. The message of expression (4.2), however, is that this approximation is also good away from the optimum. Asymptotically this is true, but typically in statistics our samples of data are 'small.' Ideally, therefore, we would like to investigate the shape of likelihood surfaces, and not just make the blanket assumption of expression (4.2). What we see will also depend on the *parameterisation* used in the model, and we shall now present an illustration of this in the following example.

Example 4.3: Reparameterisation of the logit model

When we fit the logit model of Exercise 3.10, we obtain the log-likelihood surface of Figure 4.2(a). An alternative parameterisation is to set $P(x)$, the probability of response to dose x, as: $P(x) = \left(1 + e^{-\beta(x-\mu)}\right)^{-1}$. Comparison with the previous parameterisation of $P(x) = \left(1 + e^{-(\alpha+\beta x)}\right)^{-1}$ reveals that $\mu = -\alpha/\beta$. Note that we are fitting a two-parameter model in each case, and the estimates of one set of the parameters are easily obtained from the estimates of the other set. When $x = \mu$, $P(x) = 0.5$, and μ is therefore a

Table 4.2 *The results from maximum-likelihood fitting of the model for misreporting to the fecundability data of Exercise 2.2. We note that the point estimates for the parameters, α, β and γ, satisfy the order: $\hat{\gamma} < \hat{\alpha} < \hat{\beta}$, and we comment on this in the text.*

Pill users

Parameter	Estimate	Errors/correlations				
		μ	θ	α	β	γ
μ	0.309	0.007				
θ	0.048	0.690	0.009			
α	0.342	0.086	0.066	0.029		
β	0.595	−0.003	0.008	−0.013	0.030	
γ	0.205	−0.151	−0.105	−0.367	0.052	0.044

Other contraceptive users

Parameter	Estimate	Errors/correlations				
		μ	θ	α	β	γ
μ	0.418	0.005				
θ	0.242	0.641	0.011			
α	0.346	0.036	0.002	0.018		
β	0.635	−0.019	0.051	−0.162	0.017	
γ	0.132	−0.075	−0.022	−0.329	0.054	0.031

parameter with a very useful interpretation, as it is the dose resulting in a response probability of 50%. It is sometimes denoted as the ED_{50}, that is, the dose with an expected 50% response. The log-likelihood surface now appears as shown in Figure 4.2(b). The shape of the contours suggests that the estimators $(\hat{\mu}, \hat{\beta})$ have little correlation, in direct contrast to the pair $(\hat{\alpha}, \hat{\beta})$. For further discussion, see Exercise 4.2.

Looking at surfaces is more difficult when the number of parameters in a model, d say, is greater than 2. We often have $d > 2$, and one useful approach is then to construct what we call *profiles*. Any two-parameter likelihood is a surface in three-dimensional space. If that surface is constructed on a horizontal potter's wheel, and illuminated by a light shining horizontally, then the shadow, or profile, cast on a vertical screen directly opposite the light will reveal useful features of the surface, especially maxima, in which of course we are particularly interested. Rotating the wheel will change the profile and add previously hidden information. In our context, rotation corresponds to a particular reparameterisation. Rotation is readily accomplished in MATLAB, and simple profiles from a two-parameter surface are easily obtained. In Figure 4.3, we see the result of using the MATLAB *view* command. Free rotation

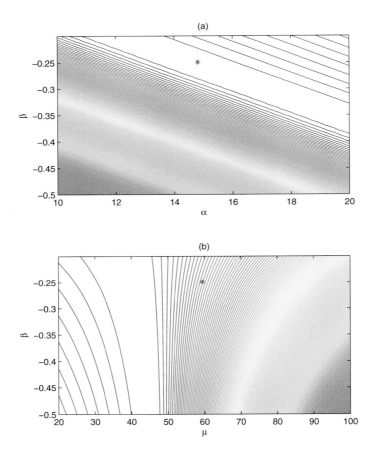

Figure 4.2 *The effect on the log-likelihood contours of changing the parameterisation of a model. Here we have the logit model with the alternative parameterisations:* (a) $P(x) = (1 + e^{-(\alpha+\beta x)})^{-1}$; (b) $P(x) = (1 + e^{-\beta(x-\mu)})^{-1}$. *In each case the location of the maximum-likelihood estimate is shown.*

of an isometric projection is easily obtained in MATLAB by clicking on the 'rotate' icon in the graphics window displaying the surface, and then using the mouse to rotate the surface.

Profiles are particularly valuable when $d > 2$. The profile log-likelihood for a particular parameter θ is a plot of the log-likelihood maximised over all of the parameters except θ, vs θ. To obtain profile log-likelihoods in MATLAB when $d > 2$ is also a straightforward procedure, using the approach which is illustrated in Figure 4.4, for $d = 2$.

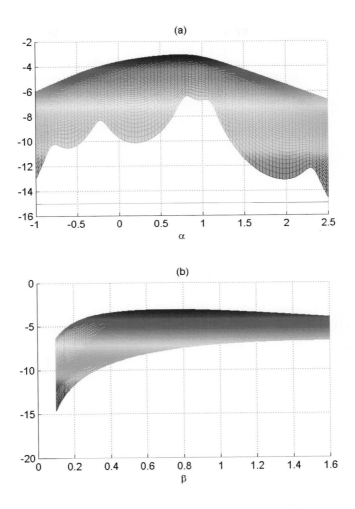

Figure 4.3 *Using the* MATLAB *command view to obtain simple profiles of a two-parameter Cauchy log-likelihood surface. We obtain the isometric projection of Figure 3.4 (a) by means of the program in Exercise 3.19. Then typing view (0,0) produces the result of (a), while typing view (90,0) produces the result of (b).*

(a)
```
% Program to maximise the beta-geometric log-likelihood
% while holding the first variable (t) fixed, and then repeating
% this over a grid of t-values in order to obtain a profile
% log-likelihood with respect to the first variable.
% The scalar 't' is global, so that the function 'beopt' can access
% the value at which 't' is fixed. 'n' is the grid size.
% Calls FMINBND, BEOPT.
%------------------------------------------------------------------
global t
 n=100; t=0;
 for i=1:n
  t=t+.01;
  x(i)=fminbnd('beopt',0.001,0.999);   % this obtains the value of the
                                       % second parameter,
                                       % which maximises the
                                       % log-likelihood when the first
                                       % parameter is fixed at t.
  y(i)=beopt(x(i));                    % this is the value for the profile
                                       % log-likelihood.
end
t1=.01:.01:1;                          % the grid of t-values for plotting
plot(x,t1);
xlabel('\mu');ylabel('\theta')         % this is the path traced in the
                                       % parameter space, as we move along
                                       % the profile log-likelihood.
plot(y,t1)                             % this is the profile log-likelihood
xlabel('- log-likelihood');ylabel('\theta')
```

(b)
```
function y=beopt(x)
%
% BEOPT obtains the beta-geometric log-likelihood
%   as a function of just the one parameter (in this case the first one)
%   so that optimisation can take place, using fminbnd, with respect to
%   that parameter alone, the other, the second in this case, being fixed.
%   This is for obtaining a profile log-likelihood.
%   The first variable, 't,' is set in the driving program, and is global.
%   Calls BEGEO, which provides the negative log-likelihood for the
%   beta-geometric model.
%------------------------------------------------------------------
global t
z=[x t];
y=begeo(z);
```

Figure 4.4 *A general approach for obtaining a profile log-likelihood, illustrated on a two-parameter beta-geometric log-likelihood. The program calls the function beopt, which in turn calls the function begeo.*

```
(c)
function y=begeo(x)
%
% BEGEO calculates the beta-geometric negative log-likelihood.
%    mu and theta are the beta-geometric parameters;
%    'data' are global, and set in the calling program
%-----------------------------------------------------------------
global data
mu=x(1);theta=x(2);
n=length(data)-1;                   % n denotes the largest number
loglik=log(mu)*data(1);             % observed, before censoring
p=mu; s=p;
for i=2:n
  p=p*(1-(mu+theta)/(1+(i-1)*theta));
                                    % this is the recursive way of
                                    % generating beta-geometric
                                    % parameters
  s=s+p;
  loglik=loglik+log(p)*data(i);
end
p=1-s;                              % the right-censored probability
n1=n+1;
y=loglik+log(p)*data(n1);
y=-y;
```

Figure 4.4 *Continued.*

4.4 Confidence regions from profiles

In Section 4.2, we saw how approximate confidence intervals can be obtained following point estimation by maximum likelihood. Now we shall construct confidence regions based on hypothesis testing, in particular on a likelihood-ratio test. We suppose again that we have a d-dimensional parameter vector $\boldsymbol{\theta}$.

Consider testing the null hypothesis, $\boldsymbol{\theta} = \boldsymbol{\theta}_0$, against the general alternative that $\boldsymbol{\theta} \neq \boldsymbol{\theta}_0$. If we denote the likelihood obtained from data $\mathbf{x}$ by $L(\boldsymbol{\theta}; \mathbf{x})$, then the test statistic for a likelihood-ratio test is

$$\lambda = \frac{L(\boldsymbol{\theta}_0; \mathbf{x})}{L(\hat{\boldsymbol{\theta}}; \mathbf{x})} \tag{4.4}$$

where $\hat{\boldsymbol{\theta}}$ is the maximum likelihood estimate of $\boldsymbol{\theta}$. If the null-hypothesis is true, then, subject to regularity conditions, asymptotically, $-2 \log \lambda$ has a χ_d^2 distribution, as we have seen from the work of Section 4.1.2. Let $\chi_{d:\alpha}^2$ denote the $\alpha\%$ point for the χ_d^2 distribution, that is, the value which is exceeded with probability $\alpha\%$ by a random variable with a χ_d^2 distribution. The likelihood-ratio test of size $\alpha\%$ then has the acceptance region given by

$$-2 \log \lambda \leq \chi_{d:\alpha}^2 \,,$$

that is, $2\{\ell(\hat{\boldsymbol{\theta}};\mathbf{x}) - \ell(\boldsymbol{\theta}_0;\mathbf{x})\} \leq \chi^2_{d:\alpha}$,

where as usual $\ell(\boldsymbol{\theta};\mathbf{x}) = \log\{L(\boldsymbol{\theta};\mathbf{x})\}$. We see therefore that a *confidence set* for $\boldsymbol{\theta}$ with *confidence coefficient* $(1-\alpha/100)$ consists of all values of $\boldsymbol{\theta}$ for which the log-likelihood $\ell(\boldsymbol{\theta};\mathbf{x})$ lies within $\frac{1}{2}\chi^2_{d:\alpha}$ of the maximum value, $\ell(\hat{\boldsymbol{\theta}};\mathbf{x})$. This is shown for the case $d=1$ in Figure 4.5, while the illustration of Figure 4.6 and Example 4.4 has $d=2$.

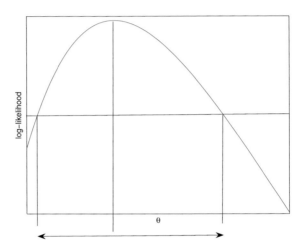

Figure 4.5 *Obtaining a likelihood-ratio based confidence interval with confidence coefficient $(1-\alpha)$ for a scalar parameter θ, by taking a suitable slice of the log-likelihood. The horizontal line is taken at a distance of $-\frac{1}{2}\chi^2_{1:\alpha}$ below the maximum.*

Example 4.4: Two-dimensional confidence regions

Consider now the data of Exercise 2.4. Shown in Figure 4.6 are several two-dimensional likelihood-ratio-based confidence regions for the parameters (δ, λ). We can see how they differ from the ellipses which would follow from using expression (4.2). This illustration is taken from Morgan et al. (2007). □

Frequently of course there are what are called *nuisance* parameters: in Example 4.4, we may just wish to focus on μ, for instance, in which case τ would then play the rôle of the nuisance parameter. The extension to the case of nuisance parameters is straightforward, following the rules of the likelihood-ratio test, as follows. Suppose we partition the parameter vector $\boldsymbol{\theta} = (\boldsymbol{\phi}, \boldsymbol{\psi})$, with $\boldsymbol{\phi}$ denoting the parameters (say d_1 of them) of interest. In Equation (4.4) the denominator remains unchanged, but in the numerator we take the maximum of $L(\boldsymbol{\theta};\mathbf{x})$, with regard to the nuisance parameters $\boldsymbol{\psi}$ in $\boldsymbol{\theta} = (\boldsymbol{\phi}_0, \boldsymbol{\psi})$; thus the $\boldsymbol{\phi}$ parameters are fixed at $\boldsymbol{\phi}_0$. The appropriate slice of the log-likelihood surface which provides the desired confidence region is then taken at a depth

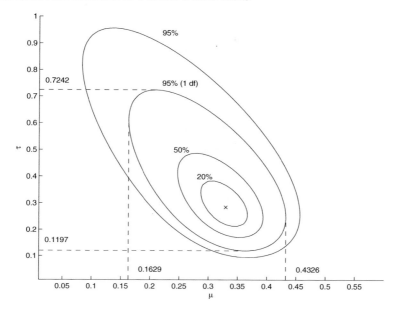

Figure 4.6 *Confidence regions resulting from a model which is an elaboration of one considered in Exercise 5.41. Shown are the 95% confidence region for a pair of the model parameters (μ, τ) together, as well as the 95% confidence-interval for the parameters taken separately. Figure taken from Morgan et al. (2007) and reproduced with permission from* Biometrics.

$\chi^2_{d_1 : \alpha}$. In Example 4.4, the resulting intervals for parameters τ and μ, taken separately, are also shown in Figure 4.6. The intervals are narrower than the appropriate projections of the two-dimensional confidence region, as has to be true — see Exercise 4.3. We see from Figure 4.6 that marginal confidence intervals or regions can be misleading when estimators appear to be highly correlated. It is partly for this reason that we often seek parameterisations of models that result in low correlations between estimators, and we revisit this issue in Section 5.2. Maximising over ψ alone is equivalent to finding the profile with respect to ϕ, and so confidence regions based on the likelihood-ratio test correspond to taking appropriate slices of profile log-likelihoods (as well as of the log-likelihood surface). Three examples of obtaining confidence intervals from profiles now follow. Empirically, evidence favours intervals and regions derived from likelihood-ratio tests, rather than from point estimates. See, for example, Lawless (2002, p.525), Williams (1986), Morgan and Freeman (1989), and Cormack (1992). This is not surprising, given that the approach based on likelihood-ratio tests explores the shape of log-likelihood surfaces away from the maximum. Traditionally it has been far simpler to invert a matrix, which is the operation required in expression (4.2), rather than to carry out the

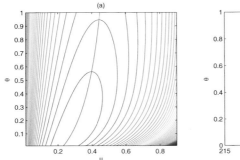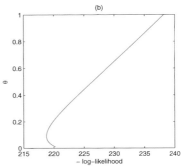

Figure 4.7 *(a) The negative log-likelihood surface for the beta-geometric model and the smokers data from Table 1.1. The minimum is at 218.77. The 95% confidence region for both parameters simultaneously is therefore the contour of value* $218.77 + \frac{1}{2}\chi^2_{2;5} = 218.77 + 2.966 = 221.77$. *Also shown is the path followed in the parameter space corresponding to the profile log-likelihood for* θ. *(b) Minus the profile log-likelihood for* θ.

range of optimisations needed to construct a profile. This is no longer true — see, e.g., Venzon and Moolgavkar (1988) — though the former approach will be faster, and is simpler from a presentational point of view, particularly when the number of parameters involved is large.

If the data contain little information on the nuisance parameters then it may be necessary to modify the profile log-likelihood. This can occur if there are many nuisance parameters, for example. For discussion, see Barndorff-Nielsen and Cox (1994, Chapter 8). An illustration is provided by Grieve (1996). We note finally that confidence regions that result from taking sections of profiles may not be connected, but rather can be made up of several disjoint parts. Illustrations are provided by Williams (1986) and in Example 4.5.

Example 4.1 continued

Continuing with error estimation for the beta-geometric model, Figure 4.7 presents the contours of the likelihood surface for the non-smokers and the 95% confidence region for both parameters simultaneously. Also shown in Figure 4.7 is the profile for the parameter θ considered separately. We can see the path traced out by the profile in the (μ, θ) space. Taking the case of a single parameter as an example, it is important to realise that profiles do not usually simply follow from taking a section of a likelihood parallel to the axis of the parameter of interest. The MATLAB function contour($\mathbf{Z}$) produces a contour plot of the matrix $\mathbf{Z}$ with automatic selection of the number of contour lines and their number. However contour $(\mathbf{Z}, \mathbf{v})$ provides contour levels specified in the vector $\mathbf{v}$, which is useful for plots such as that of Figure 4.7(a).   □

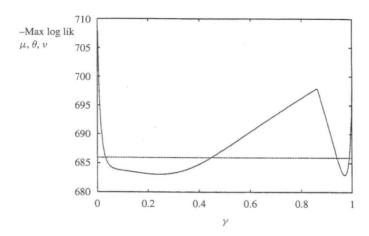

Figure 4.8 *A plot of minus the profile log-likelihood plotted against the mixing parameter γ in the mixture of a beta-binomial and a binomial distribution, for the data of Exercise 3.2. Also shown (horizontal line) is a section based on the χ_2^2 distribution to give an approximate 95% confidence interval for γ. The reason for basing the section on the χ_2^2 distribution is given in the source paper. From Brooks et al. (1997). Reproduced with permission from Biometrics.*

Example 4.5: A profile log-likelihood with two optima

Continuing the application of Exercise 3.3, in Figure 4.8 we present the profile for γ. In this case there are two alternative local optima, corresponding to two different interpretations of the mixture model. We can see that the 95% confidence interval for γ, corresponding to the section shown, is made up of two disjoint parts. The interpretation favoured by Brooks et al. (1997) is that of a mixture which is primarily beta-binomial, but contaminated by a binomial with a large probability ($\hat{p} = 0.74$), of death.   □

Example 4.6: A likelihood ridge

Here we continue the modelling introduced in Exercise 1.4. A model for avian annual mortality with complete age-dependent annual survival was proposed by Cormack (1970) and Seber (1971). The parameter set is now $(\lambda, \phi_1, \phi_2, \ldots, \phi_k)$. When this model is fitted to recoveries data of the kind illustrated in Table 1.5, but with known numbers of birds ringed in each cohort, we find that the likelihood does not possess a single maximum, but instead is maximised along a completely flat ridge. The nature of this ridge is explained by Catchpole and Morgan (1994). The profiles of Figure 4.9 reveal clearly the flatness of the likelihood surface. We discuss how a Bayesian analysis copes with the ridge to this likelihood surface in Exercise 7.22.   □

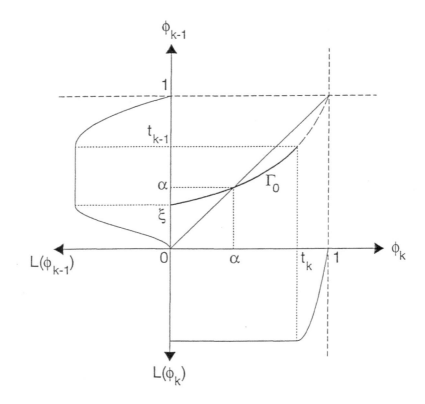

Figure 4.9 *The curve* Γ_0 *is the projection onto the* (ϕ_{k-1}, ϕ_k) *plane of a curve* Γ *in the full model parameter space which corresponds to a completely flat ridge of the likelihood surface for the bird survival model with parameters* $(\lambda, \phi_1, \phi_2, \ldots, \phi_k)$. *In the margins are shown the profile likelihoods, denoted by* L, *for* ϕ_{k-1} *and* ϕ_k, *which reveal the flatness of the likelihood surface. From Catchpole and Morgan (1994). Reproduced by permission of the Royal Statistical Society.*

4.5 Hypothesis testing in model selection

For any set of data we shall usually have a family of alternative models. An important case is when these are *nested*, so that simpler members of the family of models are the result of restrictions on the parameters of more complex models. We shall now present three examples to illustrate the nesting of models, and to show how likelihood-ratio tests and information criteria may be used for choosing between models.

Example 4.7: Digit preference

The model for digit preference for fecundability data of Example 4.2 contains the beta-geometric model ($\alpha = \beta = \gamma = 0$) which in turn contains the geometric model ($\theta = 0$). In these cases there are instances of parameters being fixed at an end of their range, and we discuss this feature later in Section 5.8. □

Example 4.8: Dose-response data over time

Consider now the data of Example 1.4. These data can be viewed as constrained and censored survival data — see for example Aranda-Ordaz (1983) and Wolynetz (1979). We can model the probability that an individual beetle dies in the time interval $(t_{j-1}, t_j]$, at dose d_i, by

$$P_{ij} = \psi(t_j, d_i) - \psi(t_{j-1}, d_i) \,,$$

for a suitable function $\psi(\ )$. For the flour beetle data, Pack and Morgan (1990) obtained a good description of the data with the function,

$$\psi(t_j, d_i) = \{(1 + e^{-(\alpha_1 + \alpha_2 \log d_i)})(1 + t_j^{-\kappa}\beta)\}^{-1} \,, \tag{4.5}$$

where the parameter set is $(\alpha_1, \alpha_2, \kappa, \beta)$. We can interpret this model as follows: the first term on the right-hand side of Equation (4.5) gives the probability of response at dose d_i, as in Exercise 2.3, and the second term is the cdf of a log-logistic distribution, producing, for susceptible individuals, the cdf of response times, *irrespective* of the dose level. This last feature is one that is sometimes imposed in a model, as in McLeish and Tosh (1990). However it arose naturally as a result of modelling the flour beetle data.

A more general approach is to embed the model of Equation (4.5) in a more complex framework, and then carry out hypothesis tests to see if simpler forms will suffice for a particular data set. For example, for the time aspect of the model, we could set the cdf of response times to be

$$F(t, d) = \left\{ \begin{array}{ll} 1 - \{1 + \lambda\eta(t, d)\}^{-1/\lambda} \,, & \lambda \neq 0 \\ 1 - \exp\{-\eta(t, d)\} \,, & \lambda = 0, \end{array} \right.$$

where $\eta(t, d) = \gamma t^\kappa$, involving parameters λ, γ and κ, but no longer depending on d.

When $\lambda = 1$ this reduces to the form in Equation (4.5), which is a representative of the *proportional odds* family of models. However when $\lambda = 0$ we get a Weibull distribution, an example of a *proportional hazards* model. Similarly, we can extend the first part of the right-hand side of Equation (4.5) in a variety of ways — see Exercise 4.6.   □

Example 4.9: Bird survival

Freeman and Morgan (1992) extended the bird survival model of Example 4.6 to allow time-dependence of both the first-year survival probability and the annual reporting probability, to produce a model with parameters: $(\{\lambda_i\}, \{\phi_{1i}\}, \phi_2, \dots, \phi_k)$. Here λ_i is the probability of rings from dead birds being reported in the ith year of the study, irrespective of when the birds were ringed, and ϕ_{1i} is the probability a first-year bird survives its first year of life when that is the ith year of the study. Time variation in either of these parameters might, for example, correspond to annual (especially winter) variation in temperature.   □

Likelihood-ratio tests provide an obvious and convenient way of selecting between the alternatives in a nested family of models. A useful omnibus benchmark for the resulting test statistics is provided by the chi-square distribution, subject to regularity conditions being satisfied, and the realisation that the chi-square result is an asymptotic one. There are alternative criteria for model-selection, and much use is also made of *information criteria*. For instance, the Akaike Information Criterion (AIC) is defined as

$$\text{AIC} = -2\log L + 2d,$$

where L is the likelihood of the fitted model, and d is the number of parameters estimated. In contrast, the Bayesian information criterion (BIC) is defined as:

$$\text{BIC} = -2\log L + d\log n,$$

where n is the number of observations. For a range of different models we can choose between them by selecting the model with the smallest value of an information criterion. This approach penalises models with large numbers of parameters, with a greater penalty for the BIC, than for the AIC. The approach makes no explicit reference to asymptotic criteria or to distributional results, and provides a very convenient and simple way of discriminating between large numbers of competing models. For a justification of this approach, see Burnham et al. (1995) and Anderson and Burnham (1998), who also discuss alternative information measures. In principle, information criteria can also be used for selecting between non-nested models. In a simulation study, Wang et al. (1996) found that the BIC slightly outperforms the AIC. Brooks et al. (1997) used both the AIC and BIC, obtaining similar conclusions to Wang et al. (1996).

We conclude this section with two examples of model-selection at work.

Example 4.2 continued

We have already seen, in Table 4.2, the results of the maximum-likelihood fit of the digit confusion model to the data of Exercise 2.2. At this stage in modelling it is always wise to make initial observations in terms of the point estimates and their standard errors. For instance, for the data sets for both the pill and other contraceptive users, it seems fairly clear (by forming crude confidence intervals) that all of the confusion parameters are significantly different from 0, for example $0.342 \pm (1.96 \times 0.029)$ does not include 0. This is, in fact, a simple instance of carrying out a Wald test, which is described later. Additionally it seems clear that in both cases, $\theta \neq 0$. Table 4.3 provides the necessary information for model selection, while fitted values are given in Table 4.4. Here we shall only use the AIC — see Exercise 4.7.

For both data sets, the AIC selects model 4, which is also the only model to provide an acceptable fit to the data, as judged by the Pearson X^2 value.

Table 4.3 *Model-selection for digit confusion: data of Exercise 2.2; d denotes the number of fitted parameters; ℓ denotes the maximised log-likelihood.*

Model	d	Pill users X^2 (df)	AIC	-2ℓ	Other contraceptive users X^2 (df)	AIC	-2ℓ
1. beta-geometric	2	224.12 (10)	5256	5252	628.42 (10)	15057	15053
2. $\alpha \neq \beta, \gamma = 0$	4	20.69 (8)	5104	5096	28.10 (8)	14646	14638
3. $\alpha = \beta, \gamma = 0$	3	43.91 (9)	5124	5118	120.15 (9)	14725	14719
4. $\alpha \neq \beta, \gamma \neq 0$	5	6.03 (7)	5092	5082	13.30 (7)	14634	14624

This therefore is in agreement with our prior views above, from forming simple confidence intervals for each parameter separately. Individual likelihood-ratio tests may be carried out of hypotheses such as $\alpha = \beta$ or $\gamma = 0$, and are all highly significant. We should note here that the point, $\alpha = \beta = \gamma = 0$, corresponding to the beta-geometric model, is on the boundary to the parameter space, which violates the regularity conditions for a likelihood ratio test. This feature is likely also to affect how we interpret information criteria. We shall discuss this non-regular feature in Section 5.8. However, in the application of this example there can be no doubt about the significance of the results. Note that for the smaller data sets of Table 1.1, both data sets are well described by model 3 of Table 4.5, with $\gamma = 0$ and $\alpha = \beta$.

An interesting feature of Table 4.5 is that, judged by X^2, both of the beta-geometric models provided an acceptable fit to the data since the 5% point for a χ^2_{10} distribution is $\chi^2_{10:5} = 18.31$. This was the conclusion of Weinberg and Gladen (1986). However, in both cases we have significantly improved the model description of the data. Note that $\chi^2_{7:10} = 12.02$ and $\chi^2_{7:5} = 14.07$, so that model 4 for other contraceptive users, judged by the X^2 value of Table 4.3, is not an especially good fit.   □

When comparing models using the AIC or likelihoods, one must not also lose sight of absolute, rather than relative, goodness of fit: the best model for a data set, selected by comparing relative model performance, may nevertheless provide a poor fit to the data.

Example 4.8 continued

When the model of Equation (4.5) is fitted to the flour beetle data we obtain the estimates of Table 4.6. Simply comparing point estimates and their standard errors across sexes suggests a sex difference, which is confirmed by a likelihood-ratio test, producing a test-statistic of 41.58, which is highly significant when referred to χ^2_4 tables. We can also test separately hypotheses that the pair of parameters (κ, β) does not vary between sexes, which is rejected, and that the pair of parameters (α_1, α_2) does not vary between sexes. The lat-

Table 4.4 *Expected values from fitting the models of Table 4.3. (i): fitted values for underlying beta-geometric; (ii): fitted values after misreporting is added.*

(i) Pill users

		Model						
		1		2		3		4
Cycle	Obs. nos.		a	b	a	b	a	b
1	383	391.3	403.7	402.3	405.4	402.3	393.3	393.3
2	267	259.0	261.4	258.5	261.7	256.4	259.4	259.4
3	209	175.1	173.8	167.8	173.6	164.3	174.9	203.2
4	86	120.8	118.3	105.8	118.0	101.2	120.4	81.6
5	49	84.8	82.2	55.5	81.9	51.0	84.4	52.0
6	122	60.5	58.3	125.7	58.0	145.3	60.1	122.8
7	23	43.8	42.0	25.1	41.7	25.7	43.5	25.4
8	30	32.1	30.7	23.5	30.5	25.6	31.9	24.1
9	14	23.8	22.8	17.1	22.6	20.2	23.7	17.7
10	11	17.9	17.1	10.8	17.0	14.2	17.8	11.2
11	2	13.6	13.0	5.2	12.9	8.0	13.5	5.4
12	43	10.4	10.0	44.5	9.9	23.5	10.3	45.6
> 12	35	41.0	40.7	32.1	40.6	36.5	40.9	32.3
Total	1284							

(ii) Users of contraceptive methods other than the pill

		Fitted values						
		Model						
		1		2		3		4
Cycle	Obs. nos.		a	b	a	b	a	b
1	1674	1669.7	1697.7	1692.4	1703.4	1690.5	1678.5	1678.5
2	790	789.6	786.1	778.3	788.2	772.3	786.3	786.3
3	480	439.9	433.6	420.0	434.2	410.9	436.5	475.2
4	206	272.0	266.7	240.2	266.6	228.8	269.6	201.7
5	108	180.7	176.8	121.1	176.4	109.9	179.1	114.0
6	263	126.6	123.8	270.1	123.3	327.2	125.6	263.1
7	54	92.4	90.3	52.2	89.9	55.3	91.8	50.5
8	56	69.7	68.2	49.9	67.7	56.7	69.3	49.7
9	21	53.9	52.8	37.3	52.4	46.8	53.7	37.7
10	33	42.7	41.8	24.3	41.4	34.7	42.6	24.8
11	8	3.4	33.7	12.0	33.4	20.5	34.3	12.4
12	130	23.1	27.6	131.9	27.3	63.1	28.2	132.4
> 12	191	214.4	214.8	184.2	209.8	197.3	218.4	187.7
Total	4014							

Table 4.5 *Model-selection for digit confusion: data of Table 1.1. d denotes the number of fitted parameters.*

Model	d	Smokers X^2 (df)	AIC	-2ℓ	Non-smokers X^2 (df)	AIC	-2ℓ
1. beta-geometric	2	15.5 (10)	442	438	12.6 (10)	1785	1781
2. $\alpha \neq \beta$	4	8.8 (8)	439	431	7.0 (8)	1783	1775
3. $\alpha = \beta$	3	9.1 (9)	437	431	6.9 (9)	1781	1775

Table 4.6 *Point estimates, together with estimates of standard error (along the diagonals) and correlation when the model of Equation (4.5) is fitted to the flour beetle data of Table 1.4.*

(i) Males

Parameter	Estimate	Errors/correlations α_1	α_2	κ	β
α_1	4.63	0.46			
α_2	3.37	0.98	0.33		
κ	2.70	−0.17	−0.15	0.14	
β	14.28	−0.11	−0.09	0.83	2.38

(ii) Females

Parameter	Estimate	Errors/correlations α_1	α_2	κ	β
α_1	2.61	0.27			
α_2	2.61	0.94	0.22		
κ	3.48	−0.05	−0.03	0.20	
β	57.05	−0.04	−0.02	0.90	14.83

ter hypothesis is also rejected, but the *parallelism* hypothesis, that α_2 is the same for both sexes, is acceptable. We conclude that males are more susceptible than females, and also have smaller mean time-to-death than females. In many insect species there is a sex difference in susceptibility to insecticide, with the females tending to be less susceptible than the males, and our findings agree with this. We note also that for the flour beetles, females are slightly heavier than males which may be a contributory explanatory factor. An example of the fitted values is given in Table 4.7. □

Model-selection is often the result of carrying out several significance tests, and we know that when this is done it increases the chance of obtaining spu-

Table 4.7 *An example of the fit of the model of Equation (4.5) to the flour beetle data of Table 1.4: doses 0.2 and 0.5 mg/cm² only. In each case we present, for males (M) and females (F), the observed numbers dead (Obs) and the numbers predicted by the model (Exp).*

	Dose $(\text{mg/cm})^2$							
	0.2				0.5			
Sex	M		F		M		F	
Day	Obs	Exp	Obs	Exp	Obs	Exp	Obs	Exp
1	3	3.0	0	0.4	5	3.2	0	0.5
2	14	14.1	2	4.2	13	15.4	4	5.0
3	24	26.1	6	11.5	24	28.3	10	13.5
4	31	33.8	14	17.7	39	36.7	16	20.8
5	35	38.1	23	21.3	43	41.4	19	25.0
6	38	40.6	26	23.2	45	44.1	20	27.3
7	40	42.1	26	24.2	46	45.7	21	28.5
8	41	43.0	26	24.8	47	46.7	25	29.2
9	41	43.5	26	25.1	47	47.3	25	29.6
10	41	43.9	26	25.3	47	47.7	25	29.8
11	41	44.2	26	25.5	47	48.0	25	30.0
12	42	44.4	26	25.6	47	48.2	26	30.1
13	43	44.6	26	25.6	47	48.4	27	30.1

rious significant results by chance. If possible, this should be allowed for by carrying out each test at a suitably 'low' level of significance. This typically involves a degree of arbitrariness, and judgement on behalf of the statistician. It is usually impossible to allow formally for the complicated process of carrying out many tests in model-selection, and impractical to be able to carry out inferences which condition upon the various tests which have been carried out. For a detailed discussion of this, see Chatfield (1995) and the references cited there. It should be realised also that many of the tests carried out during a model-selection procedure may have low power.

In choosing between models to describe data, statisticians are often parsimonious, omitting parameters which do not appear necessary in order to provide a simple description of data. However, if a model is to be used for prediction, then there may be benefits resulting from including parameters which only marginally improve the description of a data set. Issues of how models are parameterised, and of parameter-redundancy, are discussed in Sections 5.2 and 5.3 respectively. Vuong (1989) considers testing non-nested models; for an application, see Palmer et al. (2008).

4.6 Score and Wald tests

So far we have concentrated on using likelihood-ratio tests for testing hypotheses, but there are alternatives which are asymptotically equivalent under the null hypothesis and often much easier. Here we focus on *score* tests, which see much use in modern classical statistics. However, first of all we explain how an additional test, called the Wald test, operates. We know from Section 4.1.2 that in a likelihood-ratio test, a null hypothesis that $\theta = \theta_0$ is rejected at the $\alpha\%$ significance level if

$$2\{\ell(\hat{\theta}) - \ell(\theta_0)\} > \chi^2_{d:\alpha} ,$$

where d is the dimension of θ, which has maximum likelihood estimate $\hat{\theta}$, and $\chi^2_{d:\alpha}$ is as usual the $\alpha\%$ point of a χ^2_d distribution.

Rather than measure the distance between $\hat{\theta}$ and θ_0 through the values that the log-likelihood function takes for those values of θ, which is what the likelihood-ratio test does, we can simply consider the difference between the parameters, $\hat{\theta} - \theta_0$. From expression (4.2), we know that when the null hypothesis is true then asymptotically, $\hat{\theta} \sim N_d(\theta_0, \mathbf{J}^{-1}(\theta_0))$, and then, as shown in Appendix A,

$$(\hat{\theta} - \theta_0)'\mathbf{J}(\theta_0)(\hat{\theta} - \theta_0) \sim \chi^2_d.$$

We would then reject H_0 at the α % significance level if

$$(\hat{\theta} - \theta_0)'\mathbf{J}(\theta_0)(\hat{\theta} - \theta_0) > \chi^2_{d:\alpha} .$$

This is called the *Wald* test, and in fact is what we have been doing informally in this chapter, when comparing point estimates with their standard errors: in these cases we have been examining elements of θ to check whether they may be set equal to zero. We have used $\mathbf{J}(\hat{\theta})$, rather than $\mathbf{J}(\theta_0)$. Clearly the Wald test is dependent upon the particular parameterisation used, which may be a shortcoming in some cases.

Now we consider the score test. From expression (4.1), we know that asymptotically $\mathbf{U}(\theta)$ has the multivariate normal distribution,

$$\mathbf{U}(\theta) \sim N_d(\mathbf{0}, \mathbf{J}(\theta)).$$

Thus in order to test the null-hypothesis that $\theta = \theta_0$, we can form the test-statistic,

$$\mathbf{U}(\theta_0)'\mathbf{J}^{-1}(\theta_0)\mathbf{U}(\theta_0) .$$

Since, subject to regularity conditions, $\mathbf{U}(\hat{\theta}) = \mathbf{0}$, this is equivalent to forming,

$$\{\mathbf{U}(\theta_0) - \mathbf{U}(\hat{\theta})\}'\mathbf{J}^{-1}(\theta_0)\{\mathbf{U}(\theta_0) - \mathbf{U}(\hat{\theta})\} ,$$

and so we see that a score test gauges the separation of $\hat{\theta}$ and θ_0 through the values that the scores vector takes for those values of θ. It is clearly necessary also to account for the curvature of the log-likelihood surface at θ_0, and this is done through $\mathbf{J}^{-1}(\theta_0)$.

Thus the score test rejects the null hypothesis that $\boldsymbol{\theta} = \boldsymbol{\theta}_0$ at the $\alpha\%$ significance level if

$$\mathbf{U}(\boldsymbol{\theta}_0)'\mathbf{J}^{-1}(\boldsymbol{\theta}_0)\mathbf{U}(\boldsymbol{\theta}_0) > \chi^2_{d:\alpha} \,. \qquad (4.6)$$

The justification for the asymptotic chi-square distributions of the Wald and score tests has been outlined in Section 4.1.2, in terms of the observed information matrix, rather than the expected information matrix.

The great beauty of the result of (4.6) is that it does not involve $\hat{\boldsymbol{\theta}}$, and so the null-hypothesis can be tested without the maximum-likelihood estimate of $\boldsymbol{\theta}$ having to be found. Of course, if the test is significant then $\hat{\boldsymbol{\theta}}$ does need to be found. Difficulties with obtaining maximum-likelihood estimates numerically are often due to attempts to fit inappropriate models, and this is frequently avoided when score tests are used. This is particularly relevant when nuisance parameters are present, which is the case that we shall consider shortly. Of course expression (4.6) still uses the data, via the likelihood. A graphical illustration of the score and likelihood-ratio tests and how they differ is given in Figure 4.10, which is based on Buse (1982). The Wald test measures the difference between $\hat{\theta}$ and θ_0 in the direction of the θ axis.

If the parameter vector has the partitioned form $\boldsymbol{\theta} = (\boldsymbol{\phi}, \boldsymbol{\psi})'$, and we wish to test $\boldsymbol{\phi} = \boldsymbol{\phi}_0$, then we regard $\boldsymbol{\psi}$ as nuisance parameters. Let the constrained maximum likelihood estimate of $\boldsymbol{\psi}$ be $\hat{\boldsymbol{\psi}}_0$, when $\boldsymbol{\phi} = \boldsymbol{\phi}_0$. Let

$$\mathbf{J}(\boldsymbol{\phi}, \boldsymbol{\psi}) = \begin{bmatrix} \mathbf{J}_{1,1}(\boldsymbol{\phi}, \boldsymbol{\psi}), & \mathbf{J}_{1,2}(\boldsymbol{\phi}, \boldsymbol{\psi}) \\ \mathbf{J}'_{1,2}(\boldsymbol{\phi}, \boldsymbol{\psi}), & \mathbf{J}_{2,2}(\boldsymbol{\phi}, \boldsymbol{\psi}) \end{bmatrix},$$

partitioned in the same way as the parameter vector.

The score test statistic is now the scalar quantity z, given by

$$z = (\mathbf{U}(\boldsymbol{\phi}_0, \hat{\boldsymbol{\psi}}_0))'(\mathbf{J}_{1,1} - \mathbf{J}_{1,2}\mathbf{J}_{2,2}^{-1}\mathbf{J}'_{1,2})^{-1}\mathbf{U}(\boldsymbol{\phi}_0, \hat{\boldsymbol{\psi}}_0) \,, \qquad (4.7)$$

where the components of $\mathbf{J}(\boldsymbol{\phi}, \hat{\boldsymbol{\psi}})$ are calculated at $(\boldsymbol{\phi}_0, \hat{\boldsymbol{\psi}}_0)'$. Here the *only* optimisation is done with $\boldsymbol{\phi} = \boldsymbol{\phi}_0$. The scores vector $\mathbf{U}$ in Equation (4.7) only contains the scores for $\boldsymbol{\phi}$, that is the partial derivatives of the log-likelihood with respect to the elements of $\boldsymbol{\phi}$.

Note that we can also write Equation (4.7) in the form:

$$z = \mathbf{U}'\mathbf{J}^{-1}\mathbf{U} \qquad (4.8)$$

in which $\mathbf{U}$ is the complete scores vector, involving all the partial derivatives of the log-likelihood ℓ with respect to *all* of the elements of $\boldsymbol{\theta}$. This is because $(\mathbf{J}_{1,1} - \mathbf{J}_{1,2}\mathbf{J}_{2,2}^{-1}\mathbf{J}'_{1,2})^{-1}$ is the top left-hand component of the partitioned matrix $\mathbf{J}^{-1}$ (see, e.g., Morrison, 1976, p.68) and the scores vector is evaluated when $\boldsymbol{\phi} = \boldsymbol{\phi}_0$ and $\boldsymbol{\psi} = \hat{\boldsymbol{\psi}}$, which is the maximum-likelihood estimate of $\boldsymbol{\psi}$ when $\boldsymbol{\phi} = \boldsymbol{\phi}_0$. Hence subject to regularity conditions, the components of $\mathbf{U}$ corresponding to the elements of $\boldsymbol{\psi}$ are all identically equal to zero. The expression of Equation (4.8) is attractively simple if one is using a computer to evaluate the terms. Derivatives can be formed numerically or possibly from using a symbolic algebra package such as Maple — see, e.g., Catchpole and

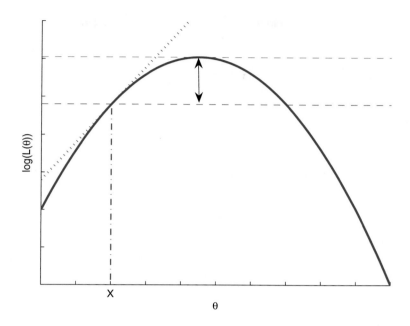

Figure 4.10 *An illustration, for scalar θ, of the key differences between the likelihood ratio test, the score test and the Wald test, which here has test statistic $(\hat{\theta} - \theta_0)^2 J(\hat{\theta})$. The likelihood-ratio test statistic is twice the distance indicated by the arrows. The score test statistic is the square of the slope of the curve of logL(θ) at θ_0, indicated by a X on the θ axis, divided by the curvature of the curve at that point.*

Morgan (1996). The MATLAB programs of Figure 4.11 illustrate the difference between Equations (4.7) and (4.8).

For quadratic log-likelihoods the three tests, score, Wald and likelihood ratio, are equivalent, as the Taylor series expansions of Section 4.1.2 are then exact. One may use the observed Fisher information instead of the expected Fisher information; Catchpole and Morgan (1996) found that the expected information performed better in a particular application. Morgan et al. (2007) showed that when the observed information matrix is used in a score test of zero-inflation, then a negative test statistic can arise. A number of the standard tests of modern statistics — such as the log-rank test (Lawless, 1992, p.417) and the Cochran-Armitage test (Morgan, 1992, p.74), have been shown to be score tests of suitable hypotheses.

We shall now consider three examples of the use of score tests.

Example 4.10: Quantal response models

The logit model has seen wide use in statistics — see for example Exercise

(a)

```
function s=longscore(x)
%
%LONGSCORE illustrates how score tests can be done using numerical
%  approximations to the derivatives that we need.
%  In this particular application, 'x' contains the maximum-
%  likelihood estimates of the beta-geometric parameters, mu and theta.
%  Calls GRAD and HESSIAN.
%_____
H=hessian('preference',x);
g=grad('preference',x);
J=inv(H);
s=g(1:2)'*(J(1:2,1:2))*g(1:2);
disp(s)
```
(b)
```
function s=score(x)
%
%SCORE calculates the score test statistic
%  of w=0, in the zero-inflated Poisson distribution. The likelihood
%  is specified by the function POIS0
%  Calls GRAD and HESSIAN.
%_____
global data
H=hessian('pois0',x)
                          % As the second component of g is
g=grad('pois0',x);        % zero, we do not have to extract
s=g'/(-H)*g;              % the relevant part of the inverse
                          % Hessian.
```
(c)
```
function y=pois0(x)
%
%POIS0 forms the negative log-likelihood for the
%  zero-inflated Poisson model, and a particular
%  data set.
%_____
data=[314 48 20 7 5 2 2 1 2 0 0 1];
n=length(data);
w=x(1);lam=x(2);
p(1)=w+(1-w)*exp(-lam);
p(2)=(1-w)*exp(-lam)*lam;
for k=2:(n-1)
  p(k+1)=p(k)*lam/k;
end
p=log(p);
y=-data*p';
```

Figure 4.11 MATLAB *programs to evaluate the score test statistic of a null hypothesis, (a) using Equation (4.7) that there is no digit preference for the smokers data of Table 1.1 – see Exercise 4.1 for the function preference which is used here; (b) that $w = 0$ in the zero-inflated Poisson distribution, using Equation (4.8). The relevant formulation of the zero-inflated Poisson distribution is given in Equation (5.1) – see (c).*

2.3. After fitting the logit model, a number of authors have employed score tests, to see whether a better fit might result from a more complex model. See, for example, Brown (1982), Kay and Little (1986), and Morgan et al. (1989). Here we shall give an example from Aranda-Ordaz (1981): the probability of response to dose d is taken as:

$$P(d) = 1 - \{1 + \lambda e^{\alpha + \beta d}\}^{-\frac{1}{\lambda}},$$

and we can see that if $\lambda = 1$ then the model reduces to the familiar logit form. We are interested in testing, for a given data set, the null hypothesis that $\lambda = 1$, and we can do this by means of a score test. What is involved is fitting only the simpler, logit model, but then conducting the algebra to produce the score test statistic of Equation (4.7). In this case, in terms of the notation of Equation (4.7), we have the correspondence: $\phi = \lambda$, $\phi_0 = 1$, $\boldsymbol{\psi} = (\alpha, \beta)'$, and the scores vector of Equation (4.7) is in fact only the partial derivative, $\dfrac{\partial \ell}{\partial \lambda}$. See Exercise 4.8 for the conclusion of this test. For more discussion of quantal response models, see Morgan (1992, 1998). □

Example 4.11: Testing for synergy

Suppose a batch of n_{ij} insects are tested with a mixture of two drugs comprising x_i units of one drug and z_j units of the other. Giltinan et al. (1988) proposed the model

$$p_{ij} = \Phi[\alpha + \beta \log\{x_i + \rho z_j + \gamma(x_i z_j)^{\frac{1}{2}}\}],$$

where p_{ij} is the probability that any insect responds to the (x_i, z_j) mixture, Φ is the cdf of the standard normal distribution, and $\alpha, \beta, \rho, \gamma$ are parameters to be estimated. In particular, γ denotes the effect of applying the drugs together, and we are therefore interested in testing whether $\gamma = 0$. Giltinan et al. (1988) showed that a score test of $\gamma = 0$ has a graphical interpretation through an added variable plot (Atkinson, 1982, p.67). This allowed an investigation of the influence of the elements of the sample on the significance of the test. □

Example 4.9 continued

Catchpole and Morgan (1994) suggested starting from a minimal model $(\lambda, \phi_1, \phi_a)$ for the analysis of recovery data from dead birds. They then used score tests to choose between models such as $(\{\lambda_i\}, \phi_1, \phi_a)$ and $(\lambda, \{\phi_{1i}\}, \phi_a)$. If one of these more complicated models was selected, they then used a further range of score tests, following fitting of the selected model, and this process could then be iterated until a situation was reached in which no score test was significant. □

We have seen that score tests do not in general involve fitting the alternative hypothesis model. However, Pregibon (1982) showed that in generalised linear models, which are discussed in Chapter 8, score tests can be readily carried out by taking just *one step* of iterated maximum likelihood for fitting the alternative hypothesis model.

4.7 Classical goodness of fit

To take just one example, the comparison of observed and fitted values of Table 4.7 demonstrated that the model is performing reasonably well. We found this also in Table 2.2, and there quantified the fit using the Pearson X^2 test. An asymptotically equivalent approach is to use the *deviance*, defined as $-2\log(L_f - L_s)$, where L_f is the likelihood of the fitted model, and L_s is the likelihood of what is called the *saturated* model, which is the model which fits the data exactly (Cf. Exercise 2.20). For the full flour beetle data, only a fraction of which is displayed in Table 4.7, the male and female deviances are, respectively, 48.39 and 49.94, each referred in this case to χ^2_{48} tables. We obtain 13 degrees of freedom from each of the single-dose multinomial distributions, and overall we lose four degrees of freedom as four parameters have been estimated. Thus the chi-square degrees of freedom are 48, obtained as $48 = (4 \times 13) - 4$. As the deviance values are close to the chi-square degrees of freedom, the deviances indicate a good match between the observed and fitted values. (The mean of a chi-square random variable with ν degrees of freedom is ν.) Comparisons between nested models are often made using deviances. In such cases L_s is unchanged, and so the comparisons are equivalent to using likelihood-ratio tests. The multiplying factor of 2 which is needed for a likelihood-ratio test is 'built-in' to the deviance. Comparisons using deviances have been found to be more reliable in general than using the deviance to provide an absolute measure of fit.

While the model fitted to fecundability data in Table 2.2 is acceptable on the basis of an X^2 test, consideration of the fitted values suggested a model with digit preference. In general we need to look carefully at various aspects of the model fit. For non-linear models, we can describe the discrepancy between observed and fitted values by means of what are called *residuals*, which may be defined in a number of different ways. For instance, for models of avian survival we can look at the individual multinomial cell contributions to the overall X^2. Consideration of the signs of contributions before they are squared can be useful for establishing whether patterns indicating model deficiency may exist.

In many cases, a model will provide only a rough approximation to reality, and so while one would want to avoid obvious deficiencies, some degree of mismatch between model and data is to be expected. In practice, therefore, goodness-of-fit criteria and their significance levels should be used as guidelines, and strict adherence to particular levels of significance should be avoided. As we have observed earlier, too many parameters in a model can result in *overfitting* the data. Additionally, a good fit may flatter a model, as is discussed in the next section.

4.8 Model selection bias

The following quotation is taken from Cormack (1994):

Data-based model-selection is difficult and dangerous. The statistical question is usually phrased in terms of the significance levels at which a new parameter should be included in the model, or an existing parameter excluded. Measures such as Akaike's criterion are advocated as objective decision rules. I suggest that biological understanding of the data set also plays a part. Reliance on simulation results is also misleading. They answer questions based on the premise that one of the selected models is true... I favour inclusion of behavioural parameters for which there is any serious evidence and which the biologist can support. I no longer seek parsimony to provide more precise intervals. Failure to reject a simplifying null hypothesis is not proof that it is true. I do seek consistency of model form over subsets ...

There are various reasons for the difficulties mentioned above. As a result of careful and detailed model selection, possibly using a combination of hypothesis testing and information criteria such as AIC, we may obtain a model which provides a good fit to a particular data set. However, the goodness-of-fit is being measured on the very data set that was used in the model-selection. As discussed in Section 2.6, if a replicate data set were available, we would not expect the model to fit that data set so well. This is a feature which is well known to students of multivariate analysis: a classical problem of multivariate analysis is to devise rules to discriminate between different types of individual (see, e.g., Young et al., 1995). The apparent error rate of a rule, obtained from applying the rule to the data that were used to construct the rule, is well known to be conservative, and to flatter the good performance of the procedure. Various ways have been devised for improving the estimation of error rates in discriminant analysis, such as using only part of the data to construct the rule, and then testing the rule on the remainder of the data. An alternative approach is to use the method of *cross-validation*, in which different subsets of the data are used in turn for constructing/testing the model. Similar procedures may be used in model assessment, if possible. An alternative approach involves fitting a range of plausible alternative models and then averaging in some way over their predictions. For discussion, see Chatfield (1995), Burnham and Anderson (2002), and Buckland et al. (1997). The Bayesian framework for modelling attributes probabilities to models, and therefore provides a natural way for averaging over different models. For discussion of this approach, see Draper (1995) and Section 7.8.

4.9 Discussion

In this chapter we have presented the basic classical tools for fitting statistical models to data. Most of these rely on asymptotic results, and therefore involve an act of faith on the part of the statistician. This is often unsatisfactory, but we can investigate further in some detail. For example expression (4.2) guarantees that asymptotically maximum-likelihood estimators will be unbiased, but usually they will be biased in standard applications. We can estimate this bias and correct for it, and an illustration of this is provided by Firth (1993).

Apart from using a computer to maximise likelihoods numerically, we have in this chapter adopted a classical approach to statistical inference, ignoring how high-speed computers can now remove an over-reliance on asymptotic assumptions. This computer-intensive work will be described in Chapters 6 and 7. However, any model-fitting procedure is likely to involve a sensible combination of the old and the new. Fundamentally one has to have confidence in the model being fitted, and this confidence will often come from prior knowledge and experience, and the practical context of the work.

4.10 Exercises

4.1 Study the MATLAB program given below for fitting a digit confusion model to fecundability data. Discuss the logistic transformation of parameters used in the program. Use the *fmax* function to obtain maximum-likelihood parameter estimators and obtain estimates of standard error and correlation for the parameter estimators. Extend the program to include confusion into the digit 3. Comment on the comparison of the estimates of μ in Tables 4.1 and 4.2.

```
function y=preference(x)
%
%PREFERENCE calculates the negative log-likelihood for
%   the fecundability model, allowing for digit preference
%   into 6 and 12 months.
%   'data' are global, and are set in the driving program.
%_____
global data
al=x(1); b1=x(2);m1=x(3); t1=x(4);
a=1/(1+exp(al));                       % logistic transformation
b=1/(1+exp(b1));                       % to keep parameters a,b,m,
m=1/(1+exp(m1));                       % in the 0-1 range.
t=exp(t1);                             % exponential transformation
                                       % to keep t positive
for i=2:17
  p(i)=p(i-1)*(1-(m+t)/(1+(i-1)*t));   % sets up the basic
end                                    % beta-geometric probabilities
s1=0;                                  % now we add the confusion
for i=1:5                              % probabilities
  p(i)=p(i)*(1-a^abs(i-6));
  s1=s1+p(i);
end
for i=7:11
  p(i)=p(i)*(1-a^abs(i-6)-b^abs(12-i));
  s1=s1+p(i);
end s=0;
for i=1:11
  s=s+p(i)*a^abs(i-6);
end
p(6)=s; s=0;
for i=7:17
  s=s+p(i)*b^abs(i-12);
end
p(12)=s;
p(13)=1-s1-p(6)-p(12);
p=log(p);
y=-data*p(1:13)';
```

4.2 For the two alternative parameterisations of the logit model of Example 4.3, estimate the correlations between the parameter estimators in each case for the data of Exercise 3.10. Comment on the results and how they tie-in with the contour plots of Figure 4.2.

4.3 Explain why the projections of the two-dimensional confidence region of Figure 4.6 are wider than the marginal confidence regions with the same confidence coefficient.

4.4 Read the paper by Venzon and Moolgavkar (1988).

4.5 Produce MATLAB code for obtaining the parameter space path of Figure 4.7.

4.6 Discuss the following extension to the first part of the right hand side of Equation (4.5):

$$a(d_i) = \begin{cases} 1 - [1 + \lambda_2 \exp\{\eta(d_i)\}]^{-1/\lambda_2} & , \quad \lambda_2 \neq 0 \\ 1 - \exp[-\exp\{\eta(d_i)\}] & , \quad \lambda_2 = 0, \end{cases}$$

where $\eta(d_i) = \alpha_1 + \alpha_2 \log(d_i)$. Show how to obtain the logit model as a special case. Note the relationship with the Aranda-Ordaz model of Example 4.10.

4.7 Repeat the model-selection of Table 4.3, using the BIC, rather than AIC.

4.8 Complete the algebra for the score test of Example 4.10. Apply this test to the data of Exercise 3.10.

4.9* A zero-inflated Poisson distribution has the probability function,

$$
\begin{aligned}
pr(X = 0) &= w + (1 - w)e^{-\lambda} \\
pr(X = k) &= \frac{(1 - w)e^{-\lambda}\lambda^k}{k!}, \quad k = 1, 2, \ldots.
\end{aligned}
$$

Study the MATLAB program given below for simulating data from a zero-inflated Poisson distribution. It makes use of MATLAB functions to simulate binomial and Poisson random variables, which are described in Chapter 5. Simulate a sample of data from this distribution, and obtain the maximum-likelihood estimate of the parameters, (λ, w). Perform both a score test and a likelihood-ratio test of the hypothesis that $w = 0$.

```
function y=zipsim(lambda,w,n)
%
%ZIPSIM simulates n realisations of a zero-inflated Poisson random
%  variable, of parameter lambda, and probability 'w' of extra
%  zeros.
%  Calls BINSIM, POISSIM
%_____
k=binsim(n,w,1);              % k is the number of extra zeros
y1=zeros(1,k);
y2=poissim(lambda,n-k);
y=[y1 y2'];
```

4.10* Suppose that the probabilities $\{\pi_i, 1 \leq i \leq t\}$ of a multinomial distribution depend on the parameter vector $\boldsymbol{\theta} = (\theta_1, \ldots, \theta_d)$. Show that the expected information matrix for $\boldsymbol{\theta}$ is given by:

$$\mathbf{J} = m\mathbf{D}'\mathbf{r}^{-1}\mathbf{D},$$

where m is the index of the multinomial distribution, $\mathbf{D}$ is a $d \times t$ matrix with (i, j)th element $\dfrac{\partial \pi_i}{\partial \theta_j}$, and $\mathbf{r}$ is a $d \times d$ diagonal matrix with diagonal elements $(\pi_1, \pi_2, \ldots, \pi_d)$. Study the MATLAB function below for producing estimates of variance and covariance in the multinomial case (Terrill, 1997). Consider how you would use this function for data from several independent multinomial distributions, as with the heron survival data of Table 1.5.

```
function I= expinf(funfcn,x,m)
%
%EXPINF Calculates numerically the expected information matrix of a
%   multinomial model. The function, 'funfcn,' gives the probability
%   matrix of the model with the rows corresponding to each
%   independent sample from the model, with corresponding multinomial
%   index m. The sum over each row equals unity.
%   x is the vector of parameter values;
%   m is the vector of multinomial indices.
%   Code due to Ted Catchpole, modified by Paul Terrill.
%-------------------------------------------------------------------
x=x(;); m=m(:);
delta=1e-6*sign(x)*x;
t=length(x);
I=zeros(t);
Delta=diag(delta); P=feval(funfcn,x); c=size(P,2); m=m(:,ones(1-c));
for i=1:t
  for j=1:t
    dr=Delta(:,i);
    ds=Delta(:,j);
    dPr=(feval(funfcn,x+dr)-feval(funfcn,x-dr))/(2*dr(i));
    dPs=(feval(funfcn,x+ds)-feval(funfcn,x-ds))/(2*ds(i));
    I(i,j)=sum(m(P>0).*dPr(P>0).*dPs(P>0)./P(P>0));
  end                    % Ref :Hunt et al (2006, p87), for use of the relationa
end                      % operator, '>'
I=(I+I')/2;
```

4.11 (Examination question, St. Andrews University.) The random variable X has the inverse Gaussian distribution, with pdf:

$$f(x) = \left(\frac{\lambda}{2\pi x^3}\right)^{\frac{1}{2}} \exp\left\{-\frac{\lambda}{2}\left(\frac{x}{\mu^2} - \frac{2}{\mu} + \frac{1}{x}\right)\right\},$$

$$0 \leq x < \infty, \quad \text{with } \lambda > 0, \ \mu > 0.$$

For the case of fixed λ, show, by calculating the first-order derivative of

$\log f(x)$ with respect to μ, that $\mathbb{E}[X] = \mu$. For the case of unknown λ and μ, calculate the expected Fisher information matrix for a single observation from the inverse Gaussian distribution. For a random sample of size n from this distribution, find the maximum likelihood estimates of the parameters λ and μ.

4.12 Verify in the model for polyspermy of Chapter 2 that if secondary fertilisations take place at rate $\mu(t)$, then the probabilities of any egg having n sperm at time t have the form:

$$p_0(t) = \exp(-\lambda t),$$
$$p_n(t) = \frac{\lambda \int_0^t \exp(-\lambda x) \exp\{-M(t-x)\} M(t-x)^{n-1} dx}{(n-1)!} \quad \text{for } n \geq 1,$$

where $M(\phi) = \int_0^\phi \mu(t) dt$.

Try to fit this model to data using maximum likelihood and MATLAB numerical integration through the *quad* function, when

$$\mu(t) = \mu_o(1 - t/\tau),$$

for a suitable constant μ_0 and $0 \leq t \leq \tau$.

4.13 We have stated that, subject to regularity conditions, evaluated at the maximum-likelihood estimator $\hat{\boldsymbol{\theta}}$, the scores vector is identically zero, that is,

$$\mathbf{U}(\hat{\boldsymbol{\theta}}) = \mathbf{0}.$$

Provide an example of when this equation does not hold.

4.14 The data following are taken from Milicer and Szczotka (1966). For a sample of 3918 Warsaw girls they record whether or not that the girls had reached menarche, which is the start of menstruation. Analyse these data by means of a logit model.

Mean age of group (years)	No. having menstruated	No. of girls
9.21	0	376
10.21	0	200
10.58	0	93
10.83	2	120
11.08	2	90
11.33	5	88
11.58	10	105
11.83	17	111
12.08	16	100
12.33	29	93
12.58	39	100
12.83	51	108
13.08	47	99
13.33	67	106
13.58	81	105
13.83	88	117
14.08	79	98
14.33	90	97
14.58	113	120
14.83	95	102
15.08	117	122
15.33	107	111
15.58	92	94
15.83	112	114
17.58	1049	1049

4.15   (Examination question, Sheffield University.) Suppose that X_1 and X_2 are independent random variables with respective exponential distributions, $Ex(a\lambda)$ and $Ex(b\lambda)$. Here $a \neq b$ are known positive constants. Find the maximum-likelihood estimators of λ and λ^{-1}. Compare and contrast the maximum-likelihood estimator of λ^{-1} and the two alternative estimators,

$$ab(X_1 + X_2)/(a + b), \quad \text{and} \quad ab(X_1 - X_2)/(b - a).$$

4.16   (Examination question, Sheffield University.) A random variable X has probability function,

$$pr(X = k) = \frac{1}{k!} \exp(\theta k - e^\theta), \quad \text{for} \quad k = 0, 1, 2, \ldots.$$

Derive the maximum-likelihood estimator of θ in terms of the members of a random sample, $(x_1, \ldots, x_n)$, of values for X.

4.17   Compare the estimated variance of the estimator of the parameter p in Example 1.1 with the estimated variance of the maximum-likelihood estimator of p.

4.18   Consider a stochastic model for avian survival, involving the parameter set $(\lambda, \phi_1, \phi_a)$. This model is obtained from setting $\phi_2 = \ldots = \phi_k = \phi_a$ in the model of Example 4.6. Construct the cell recovery probabilities for the data from 1955–1958 inclusive from Table 1.5. Deduce the form for these probabilities if the numbers of birds ringed each year are known. Discuss whether you think the estimators of ϕ_1 and ϕ_a will be substantially more precise when the ringing totals are known.

4.19   Revisit Exercise 2.1 (b) and calculate standard errors.

4.20   A bivariate parameter vector $\boldsymbol{\theta}$ can be assumed to have a known bivariate normal sampling distribution. Explain how you would construct appropriate confidence ellipsoids.

4.21   Consider how the model for misreporting, in Example 4.2, may be used to estimate the proportion of individuals who do not report the correct time to conception.

4.22*   (Morgan and North, 1985.) When quail eggs start to hatch they move through a number of distinct stages in sequence. Consider how you might devise stochastic models for such a system, and use them to analyse the data of Example 1.6.

4.23   Fit a beta-geometric model to the data of Exercise 2.1 on numbers of tests needed by car drivers.

4.24   (Examination question, University of Glasgow.)

Sharon Kennedy conducted a research project in Trinidad in July 1999 on the occurrence and distribution of larvae of the hoverfly *Diptera* on the flowering bracts of *Heliconia* plants. In a small part of her study she observed that on 20 randomly selected bracts the numbers of larvae were

$$9 \ \ 14 \ \ 3 \ \ 3 \ \ 8 \ \ 7 \ \ 7 \ \ 6 \ \ 7 \ \ 0 \ \ 6 \ \ 0 \ \ 5 \ \ 1 \ \ 3 \ \ 12 \ \ 2 \ \ 4 \ \ 0 \ \ 11.$$

Let X denote the random variable which describes the number of larvae on a randomly selected bract. We assume that X has the probability function,

$$\text{pr}(X = x) = \left(\frac{1}{1 + \mu} \right) \left(\frac{\mu}{1 + \mu} \right)^x, \quad \text{for} \quad x = 0, 1, 2, \ldots,$$

where $\mu > 0$ is the mean value of X.

Show that the log-likelihood for μ is given by

$$\ell(\mu) = 108 log(\mu) - 128 log(1 + \mu).$$

Find the maximum-likelihood estimate of μ and also the observed Fisher information.

A plot of the log-likelihood for μ is shown in Figure 4.12.

Use this graph to obtain an approximate 95% confidence interval for μ.

Compare this interval with an alternative 95% confidence interval for μ which is calculated from using the asymptotic normality of maximum likelihood estimators given in expression (4.2). Which would you prefer?

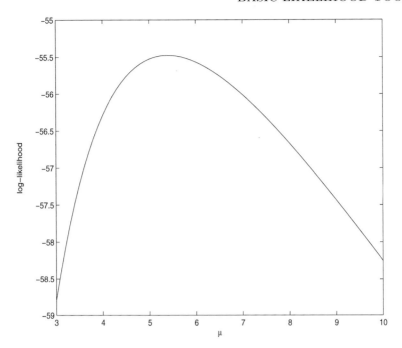

Figure 4.12 *Log-likelihood for Exercise 4.24.*

4.25 Let $x_1, \ldots, x_n$ denote a random sample from the Cauchy distribution, with single unknown parameter θ, which has pdf given by

$$f(x) = \frac{1}{\pi\{1 + (x - \theta)^2\}}, \quad -\infty < x < \infty.$$

1. Show that the Newton-Raphson iterative method for obtaining the maximum-likelihood estimate $\widehat{\theta}$ has the form below, where $\widehat{\theta}^{(m)}$ denotes the mth iterate for $\widehat{\theta}$:

$$\widehat{\theta}^{(m+1)} = \widehat{\theta}^{(m)} - \frac{\sum_{i=1}^{n} \frac{(x_i - \widehat{\theta}^{(m)})}{\{1 + (x_i - \widehat{\theta}^{(m)})^2\}}}{\sum_{i=1}^{n} \frac{(x_i - \widehat{\theta}^{(m)})^2 - 1}{\{1 + (x_i - \widehat{\theta}^{(m)})^2\}^2}}.$$

2. Fisher's expected information is $J = \frac{n}{2}$. Write down the corresponding iteration for the method of scoring.

3. A particular sample has the values: $\{0.11, 1.67, 1.01, -1.20, -2.80, -0.68, 2.28, -1.14, -7.34, -4{,}60\}$, resulting in $\widehat{\theta} = -0.6632$.

   Let $L(\theta)$ denote the likelihood. The hypothesis $\theta = 0$ may be tested by means of the score, Wald and likelihood-ratio tests. The results of these tests are shown below:

(i) $\quad \sqrt{2\log\left(\frac{L(-0.6632)}{L(0)}\right)} \quad = 1.026$

(ii) $\quad (0.6632)\sqrt{5} \qquad\qquad = 1.483$

(iii) $\quad \dfrac{\sum_{i=1}^{10}\left\{\frac{d\log f(x_i)}{d\theta}\right\}\big|_{\theta=0}}{\sqrt{5}} = 0.620$

Explain, giving your reasons, which test is used in each of these cases. Comment on the disparity between the three test results.

4.26* The data below are taken from Pierce et al. (1979), and summarise the daily mortality in groups of fish subjected to three levels of zinc concentration. Half the fish at each concentration level received one week's acclimatisation to the test aquaria, and half received two weeks' acclimatisation. There were therefore six treatment groups, and 50 fish were randomised to each group, resulting in 300 fish in total in the experiment.

| | | \multicolumn{6}{c}{Acclimatisation time} |
| | | \multicolumn{3}{c}{One week} | \multicolumn{3}{c}{Two weeks} |
Day	log zinc concentration	0.205	0.605	0.852	0.205	0.605	0.852
1		0	0	0	0	0	0
2		3	3	2	0	1	3
3		12	17	22	13	21	24
4		11	16	15	8	8	10
5		3	5	7	0	5	4
6		0	1	1	0	0	1
7		0	0	2	0	0	0
8		0	1	0	0	0	0
9, 10		0	0	0	0	0	0

The following *mixture* model is proposed for the data, in Farewell (1982): a fish is classified as either a 'long-term' survivor or a 'short-term' survivor. If **x** denotes the two-dimensional vector indicating acclimatisation time and zinc concentration, the probability that a fish with covariate **x** is a 'short-term' survivor is:

$$p(\mathbf{x}) = (1 + e^{-\boldsymbol{\beta}'\mathbf{x}})^{-1}.$$

'Long-term' survivors can be assumed not to die during the course of the experiment. For a short-term survivor the time of death, measured from the start of the experiment, will have a Weibull distribution, with pdf

$$f(t|\mathbf{x}) = \delta\lambda(\lambda t)^{\delta-1}\exp\{-(\lambda t)^{\delta}\}, \quad t \geq 0,$$

where $\lambda = \exp(-\boldsymbol{\gamma}'\mathbf{x})$. This model contains five parameters $\boldsymbol{\theta} = (\delta, \boldsymbol{\gamma}, \boldsymbol{\beta})$, the vectors $\boldsymbol{\gamma}$ and $\boldsymbol{\beta}$ containing two elements each.

(i) Write down an expression, involving $p(\mathbf{x})$ and the Weibull survivor function, for the probability that a fish with covariates $\mathbf{x}$ does not die during the course of the experiment.

(ii) Hence write down a general expression for the likelihood for a set of data of the form shown above. Indicate, in outline only, how you would proceed to obtain the maximum-likelihood estimate of $\boldsymbol{\theta}$ and associated measures of standard error.

(iii) Draw conclusions from the following estimates obtained for the above data: $\widehat{\delta} = 3.47$.

Covariate	$\widehat{\beta}$	Estimated standard error	$\widehat{\gamma}$	Estimated standard error
Acclimatisation	−0.94	0.30	−0.13	0.04
Concentration	3.59	0.56	0.17	0.08

(iv) Discuss any further analyses which you would carry out. Note that a version of this question also appears in Morgan (1992, p.134).

4.27 Explain how you would use geometric and beta-geometric models for the time to conception in order to produce statistical tests of the effect of a woman smoking on the probability of conception.

4.28 (Examination question, University of Glasgow.)

In an industrial bottling plant, labels are glued to the bottles at the final stage of the process. Unfortunately, the shape of the bottles used in the plant sometimes leads to bottles not being held completely straight by the labelling machine, with the result that the label is sometimes glued at a slight angle. A quality control inspector has taken 20 samples, each of 25 bottles, at various times throughout the day. The numbers of poorly labelled bottles are shown below.

No. of poorly labelled bottles	0	1	2	3	4	5	6
Frequency		1	3	3	6	5	1	1

Assume that the number of poorly labelled bottles in each batch of 25 can be modelled by a $Bin(25, \theta)$ random variable.

1. Write down the likelihood and obtain the maximum likelihood estimate $\hat{\theta}$.

2. Explain how to construct an approximate 95% likelihood-ratio-based confidence interval for θ, and describe a Newton-Raphson procedure to determine the end-points of this interval.

4.29 The two-parameter Cauchy distribution has pdf given by

$$f(x) = \frac{\beta}{\pi\{\beta^2 + (x - \alpha)^2\}}, \quad \text{for} \quad -\infty < x < \infty.$$

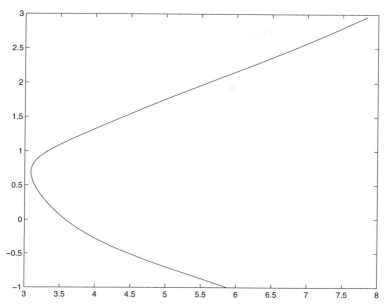

Figure 4.13 *Profile log-likelihood for Exercise 4.29.*

Corresponding to the case of a random sample $x = \{1.09, -0.23, 0.79, 2.31,$ $-0.81\}$, we obtain the graphs of Figures 4.13, 4.14, and 4.15. The first of these graphs is a profile log-likelihood; the second is a section of the log-likelihood surface corresponding to $\beta = 0.2$, and the third is the path traversed in the parameter space corresponding to the profile log-likelihood.

Explain the difference between Figures 4.13 and 4.14, and explain the relevance of the path of Figure 4.15 to the profile of Figure 4.13. Describe how you would obtain a confidence interval for α from the profile of Figure 4.13.

4.30   (Examination question, University of Glasgow.)

An investigation was carried out into the effect of artificial playing pitches on the results of certain professional English soccer teams. In particular, the results for Luton Town on their own pitch over the periods 1980–1985 (grass pitch) and 1986–1990 (plastic pitch) are given. The performance of the team is measured as either a Won, Drawn, or Lost game.

	Won	Drawn	Lost
Grass	61	31	32
Plastic	65	34	18

It is desired to fit a single-parameter multinomial model to the plastic pitch results with parameter θ and

$$p(\text{win}) = 1 - \theta, \quad p(\text{draw}) = \theta - \theta^2, \quad p(\text{lose}) = \theta^2.$$

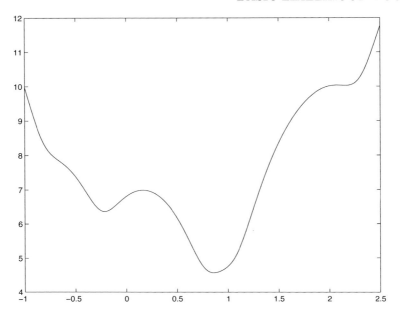

Figure 4.14 *Section of log-likelihood surface, when $\beta = 0.2$. For Exercise 4.29.*

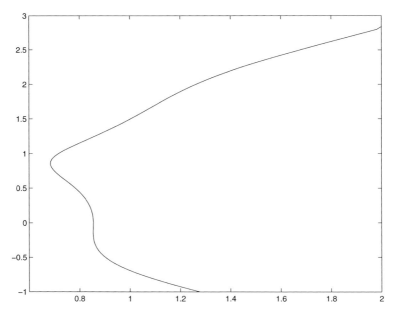

Figure 4.15 *Path traversed in the parameter space corresponding to the profile log-likelihood. For Exercise 4.29.*

If there are observed counts of x_1, x_2, x_3 for the numbers of wins, draws and losses respectively, out of n games, write down the likelihood, and obtain the maximum-likelihood estimate for θ.

Fit the model to the plastic pitch data by maximum-likelihood estimation, and carry out a suitable test for goodness of fit.

Fit the same model to the grass pitch results, and to the data for both types of pitch taken together. Calculate the log-likelihood in all three cases (grass, plastic, combined) and hence perform a likelihood-ratio test of the hypothesis that $\theta_g = \theta_p$, where θ_g and θ_p are respectively the values of θ for the grass and plastic pitches. What is flawed in this comparison?

4.31 (Continuation of Exercise 2.26)

An experiment took place in St. Andrews University, in which groups of golf tees were placed randomly in a field, and records were kept of which groups were observed on each of T sampling occasions; see Borchers et al. (2002, p.13). If f_i denotes the number of groups that were found i times, then the resulting data are shown below.

f_1	f_2	f_3	f_4	f_5	f_6	f_7	f_8
46	28	21	13	23	14	6	11

The purpose of the study was to simulate a standard ecological investigation, aimed at estimating animal abundance. For a study of this kind, the likelihood in general consists of components such as:

$$\psi_i = \left\{ \binom{T}{i} p^i (1-p)^{T-i} \right\}^{f_i},$$

where p denotes the probability that any group is found on any of the sampling occasions.

Explain how you would modify the binomial model in the expression for ψ_i to account for heterogeneity in the probability of finding groups of golf tees, and devise the modified form of ψ_i in terms of beta functions.

Five distinct models have been fitted to the St. Andrews data. Each model is given a name in the table below, and the table presents the value of minus the maximum log-likelihood for each model, the X^2 goodness-of-fit statistic and the number, d, of parameters in the model.

Model	−Max log-lik	X^2	d
Bin	104.9	923.2	2
Beta-Bin	22.6	8.4	3
LNB	22.6	8.8	3
2 Bins	27.3	21.8	4
Bin + Beta-Bin	22.2	8.1	5

The number of cells for each of the X^2 tests was 8. Construct the AIC values for the 5 models.

Explain which model or models you believe is/are the best for the St. Andrews data. Note that in order to answer the last part of the question it is not necessary to understand how each model is constructed. For further discussion, see Morgan and Ridout (2008).

4.32   The pdf of a stable distribution with fixed parameter $\alpha = 1/2$ exists in closed form, and is given by

$$f(x;c) = \frac{c}{(2\pi x^3)^{1/2}} exp\left(\frac{-c^2}{2x}\right), \qquad \text{for} \quad x > 0, c > 0,$$

where c is the only free parameter of the distribution.

Write down the likelihood corresponding to a random sample, $\{x_i, i = 1, \ldots, n\}$. If ℓ denotes the log-likelihood, using the fact that

$$\mathbb{E}\left[\frac{d\ell}{dc}\right] = 0,$$

or otherwise, derive the asymptotic distribution of the maximum-likelihood estimator $\hat{c}$. For further discussion, see Besbeas and Morgan (2004).

4.33   (Examination question, University of Glasgow.)

During an investigation into the possible effect of the chemical DES on the development of human cells, a geneticist grew independent samples of human cells in tissue culture at five different dose levels and counted the number of chromosomes in each cell at metaphase. Each of the cells was then classified as being Abnormal or Normal. The following data were obtained.

Dose level (μg./dl)	Number of abnormal cells	Sample size
5.0	21	99
10.0	25	99
15.0	31	100
20.0	37	97
25.0	42	99

A logistic model, based on a binomial distribution, is fitted to these data, in terms of dose level (and not its logarithm).

Write down the likelihood function. The maximum-likelihood estimates of the parameters and the inverse observed Fisher information matrix are given below, in standard notation:

$$\hat{\alpha} = -1.58023, \qquad \hat{\beta} = 0.05221,$$

$$\hat{\mathbf{I}}^{-1} = \begin{bmatrix} 0.0601345 & -0.0031698 \\ -0.0031698 & 0.0001990 \end{bmatrix}.$$

Construct a 95% confidence interval for the dose level of DES at which 25% of the cells will be abnormal, based on the asymptotic normality of maximum-likelihood estimators.

Construct a corresponding confidence interval for the proportion of abnormal cells likely to be found at a dose level of 17 μg./dl.

4.34 Show that the score function for comparing a Weibull distribution with an exponential distribution has the form

$$u = d + \sum_u logx_i - d\frac{\sum x_i logx_i}{\sum x_i},$$

where d is the number of uncensored terms in the sample and $\sum_u$ denotes summation for only uncensored terms.

Complete the score test statistic.

4.35 A batch of n eggs is infected by bacteria, and the number of bacteria per egg is thought to be described by a Poisson distribution, with mean μ. Write down the likelihood if it is observed that r eggs are free from infection. Obtain the maximum-likelihood estimator for μ, and show that its variance is approximately given by

$$\text{Var}(\hat{\mu} \approx \frac{(e^{\hat{\mu}} - 1)}{n}.$$

4.36 With reference to the fecundity illustration of Section 2.2, show that if the geometric model is appropriate, and that there are n couples, with conception times censored at r cycles, then

$$\mathbb{E}\left[\frac{d^2\ell}{dp^2}\right] = -\frac{n\{1 - (1 - p)^r\}}{p^2(1 - p)},$$

where ℓ denotes the log-likelihood. Compare the asymptotic variances of two estimators of p, the maximum-likelihood estimator and $\tilde{p}$, where $\tilde{p} = n_1/n$, used in Example 1.1.

CHAPTER 5

General Principles

5.1 Introduction

In this chapter we consider a wide range of aspects of model-fitting which have not yet been discussed. In most cases, we develop *general principles*, such as the δ-method, a test for parameter-redundancy, the EM (expectation-maximisation) algorithm, etc. We also discuss the breakdown of the customary regularity conditions, and present several methods of parameter-estimation which are alternatives to the method of maximum likelihood, such as the method of minimum chi-square, and the method of empirical transforms. The Bayesian approach to model-fitting is so widely applicable, that we defer discussion of this important approach until Chapter 7, which is devoted entirely to Bayesian statistics and modern computational methods for conducting Bayesian inference.

5.2 Parameterisation

Consider the exponential pdf, $f(x) = \lambda e^{-\lambda x}$. We may alternatively write $f(x) = \frac{1}{\mu} e^{-\frac{x}{\mu}}$. This is a *re-parameterisation* of the pdf. It does not change the number of parameters in the pdf, but changes the parameters through a one-to-one transformation. Other examples of reparameterisation have been encountered in Sections 2.3 and 4.3. It may be that changing the way a model is parameterised can be beneficial, and we shall now examine aspects of re-parameterisation.

5.2.1 Ways of simplifying optimisation

We start by looking at two examples which both involve two parameters, but where maximum-likelihood estimation of the parameters only requires a one-dimensional search.

Example 5.1: Zero inflation

Zero inflation was introduced in Exercises 1.3 and 4.9. The data of Table 5.1 describe the distribution of stillbirths in 402 litters of New Zealand white rabbits.

We are not, however, provided with litter sizes. If we had litter sizes we may have used a binomial or a beta-binomial model for the data, as has been done previously — see Exercise 3.3. A natural model for the data of

123

Table 5.1 *Distribution of stillbirths in 402 litters of New Zealand white rabbits.*

No. of stillbirths:	0	1	2	3	4	5	6	7	8	9	10	11
No. of litters:	314	48	20	7	5	2	2	1	2	0	0	1

Table 5.1 is a Poisson distribution, as the recorded data are non-negative, discrete and unbounded. However there are far too many 0's in the data for a Poisson distribution to provide a good fit without modification. The obvious modification is to set

$$
\begin{aligned}
pr(Y = 0) &= w + (1 - w)p_0 \\
pr(Y = i) &= (1 - w)p_i , \qquad \text{for } i > 0,
\end{aligned}
\tag{5.1}
$$

where Y is the number of stillbirths, w is the probability which inflates the zero response category and $\{p_i\}$ is a standard Poisson distribution. This is the zero-inflated Poisson distribution of Exercise 4.9. We therefore have a mixture model — we either obtain a zero response with probability w, or with probability $(1 - w)$ a Poisson response occurs, which could also of course result in a zero response. This model may be justified in practice if fundamentally the population of rabbits was inhomogeneous, with one type which never produced stillbirths, and another which might.

The likelihood under model (5.1) is given by

$$
L(w, \lambda) = \prod_{i=0}^{\infty} pr(Y = i)^{n_i},
$$

where n_i is the number of occurrences of i in the data, $i \geq 0$, and λ is the mean of the Poisson distribution. Substituting from Equation (5.1) we obtain the explicit form:

$$
L(w, \lambda) = \{w + (1 - w)e^{-\lambda}\}^{n_0} \prod_{i=1}^{\infty} \left\{ \frac{(1 - w)e^{-\lambda}\lambda^i}{i!} \right\}^{n_i}.
\tag{5.2}
$$

At first sight it may appear that iteration is needed in order to obtain the maximum likelihood estimates, $(\widehat{w}, \widehat{\lambda})$. However, (Ridout, pers.com.) we may re-parameterise the model by setting

$$
w + (1 - w)e^{-\lambda} = 1 - \theta,
$$
$$
\text{so that} \quad \theta = (1 - w)(1 - e^{-\lambda}).
$$

We now express the likelihood as a function of θ and λ, as follows:

$$L(\theta, \lambda) = (1-\theta)^{n_0} \left(\frac{\theta}{1-e^{-\lambda}}\right)^{\sum_{i=1}^{\infty} n_i} \prod_{i=1}^{\infty} \left(\frac{e^{-\lambda}\lambda^i}{i!}\right)^{n_i}$$

$$= (1-\theta)^{n_0} \theta^{\sum_{i=1}^{\infty} n_i} \left(1-e^{-\lambda}\right)^{-\sum_{i=1}^{\infty} n_i} \prod_{i=1}^{\infty} \left(\frac{e^{-\lambda}\lambda^i}{i!}\right)^{n_i}.$$

(5.3)

We can see that the likelihood factorises into two components, and that the one involving θ is of binomial form with the explicit maximum-likelihood estimate:

$$\widehat{\theta} = \left(\frac{\sum_{i=1}^{\infty} n_i}{\sum_{i=0}^{\infty} n_i}\right).$$

Iteration is still needed in order to obtain $\widehat{\lambda}$ (Exercise 5.1), but the problem of maximising the likelihood has been reduced from two-dimensions to one-dimension. The parameter θ is the probability that the random variable $Y > 0$, and we can see that the estimate $\widehat{\theta}$ is simply the proportion of litters with stillbirths, as one might expect. □

In the re-parameterisation of Example 5.1, there is a single parameter for the probability of a zero. The probability $pr(Y = i)$ is expressed as:

$$pr(Y = i) = \frac{(1-w)e^{-\lambda}\lambda^i}{i!} \qquad \text{for } i > 0$$

$$= \theta \frac{e^{-\lambda}\lambda^i}{i!(1-e^{-\lambda})},$$

i.e., $pr(Y = i)$ is the product of the probability that the response is not zero (θ) and then the conditional probability that it is i, which is a term from the zero-*truncated* Poisson distribution:

$$pr(X = i) = \frac{e^{-\lambda}\lambda^i}{i!(1-e^{-\lambda})}, \qquad i \geq 1.$$

We shall return to the zero-truncated Poisson distribution in Example 5.8.

Example 5.2: Fitting the Weibull distribution

The Weibull distribution has been encountered in Example 4.8. It is a convenient model for survival time data when the times are continuous quantities, and not measured in years, as is often the case with wild animals. The pdf for survival times under the Weibull model has the form:

$$f_T(t) = \kappa\rho(\rho t)^{\kappa-1} \exp\{-(\rho t)^\kappa\}, \quad \text{for } t \geq 0.$$

We see, therefore, that the Weibull model has two parameters, κ and ρ, and that when $\kappa = 1$, it reduces to the exponential distribution.

Suppose we observe uncensored times: $t_1, t_2, \ldots, t_m$, and censored times, $t_{m+1}, t_{m+2}, \ldots, t_n$. As the survivor function of the Weibull distribution has the form $Pr(T > t) = S(t) = \exp\{-(\rho t)^\kappa\}$, then the likelihood is:

$$L(\rho, \kappa) = (\kappa\rho)^m \prod_{i=1}^{m} (\rho t_i)^{\kappa-1} \exp\{-(\rho t_i)^\kappa\} \prod_{j=m+1}^{n} \exp\{-(\rho t_j)^\kappa\}.$$

Hence the log-likelihood is given by:

$$\ell(\rho, \kappa) = m \log \kappa + m\kappa \log \rho + (\kappa - 1) \sum_{i=1}^{m} \log t_i - \rho^\kappa \sum_{i=1}^{n} t_i^\kappa.$$

In order to obtain maximum likelihood estimates we need to solve the two equations, $\dfrac{\partial \ell}{\partial \rho} = 0$ and $\dfrac{\partial \ell}{\partial \kappa} = 0$, where,

$$\frac{\partial \ell}{\partial \rho} = \frac{m\kappa}{\rho} - \kappa\rho^{\kappa-1} \sum_{i=1}^{n} t_i^\kappa$$

$$\frac{\partial \ell}{\partial \kappa} = \frac{m}{\kappa} + m \log \rho + \sum_{i=1}^{m} \log t_i - \rho^\kappa \sum_{i=1}^{n} t_i^\kappa \log(\rho t_i).$$

$$(5.4)$$

We note that from the first of Equations (5.4), when $\dfrac{\partial \ell}{\partial \rho} = 0$ we can solve explicitly for ρ in terms of κ:

$$\hat{\rho} = \left(\frac{m}{\sum_{i=1}^{n} t_i^\kappa} \right)^{\frac{1}{\kappa}}. \tag{5.5}$$

We can then substitute for $\hat{\rho}$ in $\dfrac{\partial \ell}{\partial \kappa} = 0$, resulting in a single equation to be solved numerically to obtain $\hat{\kappa}$, which will then also give $\hat{\rho}$ from Equation (5.5). Thus here too, as in Example 5.1, a two-parameter model is fitted by maximum-likelihood following a one-dimensional numerical search for a function maximum — see Exercise 5.4. □

The result of Example 5.2 is not a consequence of a reparameterisation, but arises simply because when the random variable T has the given Weibull distribution, the random variable T^κ has the exponential distribution $Ex(\rho^\kappa)$ — see Exercise 5.4. Thus for fixed κ, the maximum-likelihood estimate of ρ^κ has the well-known form resulting from Equation (5.5), which is the continuous-time analogue of the geometric distribution result of Equation (2.1).

It is interesting to note that the solution of Equation (5.5) provides an easy way of obtaining the profile likelihood with respect to κ, which then may be used to provide $\hat{\kappa}$. This is a particular illustration of *nested optimisation*: likelihood maximisation may be relatively simple when certain parameters are treated as constant. The optimisation can then proceed in a nested fashion, by first maximising with regard to the parameters which are not regarded as constant and then maximising over the parameters which had been treated as constant — see Exercise 5.5.

We see from Example 5.1 that judicious reparameterisation of a model can be most advantageous, and other examples have been provided earlier in the book. For example, in a beta-geometric distribution it is convenient to repa- rameterise from (α, β), the pair of parameters describing the beta distribution, to (μ, θ), where μ is the mean of the beta-distribution, and θ is proportional to the variance of the beta-distribution. Another example arises in the logit model, where we changed the parameterisation from the location and scale pair of parameters (α, β) to the pair (μ, β), where the parameter μ was the ED_{50} dose, with a direct interpretation.

In the case of the logit model, the correlation between $\widehat{\mu}$ and $\widehat{\beta}$ is far less than the correlation between $\widehat{\alpha}$ and $\widehat{\beta}$. A similar example for quantal response data is given in Exercise 5.3. We now consider an *automatic* procedure which may, in certain cases, be useful for producing uncorrelated parameter estimators.

5.2.2 *Orthogonal parameter estimators

The work of this section is due to Cox and Reid (1987). Suppose that the parameter vector $\boldsymbol{\theta}$ is partitioned into the two vectors, $\boldsymbol{\theta}_1$ and $\boldsymbol{\theta}_2$, of lengths p_1 and p_2 respectively. We define $\boldsymbol{\theta}_1$ to be *orthogonal* to $\boldsymbol{\theta}_2$ if the elements of the expected information matrix satisfy:

$$\mathbb{E}\left[\frac{\partial \ell}{\partial \theta_s}\frac{\partial \ell}{\partial \theta_t}\right] \equiv - \mathbb{E}\left[\frac{\partial^2 \ell}{\partial \theta_s \partial \theta_t}\right] = 0$$

for $1 \leq s \leq p_1, \quad p_1 + 1 \leq t \leq p_1 + p_2$.

Note that $\mathbb{E}\left[\dfrac{\partial \ell}{\partial \theta_i}\right] = 0$ for all i, as discussed in Chapter 4, so that orthog- onality is equivalent to setting the relevant correlations equal to zero.

It is not usually possible to find a set of parameters that are orthogonal, but progress can be made in the special but useful case of one parameter of interest, ψ, say, being orthogonal to the remaining model parameters, say $\lambda_1, \ldots, \lambda_q$, which we regard as nuisance parameters in this context.

Suppose that initially the likelihood is specified in terms of the parameters: $(\psi, \phi_1, \ldots, \phi_q)$.

We set

$$\begin{aligned} \phi_1 &= \phi_1(\psi, \boldsymbol{\lambda}), \\ \phi_2 &= \phi_2(\psi, \boldsymbol{\lambda}), \\ &\vdots \\ \phi_q &= \phi_q(\psi, \boldsymbol{\lambda}). \end{aligned}$$

Thus we may write the log-likelihood as:

$$\ell(\psi, \boldsymbol{\lambda}) = \ell^*\{\psi, \phi_1(\psi, \boldsymbol{\lambda}), \ldots, \phi_q(\psi, \boldsymbol{\lambda})\},$$

so that ℓ^* is a function of ψ and $\boldsymbol{\phi}$. Then,

$$\frac{\partial \ell}{\partial \psi} = \frac{\partial \ell^*}{\partial \psi} + \sum_{r=1}^{q} \frac{\partial \ell^*}{\partial \phi_r}\frac{\partial \phi_r}{\partial \psi},$$

and

$$\frac{\partial^2 \ell}{\partial \psi \partial \lambda_t} = \sum_{s=1}^{q} \frac{\partial^2 \ell^*}{\partial \psi \partial \phi_s} \frac{\partial \phi_s}{\partial \lambda_t} + \sum_{s=1}^{q} \sum_{r=1}^{q} \frac{\partial^2 \ell^*}{\partial \phi_r \partial \phi_s} \frac{\partial \phi_s}{\partial \lambda_t} \frac{\partial \phi_r}{\partial \psi} + \sum_{r=1}^{q} \frac{\partial \ell^*}{\partial \phi_r} \frac{\partial^2 \phi_r}{\partial \psi \partial \lambda_t}.$$

We now take expectations.

Because $\mathbb{E}\left[\dfrac{\partial \ell^*}{\partial \phi_r}\right] = 0$, in order to obtain $\mathbb{E}\left[\dfrac{\partial^2 \ell}{\partial \psi \partial \lambda_t}\right] = 0$, we need:

$$\sum_{s=1}^{q} \frac{\partial \phi_s}{\partial \lambda_t} \left(i^*_{\psi \phi_s} + \sum_{r=1}^{q} i^*_{\phi_r \phi_s} \frac{\partial \phi_r}{\partial \psi} \right) = 0, \qquad 1 \le t \le q,$$

where $i^*_{\alpha\beta} = \mathbb{E}\left[\dfrac{\partial^2 \ell^*}{\partial \alpha \partial \beta}\right]$.

It is sufficient that we have

$$\sum_{r=1}^{q} i^*_{\phi_r \phi_s} \frac{\partial \phi_r}{\partial \psi} = -i^*_{\psi \phi_s}, \qquad \text{for } 1 \le s \le q. \tag{5.6}$$

Thus solution of Equation (5.6) determines how ϕ depends on ψ, but how ϕ depends on λ is arbitrary.

Example 5.3: Reparameterising exponential distributions

The random variables Y_1 and Y_2 are respectively, $Ex(\phi^{-1})$ and $Ex(\psi^{-1}\phi^{-1})$. Suppose we observe $Y_1 = y$, and $Y_2 = y_2$; then

$$\ell(\phi, \psi) = \log\left\{ \frac{1}{\phi} e^{-\frac{y_1}{\phi}} \frac{1}{\phi\psi} e^{-\frac{y_2}{\phi\psi}} \right\},$$

i.e., $\ell(\phi, \psi) = -2\log\phi - \dfrac{y_1}{\phi} - \log\psi - \dfrac{y_2}{\phi\psi}$, and we obtain the partial derivatives:

$$\frac{\partial \ell}{\partial \phi} = -\frac{2}{\phi} + \frac{y_1}{\phi^2} + \frac{y_2}{\phi^2 \psi},$$

$$\frac{\partial^2 \ell}{\partial \phi^2} = \frac{2}{\phi^2} - \frac{2y_1}{\phi^3} - \frac{2y_2}{\phi^3 \psi},$$

$$\frac{\partial^2 \ell}{\partial \phi \partial \psi} = \frac{y_2}{\phi^2 \psi^2}.$$

Hence, $\mathbb{E}\left[\dfrac{\partial^2 \ell}{\partial \phi^2}\right] = -\dfrac{2}{\phi^2}$; $\qquad \mathbb{E}\left[\dfrac{\partial^2 \ell}{\partial \phi \partial \psi}\right] = -\dfrac{1}{\phi\psi}$.

Equation (5.6) then becomes:

$$\frac{2}{\phi^2} \frac{\partial \phi}{\partial \psi} = -\frac{1}{\phi\psi},$$

with solution, $\phi\psi^{\frac{1}{2}} = a(\lambda)$, for any function $a(\ )$. For example, if we take $a(\lambda) = \lambda$, then the orthogonal reparameterisation of this problem gives:

$$Y_1 \sim Ex(\lambda^{-1}\psi^{\frac{1}{2}}), \quad Y_2 \sim Ex(\lambda^{-1}\psi^{-\frac{1}{2}}).$$

$\square$

5.2.3 *Curvature

For a one-dimensional problem, and a likelihood which is a function of a single parameter, the curvature of the graph of the log-likelihood is related to the second-order derivative of the log-likelihood with respect to the parameter. In higher dimensions, curvature has two orthogonal components, the *intrinsic* curvature and the *parameter-effects* curvature. Changing the way a model is parameterised changes the parameter-effects curvature. It is desirable that the parameter-effects curvature is small, and this can often follow from a search for orthogonal parameter estimators. For detailed discussion, see Seber and Wild (1989, Chapter 4), Venables and Ripley (2002, p.237), and Ratkowski (1983, 1990).

5.2.4 The δ method

Maximum likelihood parameter estimates enjoy an *invariance* property. In the case of the exponential pdf at the start of Section 5.2, if $\widehat{\lambda}$ is the maximum-likelihood estimate of λ, then the maximum-likelihood estimate of μ when $\mu = \frac{1}{\lambda}$ is given by $\widehat{\mu} = \frac{1}{\widehat{\lambda}}$. Similarly, in the logit modelling of dose-response data, the maximum-likelihood estimate of the ED_{50}, μ, is given by:

$$\widehat{\mu} = -\widehat{\alpha}/\widehat{\beta},$$

where $\widehat{\alpha}$ and $\widehat{\beta}$ are the maximum likelihood estimates of α and β respectively, and $\mu = -\alpha/\beta$.

In general terms, if the parameterisation of a model is changed from $\boldsymbol{\theta}$ to $\boldsymbol{\eta}$, say, through a suitable transformation, $\boldsymbol{\eta} = \psi(\boldsymbol{\theta})$, then the corresponding maximum likelihood estimates are related by:

$$\widehat{\boldsymbol{\eta}} = \psi(\widehat{\boldsymbol{\theta}}).$$

If we are interested in a scalar function, $\eta = \psi(\boldsymbol{\theta})$, then a key issue is how we can approximate to the variance of $\widehat{\eta}$ when we already have an estimate, $\boldsymbol{S}$, say of the dispersion (i.e., variance-covariance) matrix for $\widehat{\boldsymbol{\theta}}$. For instance we may, from Equation (4.2), use $\boldsymbol{S} = \boldsymbol{J}^{-1}(\widehat{\boldsymbol{\theta}})$. Typically we would as usual be assuming that $\widehat{\eta}$ is asymptotically unbiased and normally distributed. The answer is supplied very simply and neatly by the so-called δ-method:

$$\text{Var}(\widehat{\eta}) \approx \mathbf{a}'\boldsymbol{S}\mathbf{a} \tag{5.7}$$

where $\mathbf{a}' = \left(\dfrac{\partial \eta}{\partial \theta_1}, \dfrac{\partial \eta}{\partial \theta_2}, \dots, \dfrac{\partial \eta}{\partial \theta_d} \right)$, evaluated at $\widehat{\boldsymbol{\theta}}$. The detailed derivation of the δ-method and explanation of how it extends to the case of a vector $\boldsymbol{\eta}$ are given by Bishop et al. (1975, p.486).

Heuristically, consider the simple case of $\eta = \psi(\theta)$. By once again taking a truncated Taylor series expansion, we can write

$$\widehat{\eta} = \psi(\widehat{\theta}) \approx \psi(\mu) + (\widehat{\theta} - \mu)\psi'(\mu),$$

where $\mu = \mathbb{E}[\widehat{\theta}]$.

Thus $\mathbb{E}[\psi(\widehat{\theta})] \approx \psi(\mathbb{E}[\widehat{\theta}])$, and

$$\{\psi(\widehat{\theta}) - \psi(\mu)\}^2 \approx (\widehat{\theta} - \mu)^2 \psi'(\mu)^2,$$

so that what we require, $\mathrm{Var}(\psi(\widehat{\theta}))$, is approximately given by:

$$\mathrm{Var}(\psi(\widehat{\theta})) \approx \mathbb{E}[\{\psi(\widehat{\theta}) - \psi(\mu)\}^2] \approx \mathrm{Var}(\widehat{\theta})\psi'(\mu)^2,$$

which is the univariate version of expression (5.7), when we estimate μ by $\widehat{\theta}$.

Example 5.4: ED$_{50}$ estimation

As we saw in Example 4.3, the logit model for dose-response data can be parameterised as either $P(d) = 1/(1 + e^{-(\alpha + \beta d)})$, or $P(d) = 1/(1 + e^{-\beta(d-\mu)})$, where μ is the ED$_{50}$ — when $d = \mu$, $P(d) = 0.5$. If the second parameterisation is used, we obtain a direct maximum-likelihood estimate of μ and of $\mathrm{Var}(\widehat{\mu})$. If the first parameterisation is used then the maximum-likelihood estimate of μ is obtained as

$$\widehat{\mu} = -\frac{\widehat{\alpha}}{\widehat{\beta}},$$

and by the δ-method, from expression (5.7),

$$\mathrm{Var}(\widehat{\mu}) \approx \left(-\frac{1}{\widehat{\beta}}, \frac{\widehat{\alpha}}{\widehat{\beta}^2}\right) \mathbf{S} \begin{pmatrix} -\frac{1}{\widehat{\beta}} \\ \frac{\widehat{\alpha}}{\widehat{\beta}^2} \end{pmatrix},$$

where $\mathbf{S}$ is the estimated dispersion matrix of $(\widehat{\alpha}, \widehat{\beta})$. See Exercise 5.7 for a continuation of this example. □

5.3 *Parameter redundancy

Later in this chapter (in Sections 5.6.4, 5.7.4, and 5.8) we shall be discussing mixture models, and how they may be fitted to data. The simplest example of a mixture is when we can write a pdf $f(x)$ in the form:

$$f(x) = \alpha f_1(x) + (1 - \alpha)f_2(x),$$

where α is a probability and $f_1(x)$, $f_2(x)$ are also pdfs. If we try to fit such a model to data which arise from $f_1(x)$, then we would expect the likelihood to provide essentially no information on the parameters of $f_2(x)$. Some models may not just result in a likelihood surface which is fairly flat in certain directions, as in this case, but they always result in a likelihood surface which is in part *completely* flat — possibly along a ridge, for example. A trivial illustration of this is if one tried to estimate the parameters (ψ, ϕ) of Example 5.3, using only observations on the single exponential random variable Y_2. In this case the parameters only occur as the product, $\psi\phi$, and, without information on Y_1, it is only that product which may be estimated using the method of maximum-likelihood. In this section we provide a test for whether a model contains so many parameters that it is not possible to estimate all of them.

5.3.1 Too many parameters in a model

The topic of parameter redundancy is a general one, of relevance to all stochastic modelling. We shall introduce the issues through a particular model from statistical ecology.

Example 5.5: The Cormack-Seber model for bird survival

An attractive model for recovery data, similar to that of Table 1.5, is one with a separate annual survival probability for each year of a bird's life, and a single probability, λ, for the finding and reporting of a ring from a previously marked dead bird. The full set of model parameters for a k-year study is then: $\{\lambda, \phi_1, \phi_2, \ldots, \phi_k\}$, where ϕ_i is the probability that a bird survives its ith year of life, conditional on it having survived $(i-1)$ years. This model was considered in Example 4.6.

Given that a bird is ringed in year i, the probability that it is reported dead in year $(i+j)$ is $\lambda \phi_1 \ldots \phi_j (1 - \phi_{j+1})$, and the probability that it is never recovered is

$$1 - \lambda(1 - \phi_1 \cdots \phi_{k+1-i}), \qquad \text{for } 1 \leq i \leq k, \text{ and } 0 \leq j \leq k - i.$$

If we assume that birds behave independently, an assumption which may be violated in practice, then the likelihood is of product-multinomial form, i.e., it is the product of several multinomials:

$$L(\lambda, \boldsymbol{\phi}) \propto \prod_{i=1}^{k} \{1 - \lambda(1 - \phi_1 \cdots \phi_{k+1-i})\}^{F_i} \prod_{j=0}^{k-i} \{\lambda \phi_1 \cdots \phi_j (1 - \phi_{j+1})\}^{m_{i,i+j}}.$$

Here we have omitted the term involving multinomial coefficients, as it does not involve the model parameters. Additionally,

$$F_i = R_i - \sum_{j=0}^{k-i} m_{i,i+j}, \qquad 1 \leq i \leq k,$$

R_i is the number of birds ringed in the ith cohort, $1 \leq i \leq k$ and m_{ij} of these are found and reported dead in the jth year. Thus F_i is the number of birds ringed in year i that have not been recovered at the end of the experiment. We may note that in Table 1.5, the values for R_i are not presented, and so a conditional analysis of the data would be necessary in that case.

There is a problem with now simply taking the likelihood $L(\lambda, \boldsymbol{\phi})$, and trying to maximise L with respect to the $(k + 1)$ parameters, $(\lambda, \boldsymbol{\phi})$. The problem is that the likelihood surface possesses a completely flat ridge, the precise nature and location of which are described in Catchpole and Morgan (1994). Two profile log-likelihoods have been shown in Figure 4.9, and indicate the extent of the ridge. □

In a model such as that of Example 5.4, no matter how much data we collect, we cannot estimate all of the parameters in the model. We say that the model is parameter-redundant. Quite often it is obvious that the way we specify a model, although perfectly natural and potentially useful, will result in parameter-redundancy. We have seen an example of this in our earlier discussion of observing only the random variable Y_2 of Example 5.3. However in many instances parameter-redundancy is subtle and not at all obvious; Example 5.4 provides a classical illustration of this.

The questions that need answering include the following:

(i) How can I tell whether a particular model is parameter-redundant?

(ii) If a model is not parameter-redundant, might it nevertheless be rendered parameter-redundant by the presence of insufficient or missing data?

(iii) If a model is parameter-redundant, what is its *deficiency*? The deficiency of a model is the number of parameters we are not able to estimate from the data. In the case of Example 5.4, we can estimate k of the $(k + 1)$ parameters, as we shall now see, and so it has deficiency 1.

Example 5.5 continued

Suppose now that we re-parameterise by setting

$$\tau_1 \cdots \tau_i = 1 - \lambda(1 - \phi_1 \cdots \phi_i), \qquad 1 \le i \le k.$$

Thus we move from the set of $(k+1)$ parameters, $(\lambda, \phi_1, \ldots, \phi_k)$, to the reduced set of k parameters, $(\tau_1, \ldots, \tau_k)$. If we had a situation in which $\lambda = 1$, i.e., all dead animals were found and reported, then the $\{\phi_j\}$ would become the $\{\tau_j\}$. Intensive studies of a population of Soay sheep on the remote island of Hirta in the St. Kilda archipelago off the west coast of Scotland approximate the $\lambda = 1$ situation when large numbers of ecologists are employed — see Catchpole et al. (2000).

It was shown by Catchpole and Morgan (1991) that there are explicit maximum-likelihood estimators of the $\{\tau_i\}$, given by

$$\widehat{\tau}_j = 1 - \frac{D_j}{U_j}, \qquad 1 \le j \le k,$$

where $D_j = \sum_{i=1}^{k-j+1} m_{i,i+j-1}$ is the number of ringed birds which die and are recovered in their jth year of ringing, and

$$U_j = \sum_{i=j}^{k} (D_i + F_{k+1-i})$$

is the number which we may regard as 'available for recovery' in that year, in that they have not been recovered in previous years.

Thus the deficiency of this model is 1, as we cannot estimate $(k + 1)$ parameters from the data, but we can estimate k parameters.   □

5.3.2 A test using symbolic algebra

(Catchpole and Morgan, 1997.) First, consider taking a random sample, $y_1, \ldots, y_n$, from an *exponential family*, with density functions or probability functions

$f(y_i)$ given by:

$$\log f(y_i) = \frac{y_i \xi_i - b(\xi_i)}{\gamma} + c(y_i, \gamma), \quad \text{for } 1 \le i \le n$$

for functions $b(\ )$ and $c(\ )$, where the scale parameter γ is known, and

$$\mathbb{E}[Y_i] = \mu_i = b'(\xi_i), \quad \text{for } 1 \le i \le n$$

for parameters $\{\xi_i\}$. Here Y_i is the random variable taking the value y_i. This is an important general class of distributions, which not only includes pdfs such as the normal and exponential, but also extends to discrete probability functions such as the Poisson. We shall encounter the exponential family again in Chapter 8, when we discuss generalised linear models.

Thus the log-likelihood is given by

$$\ell = \sum_{i=1}^{n} \left[\{y_i \xi_i - b(\xi_i)\}/\gamma + c(y_i, \gamma) \right].$$

We suppose now that $\boldsymbol{\mu} = (\mu_1, \ldots, \mu_n)$ is specified in terms of a d-dimensional parameter vector $\boldsymbol{\theta}$, so that the elements of the scores vector are given by:

$$\frac{\partial \ell}{\partial \theta_s} = \gamma^{-1} \sum_{i=1}^{n} (y_i - \mu_i) \frac{\partial \xi_i}{\partial \theta_s}, \quad 1 \le s \le d.$$

The scores vector $\mathbf{U} = \left(\dfrac{\partial \ell}{\partial \theta_1}, \ldots, \dfrac{\partial \ell}{\partial \theta_d} \right)'$ can therefore be written as

$$\mathbf{U} = \gamma^{-1} \mathbf{D}(\mathbf{y} - \boldsymbol{\mu}) \tag{5.8}$$

where $\mathbf{D}$ is the *derivative matrix* whose elements are

$$d_{s_i} = \frac{\partial \mu_i}{\partial \theta_s}, \quad 1 \le i \le n, \quad 1 \le s \le d.$$

The model is said to be parameter-redundant if it can be expressed in terms of a smaller parameter vector $\boldsymbol{\beta}$ of dimension $q < d$.

From Chapter 4, we know that the expected Fisher information matrix is given by $\mathbf{J} = \mathbb{E}[\mathbf{U}\mathbf{U}']$, and hence from Equation (5.8)

$$\gamma^2 \mathbf{J} = \mathbf{D}\boldsymbol{\Sigma}\mathbf{D}', \tag{5.9}$$

where $\boldsymbol{\Sigma}$ is the variance-covariance matrix of $\mathbf{y}$.

We should expect the matrix $\mathbf{D}$ to play a key role in the redundancy status of the model, as it relates to the transformation from the $\boldsymbol{\mu}$-parameterisation to the $\boldsymbol{\theta}$-parameterisation. Similarly we should expect the redundancy status of the model to depend on the rank of the Fisher expected information matrix $\mathbf{J}$. As $\boldsymbol{\Sigma}$ is positive-definite, then Equation (5.9) shows that $\mathbf{J}$ and $\mathbf{D}$ have the same rank. It is simpler to work with $\mathbf{D}$, rather than $\mathbf{J}$, and, it is shown in Catchpole and Morgan (1997) that a model is parameter-redundant if and only if $\mathbf{D}$ is symbolically rank-deficient, that is, there exists a vector function

$\boldsymbol{\alpha}(\boldsymbol{\theta})$, non-zero for all $\boldsymbol{\theta}$, such that

$$\boldsymbol{\alpha}(\boldsymbol{\theta})'\mathbf{D}(\boldsymbol{\theta}) = 0 \tag{5.10}$$

for all $\boldsymbol{\theta}$. Thus in order to test a model for parameter redundancy, we form the derivative matrix $\mathbf{D}$ and determine its row rank. If it is of full rank then Equation (5.10) has no suitable solution for $\boldsymbol{\alpha}$. Both obtaining $\mathbf{D}$ and checking its row rank are easily done in a symbolic algebra package such as Maple, and Catchpole (1999) provides the required computer programs. We shall now consider an example involving multinomial distributions, for which the results of Equation (5.10) also hold (Catchpole and Morgan, 1997; see Exercise 5.34 below).

Example 5.6: A model of deficiency 1

Consider a bird ring-recovery model with 2 years of ringing and 4 years of recoveries, and probability matrix,

$$\mathbf{P} = \left[\begin{array}{cccc} (1-\phi_{11})\lambda_1 & \phi_{11}(1-\phi_2)\lambda_a & \phi_{11}\phi_2(1-\phi_3)\lambda_a & \phi_{11}\phi_2\phi_3(1-\phi_4)\lambda_a \\ 0, & (1-\phi_{12})\lambda_1 & \phi_{12}(1-\phi_2)\lambda_a & \phi_{12}\phi_2(1-\phi_3)\lambda_a \end{array} \right].$$

In each row we have excluded the probability of birds not being reported dead by the end of the study. In this model, first-year birds survive their first year of life with probability ϕ_{11}, for year 1 of the study, and ϕ_{12} for year 2 of the study. Older birds survive their ith year of life with probability ϕ_i, $2 \leq i \leq 4$. Dead first-year birds are found and reported dead with probability λ_1, while the corresponding rate for older birds is λ_a. First-year birds often behave differently from older birds, and this can affect the reporting rate of dead birds. This model has seven parameters, and from the two multinomial distributions which make up the probability matrix $\mathbf{P}$ there are seven degrees of freedom. The question now to be answered is whether the parameters are so combined in the cells of $\mathbf{P}$ that they cannot all be estimated. The answer lies in whether we can find solutions to Equation (5.10).

It is shown in Catchpole and Morgan (1997) that in order to determine the deficiency of a model, it is not necessary to complete the two multinomial distributions by adding an extra cell to each row to correspond to the birds never found. Additionally, we may form $\mathbf{D}$ after taking logarithms of the cell probabilities. This results in the derivative matrix:

$$\mathbf{D} = \left[\begin{array}{ccccccc} -(1-\phi_{11})^{-1} & \phi_{11}^{-1} & \phi_{11}^{-1} & \phi_{11}^{-1} & 0 & 0 & 0 \\ 0 & 0 & 0 & 0 & -(1-\phi_{12})^{-1} & \phi_{12}^{-1} & \phi_{12}^{-1} \\ 0 & -(1-\phi_2)^{-1} & \phi_2^{-1} & \phi_2^{-1} & 0 & -(1-\phi_2)^{-1} & \phi_2^{-1} \\ 0 & 0 & -(1-\phi_3)^{-1} & \phi_3^{-1} & 0 & 0 & -(1-\phi_3)^{-1} \\ 0 & 0 & 0 & -(1-\phi_4)^{-1} & 0 & 0 & 0 \\ \lambda_1^{-1} & 0 & 0 & 0 & \lambda_1^{-1} & 0 & 0 \\ 0 & \lambda_a^{-1} & \lambda_a^{-1} & \lambda_a^{-1} & 0 & \lambda_a^{-1} & \lambda_a^{-1} \end{array} \right],$$

corresponding to the order of differentiation: $\phi_{11}, \phi_{12}, \phi_2, \phi_3, \phi_4, \lambda_1, \lambda_a$.

In this case the equation $\boldsymbol{\alpha}'\mathbf{D} = \mathbf{0}$ has a single solution. Thus only six parameters may be estimated by maximum-likelihood. $\square$

5.3.3 Determining the estimable parameters

When a model is parameter-redundant, we are interested in whether we might still be able to obtain a maximum-likelihood estimate of any of the original model parameters. The answer is provided by Catchpole et al. (1998), who show that a particular parameter θ_r is estimable if and only if $\alpha_{r_j} = 0$ for all vector functions $\boldsymbol{\alpha}_j(\boldsymbol{\theta})$ which satisfy Equation (5.10).

Example 5.7 continued

We already know that we cannot estimate all seven of the parameters of the model of Example 5.6. We now investigate what may be estimated. The single solution to Equation (5.10) is of the form:

$$\boldsymbol{\alpha}' \propto [0, 0, (1 - \phi_2), (1 - \phi_3)/\phi_2, (1 - \phi_4)/(\phi_2\phi_3), 0, \lambda_a].$$

Thus the parameters ϕ_{11}, ϕ_{12} and λ_1 are estimable. It can be shown that three additional parameters which may be estimated are:

$$\kappa_1 = \lambda_a(1 - \phi_2), \quad \kappa_2 = \phi_2(1 - \phi_3)/(1 - \phi_2), \quad \kappa_3 = \phi_3(1 - \phi_4)/(1 - \phi_3)$$

— see Catchpole et al. (1998b).

This example provides an interesting contrast to the Cormack-Seber model of Example 5.4. In that case the model is also of deficiency 1, but the set of parameters that may be estimated is $\{\tau_i\}$, given earlier, which does not include any of the original parameters. For this model the single solution $\boldsymbol{\alpha}$ to Equation (5.10) has no zero components.   □

Questions of parameter redundancy arise in many different areas of statistics. An application to compartment modelling is presented in Exercise 5.30. Our discussion of parameter-redundancy has focussed on the model alone. Catchpole and Morgan (2000) show how the test of Equation (5.10) is easily extended to deal with the effect of missing data. Recent developments involve matrix decompositions of the derivative matrix, which help to refine issues of rank-deficiency; see Gimenez et al. (2003) and Cole and Morgan (2007). Note also that our discussion has been restricted to the case of parameter estimation by the method of maximum-likelihood. As we shall see in Chapter 7, the Bayesian approach can result in different conclusions.

5.4 Boundary estimates

A likelihood such as that of Example 5.4 contains parameters which are probabilities. Naive attempts at optimising functions with parameters which have restricted ranges can result in an optimum being located in an inadmissible region of the parameter space, which is generally unacceptable. The likelihood is then maximised on a boundary to the allowed parameter space. In such a case it is possible to use an optimisation routine which allows bounds to be placed on the parameters. An example of this is provided by the MATLAB routine *fmincon* which is available in the MATLAB optimisation toolbox. It uses a sequential quadratic programming method, which was mentioned for standard deterministic optimisation in Example 3.2. An added attraction of *fmincon* is that it provides an estimate of the Hessian matrix. However experience suggests that this should be checked in practice; cf. the comments made in Section 3.2.

Alternatively, we can use suitable re-parameterisations. For example, a probability p can be written as:

$$p = \frac{1}{1 + e^{-\alpha}} \ .$$

If optimisation then takes place with regard to α, rather than p, then α can range without restriction and p remains in range. This approach was used in the MATLAB function *preference* of Exercise 4.1. See Exercise 5.2 for discussion of alternatives. Another way to avoid part of the parameter space is to penalise the likelihood in the appropriate region. However this approach may prevent optimisation routines from finding optima close to a boundary — see Morgan and Ridout (2008).

If a model has many parameters, it is possible to miss a single parameter being estimated on the boundary of the parameter space. When a boundary estimate occurs, it naturally casts doubt on the use and validity of estimates of standard error resulting from inverting information matrices. Parameter redundancy leads to likelihoods with flat features, which can result in practice in boundary optimisation and estimates of correlation between parameter estimators which are close to ±1. For this reason it is always important to check estimates of correlations between parameter estimators.

5.5 Regression and influence

5.5.1 Standardising variables

The basic idea of regression is to relate a variable y to a set of explanatory variables $\mathbf{x}$. In linear regression we have the equation, $y = \boldsymbol{\alpha}'\mathbf{x} + \varepsilon$, where ε has a suitable distribution, and $\boldsymbol{\alpha}$ is a set of parameters to be estimated. An interesting problem here concerns choosing the appropriate elements of $\mathbf{x}$ to include in the model. We have seen that the basic regression idea extends, for instance, to logistic regression, in which a probability $p = 1/(1 + e^{-\boldsymbol{\alpha}'\mathbf{x}})$, and further extensions form the topic of Chapter 8.

In simple linear regression with a single x-variable, it is well known that if we replace

$$y = \alpha + \beta x + \varepsilon \quad \text{by}$$
$$y = \gamma + \beta(x - \overline{x}) + \varepsilon,$$

where $\overline{x}$ is the sample mean of the x-variables, so that $(\gamma - \beta\,\overline{x}) = \alpha$, then the estimators $\widehat{\gamma}$ and $\widehat{\beta}$ have zero correlation — see for example, Wetherill (1981, p.32). In non-linear models, such as logistic regression, it is also important to standardise x-variables before model fitting, so that each transformed variable has unit sample variance and has been *centred* so that it has zero sample mean. The effect of this transformation will not, as in the linear case, produce estimators with exactly zero correlation, but it will greatly reduce the correlations between parameter estimators and thereby facilitate the model-fitting process by maximum-likelihood — see Marquardt (1980), and Exercise 5.3. Thus in the regression context we have a simple way of trying to obtain orthogonal parameter estimators, the topic of Section 5.2.2. Although this procedure should be standard, it is often forgotten, even by experienced statisticians.

5.5.2 Influence

Individual variables and data points can exert appreciable influence on the results of a statistical analysis. This is well documented in areas such as principal component analysis (Pack et al., 1988) and linear regression (Cook, 1986). Pregibon (1981) used an approximate linearisation, resulting from taking just one step of a Newton-Raphson iteration for obtaining maximum-likelihood estimates, in order to obtain influence results for logistic regression. This work has important consequences for all non-linear modelling, but it currently sees little use. An exception was encountered in Example 4.11, in which Giltinan et al. (1988) were able to base diagnostics on the results of a score test. In any analysis it is important to consider whether parts of the data are particularly influential. For example, pyrethrins, which are presumed to cause the mortalities of Table 4.7, are plant-based, fast-acting insecticides. Consequently, in spite of careful attention to minimising natural mortality in the experiment, the late deaths observed in Table 1.4 are almost certainly not due to the insecticide. Due to their extreme position, we might expect them to exert some influence on the parameter estimates. In another illustration, Freeman and Morgan (1990) demonstrated how the results of a model for bird survival were dependent on particular aspects of the data. Influence considerations can be especially important when data are sparse. Formal study of influence may be done through the *influence function* (Cox and Hinkley, 1986, p.345). Influence functions may also be used to calculate the asymptotic variances of parameter estimators.

5.6 The EM algorithm

5.6.1 Introduction

Let us start by considering five statistical modelling situations:

(i) A multinomial model with a pooled cell, that is to say, we do not observe the components of two of the multinomial cells, but only the total occupancy of those two cells.

(ii) Abbott's formula for natural mortality in a dose-response experiment.

Here the probability of response to dose level d of a substance is written as:

$$P(d) = \lambda + (1 - \lambda)F(d),$$

where $F(\ )$ is a cdf and λ is a probability of natural response, that is of death which is not due to the effect of the substance. Note that in this formulation we would expect $F(0) = 0$, which would arise in a logit model with regression on log dose, for example.

(iii) The mover-stayer model for the movement of individuals between levels in a hierarchy, such as grades of employment or states of a disease. The transition probability, q_{ij}, is the probability of moving from state i to state j. Here the transition probabilities are of the form:

$$q_{ij} = (1 - s_i)p_j \qquad \text{for } i \neq j$$
$$q_{ii} = (1 - s_i)p_i + s_i$$

where $\{p_i\}$ forms a probability distribution, and s_i is a probability for all i; s_i is the probability of being a 'stayer' while p_i is the probability of moving to state i, given that a move is to be made, with probability $1 - s_i$.

(iv) A mixture model: here a pdf $f(x)$ is written as

$$f(x) = \alpha f_1(x) + (1 - \alpha)f_2(x).$$

This might for example be a suitable model to describe the distribution of heights in a population of men and women. Both $f_1(x)$ and $f_2(x)$ are pdfs, and α is the mixing probability.

(v) Censored survival data: this is the common situation in which a fraction of survival times are not known exactly, but are known to lie in an interval. The data of Table 1.1 are partially censored.

These examples are evidently quite varied, and apparently quite different. The power of the so-called EM (expectation-maximisation) algorithm is that it may be applied to all of these examples, and in each case simplifies to some extent the process of model-fitting using maximum-likelihood. The basic idea of the EM algorithm is elegant, and easy to understand, as we shall now see.

5.6.2 Two examples

Example 5.7: An example from genetics (Rao, 1973, pp.368–369)

This particular example is used by Dempster et al. (1977) to introduce the EM algorithm, and is often cited by subsequent authors.

A model from genetics results in a multinomial distribution with cell probabilities:

$$\left[\frac{1}{2} + \frac{\pi}{4} \quad \frac{1}{4}(1 - \pi) \quad \frac{1}{4}(1 - \pi) \quad \frac{\pi}{4}\right].$$

Suppose we observe the multinomial data $\mathbf{r} = [r_1 \ldots r_4]$. The likelihood is given by

$$L(\pi) \propto \left(\frac{1}{2} + \frac{\pi}{4}\right)^{r_1} \left\{\frac{1}{4}(1 - \pi)\right\}^{r_2+r_3} \left(\frac{\pi}{4}\right)^{r_4},$$

so that the log-likelihood is

$$\ell(\pi) = \text{constant} + r_1 \log \left(\frac{1}{2} + \frac{\pi}{4} \right) + (r_2 + r_3) \log(1 - \pi) + r_4 \log \pi.$$

In search of the maximum-likelihood estimator $\widehat{\pi}$ we set $\dfrac{d\ell}{d\pi} = 0$. This gives the equation,

$$\frac{r_1}{2 + \pi} - \frac{(r_2 + r_3)}{1 - \pi} + \frac{r_4}{\pi} = 0. \tag{5.11}$$

When we try to solve this equation for π, we obtain a quadratic equation, the positive root of which gives $\widehat{\pi}$.

The approach of the EM algorithm is to observe that we may regard the first of the multinomial cells as resulting from pooling two cells, one with probability $\frac{1}{2}$, with corresponding data value s, say, which is not known, and the other with probability $\frac{\pi}{4}$, corresponding to data value $r_1 - s$.

Of course we do not observe s, but if we did then the likelihood would be of the binomial form:

$$L(\pi) \propto \pi^{r_1 - s + r_4} (1 - \pi)^{r_2 + r_3},$$

resulting in the explicit maximum likelihood estimate

$$\widehat{\pi} = (r_1 - s + r_4)/(r_1 + r_2 + r_3 + r_4 - s) . \tag{5.12}$$

By inventing the value s we have simplified the likelihood, removing the $+$ sign, which complicates things; note that in this example it is the $+$ sign that we want to remove, and the $-$ sign does not present a problem. The really neat idea which is the basis of the EM algorithm is to observe that an initial guess at s will provide a value $\widehat{\pi}$ from Equation (5.12), which can then be used to give a better estimate of s, as follows.

If we just focus at how r_1 is divided into s and $r_1 - s$, we can see that

$$s \sim \text{Bin} \left(r_1, \left(\frac{\frac{1}{2}}{\frac{1}{2} + \frac{\pi}{4}} \right) \right),$$

and so we can estimate s from π by using the mean of this binomial distribution. Note that $\left(\frac{1}{2} + \frac{\pi}{4} \right)$ is the probability of obtaining a value in the first cell of the multinomial distribution. By focusing on the first cell only, we have to condition on this event, which is why the term $\left(\frac{1}{2} + \frac{\pi}{4} \right)$ appears in the denominator of the binomial distribution. We can now see that this process can be iterated, as we move from s to π to s, etc., successively updating our estimates in sequence.

A real data set has $\mathbf{r} = (125 \ \ 18 \ \ 20 \ \ 34)$. If our first estimate of s is $s^{(0)}$, then from Equation (5.12), our next estimate of π is

$$\pi^{(1)} = (159 - s^{(0)})/(197 - s^{(0)}),$$

and then from using the mean of the above binomial distribution,

$$s^{(1)} = 125 \left(\frac{\frac{1}{2}}{\frac{1}{2} + \frac{1}{4}\pi^{(1)}} \right) .$$

We then form the sequence $\ldots \pi^{(m)}, s^{(m)}, \pi^{(m+1)} \ldots$, where

$$s^{(m)} = \left(\frac{250}{2 + \pi^{(m)}} \right),$$

leading to:
$$\pi^{(m+1)} = \frac{159 - \left(\frac{250}{2+\pi^{(m)}} \right)}{197 - \left(\frac{250}{2+\pi^{(m)}} \right)},$$

that is

$$\pi^{(m+1)} = \frac{68 + 159\pi^{(m)}}{144 + 197\pi^{(m)}}, \quad \text{etc.} \tag{5.13}$$

We can see now that Equation (5.13) provides an iterative solution to the quadratic equation which results from setting $\pi^{(m)} = \pi^{(m+1)} = \pi$, and which is identical to the quadratic equation resulting from Equation (5.11). For the above real data we obtain the estimate, $\widehat{\pi} = 0.627$. $\qquad \square$

Example 5.8: The zero-truncated Poisson distribution

We encountered the zero-truncated Poisson distribution when we discussed the zero-inflated Poisson distribution in Example 5.1. For the zero-truncated case, the probability function is:

$$pr(X = i) = \frac{e^{-\lambda}\lambda^i}{i!(1 - e^{-\lambda})}, \qquad \text{for } i \geq 1,$$

that is

$$pr(X = i) = \frac{\lambda^i}{i!(e^\lambda - 1)}, \qquad \text{for } i \geq 1.$$

Suppose we observe i, n_i times, for $i \geq 1$. Then we can write the log-likelihood as:

$$\ell(\lambda) = \text{constant} + (\log \lambda) \sum_{i=1}^{\infty} i\, n_i - \{\log(e^\lambda - 1)\} \sum_{i=1}^{\infty} n_i.$$

From setting $\frac{d\ell}{d\lambda} = 0$, we can see that the maximum likelihood estimate is given by the solution to the equation:

$$1 - e^{-\widehat{\lambda}} = \left(\frac{\sum_{i=1}^{\infty} n_i}{\sum_{i=1}^{\infty} i n_i} \right) \widehat{\lambda}. \tag{5.14}$$

We can obtain $\widehat{\lambda}$ by estimating the point of intersection of graphs of $y = 1 - e^{-\lambda}$ and $y = \{(\sum_{i=1}^{\infty} n_i) / (\sum_{i=1}^{\infty} i\, n_i)\} \lambda$, each plotted against λ.

We can also of course maximise ℓ explicitly by an iterative search, for example, by using the MATLAB command *fminbnd* — see Figure 5.1(a).

An alternative approach is provided by the EM algorithm, as illustrated by the MATLAB program of Figure 5.1(b). In Example 5.1 there was missing information on how a pooled multinomial cell was constituted. In the case of this example we are effectively missing the zero cell of the Poisson distribution. If we assume a value for this cell, say s, then we can readily

(a)
```
function y=zeropo(lambda)
%
%ZEROPO calculates the negative log-likelihood corresponding
% to a zero-truncated Poisson distribution, of parameter 'lambda'.
% 'data' are a global row vector, and declared in the driving program.
%-------------------------------------------------------------------
global data
n=length(data);            % n is the largest value in the sample
p(1)=exp(-lambda);         % this is pr(0)
for i=2:n+1
  p(i)=p(i-1)*lambda/(i-1); % iterative formation of the
end                        % Poisson probabilities;
p=p/(1-p(1));              % these are the truncated Poisson
y=-[0 data]*(log(p))';     % probabilities.
```

(b)
```
% Program to take a data set
% from a zero-truncated Poisson distribution,
% and then   estimate the Poisson parameter using the
% EM algorithm. For illustration, we show the result of
% 'nit' iterations.
%-------------------------------------------------------------
nit=10;
data(2:14)=[7 9 13 16 19 24 27 33 18 14 9 4 2];
data(1)=1;                 % this is the first E-step
z=0:1:13;
for i=1:nit
  w=sum(data);
  n=z*data';
  mean=n/w;                % this is the maximisation stage
  e1=exp(-mean);
  data(1)=e1*w;            % this is the next E-stage
  mean
end
```

Figure 5.1 *(a) A* MATLAB *function to specify the log-likelihood for a set of data from a zero-truncated Poisson distribution. (b) A* MATLAB *program to produce the maximum-likelihood estimate of the parameter in a zero-truncated Poisson distribution, using the EM algorithm.*

form the maximum-likelihood estimate of λ using the well-known result that $\widehat{\lambda} = \left(\sum_{i=1}^{\infty} i\, n_i \right) / \left(s + \sum_{i=1}^{\infty} n_i \right)$.

Using this value we can now form a new estimate for the zero cell, which is $\left(s + \sum_{i=1}^{\infty} n_i \right) e^{-\widehat{\lambda}}$, leading to a new estimate of $\widehat{\lambda}$, etc. We obtain the maximum-likelihood estimate, $\widehat{\lambda} = 6.6323$, after just 3 iterations of the algorithm, by which time it has converged.   □

5.6.3 General theory

We know from our consideration of the Newton-Raphson iterative method that iterative procedures do not necessarily converge. What is attractive about the EM algorithm is that in quite general cases the procedure results in a sequence of values for the likelihood which increase, and also converge to a stationary value for the likelihood.

Suppose that we write the likelihood as

$$L(\boldsymbol{\theta}; \mathbf{x}) = f(\mathbf{x}|\boldsymbol{\theta}),$$

where $\boldsymbol{\theta}$ is the parameter vector and $\mathbf{x}$ denotes *incomplete* data. If we complete $\mathbf{x}$, to produce $\mathbf{y}$, say, in which the incomplete data $\mathbf{x}$ have been augmented in some way, then we write the likelihood based on $\mathbf{y}$ as $g(\mathbf{y}|\boldsymbol{\theta})$. Suppose we start the iteration from $\boldsymbol{\theta}^{(0)}$, then we shall generate a sequence of iterates $\{\boldsymbol{\theta}^{(m)}\}$, in which each iteration is a *double* step. The first step involves forming:

$$\mathbb{E}[\log g(\mathbf{y}|\boldsymbol{\theta})|\mathbf{x}, \boldsymbol{\theta}^{(m)}] = Q(\boldsymbol{\theta}, \boldsymbol{\theta}^{(m)}) , \quad \text{say.}$$

This is the E-step, where E denotes expectation. At the M-step we find the value of $\boldsymbol{\theta}$, $\boldsymbol{\theta}^{(m+1)}$, which maximises $Q(\boldsymbol{\theta}, \boldsymbol{\theta}^{(m)})$; thus M stands for maximisation.

It can be shown (e.g., McLachlan and Krishnan, 1997, p.83) that the EM algorithm outlined here is *monotonic*, in that $L(\boldsymbol{\theta}^{(m+1)}) \geq L(\boldsymbol{\theta}^{(m)})$. In most cases the EM algorithm converges to a stationary value of the likelihood surface, which is usually a local maximum. As with any deterministic optimisation method, if the likelihood surface has a number of local optima, then the EM algorithm may converge to one of them. There is no guarantee that the EM algorithm will produce a global optimum, and it should be repeated for different values of $\boldsymbol{\theta}^{(0)}$, as discussed in Chapter 3 for general deterministic search methods. The formal conditions which need to be satisfied so that the EM algorithm converges to a stationary value for L are given by Wu (1983).

Example 5.7 continued

In this example we have a scalar parameter $\theta = \pi$, the incomplete data, which is effectively trinomial, $\mathbf{x} = [125\ 38\ 34]$ and the completed vector, $\mathbf{y} = [s\ 125\ 38\ 34]$. We already know that

$$f(\mathbf{x}|\pi) = \left(\frac{1}{2} + \frac{\pi}{4}\right)^{125} \left\{\frac{1}{4}(1 - \pi)\right\}^{38} \left(\frac{\pi}{4}\right)^{34},$$

and for $\mathbf{y}$ we can write

$$g(\mathbf{y}|\pi) = \left(\frac{1}{2}\right)^{s} \left(\frac{\pi}{4}\right)^{125-s} \left\{\frac{1}{4}(1 - \pi)\right\}^{38} \left(\frac{\pi}{4}\right)^{34} , \text{ so that}$$

$$\log(g) = \text{ constant } + (125 - s) \log \pi + 38 \log(1 - \pi) + 34 \log \pi.$$

Hence in this example, in order to form $\mathbb{E}[\log g(\mathbf{y}|\pi)|\mathbf{x}, \pi^{(m)}]$, all we require is

$\mathbb{E}[s|\mathbf{x}, \pi^{(m)}]$, and we obtain

$$Q(\pi, \pi^{(m)}) = 125 \left(\frac{\frac{\pi^{(m)}}{4}}{\frac{1}{2} + \frac{\pi^{(m)}}{4}} \right) \log \pi + 38 \log(1 - \pi) + 34 \log \pi . \qquad (5.15)$$

We can see that we now obtain $\pi^{(m+1)}$ as a straightforward binomial optimisation of the function of Equation (5.15). Note that as we condition on $\mathbf{x}$, the appropriate expectation is the conditional binomial used in Equation (5.15). It would have been wrong to have used $\pi^{(m)}/4$ instead of $\left(\frac{\pi^{(m)}}{4} \right) / \left(\frac{1}{2} + \frac{\pi^{(m)}}{4} \right)$.
A pitfall to avoid when using the EM algorithm in practice is carrying out the expectation step incorrectly. It is very easy to get this wrong, resulting in an 'iteration' which does not change the parameter values.   □

In the last example, it was not necessary to do the EM algorithm, as we have an explicit solution for $\hat{\pi}$. So why did we use the EM algorithm? Of course this example has been used for illustration, but the question can be put more generally: why should we want to use the EM algorithm when a range of optimisation methods are available, as we have seen in Chapter 3? The answer lies in the attractive simplicity of the EM algorithm. In the last example we were able to use a sequence of simple explicit maximisations based on 'completed' data in order to maximise a more complex likelihood which arose due to missing data. Although the EM algorithm was iterative, we did not need to use one of the numerical methods of Chapter 3.

The maximisation at the M-step may well not be as simple as in the last example. In such a case one may just seek $\boldsymbol{\theta}^{(m+1)}$ to increase Q, that is to obtain a $\boldsymbol{\theta}^{(m+1)}$ so that

$$Q(\boldsymbol{\theta}^{(m+1)}, \boldsymbol{\theta}^{(m)}) \geq Q(\boldsymbol{\theta}^{(m)}, \boldsymbol{\theta}^{(m)}). \qquad (5.16)$$

The resulting algorithm is called the GEM algorithm (where G stands for 'generalised'), and it enjoys the same monotonicity and convergence properties as the EM algorithm. One way of obtaining a GEM algorithm is to modify the result of taking just one step of a Newton-Raphson iteration, as described by Rai and Matthews (1993). A suitable GEM algorithm can be appreciably faster than the corresponding EM algorithm as less time is spent at each M-step. First-time users of the EM algorithm are usually surprised by how slow it can be to converge. Speed of convergence is a function of how much data are missing — the closer the data are to being complete, the faster we expect the algorithm to work. Much research has been devoted to ways of speeding up the EM algorithm, but this is at the expense of the loss of simplicity of the algorithm — see McLachlan and Krishnan (1997, Chapter 4). We note also that in common with many methods of likelihood optimisation, once a maximum-likelihood estimate of $\boldsymbol{\theta}$ has been obtained, it is still necessary to obtain error estimates and estimators of the correlation between parameter estimators. Louis (1982) showed how this can be done using the observed Fisher information matrix. Furthermore, Oakes (1999) demonstrates that it is very simple to obtain the first- and second-order derivatives of the log-likelihood, $\log L(\boldsymbol{\theta}; \mathbf{x})$, from appropriate derivatives of Q. Thus the standard Newton-Raphson method for maximising $L(\boldsymbol{\theta}; \mathbf{x})$ can be performed in terms of Q.

The expectation step of the EM algorithm typically involves forming an integral, which may also be difficult. While this may be tackled using numerical integration, or Gibbs sampling, which is discussed in Chapter 7, one possibility is to use Monte Carlo EM (which is denoted by MCEM), as suggested by Wei and Tanner (1990). The idea of MCEM is, very simply, to use simulation to estimate the expectation. We can appreciate what is involved through the following example.

Example 5.7 concluded

In order to form $\mathbb{E}[\log(g)]$ we may take a random sample of size n from the binomial distribution, $Bin\left(125, \left(\frac{\pi^{(m)}}{4}\right) \Big/ \left(\frac{1}{2} + \frac{\pi^{(m)}}{4}\right)\right)$, resulting in values $(s_1, \ldots, s_n)$. Then instead of using $125\left\{\frac{\pi^{(m)}}{4} \Big/ \left(\frac{1}{2} + \frac{\pi^{(m)}}{4}\right)\right\}$ to estimate $s^{(m)}$, we can use the sample mean: $\bar{s} = \frac{\sum_{i=1}^{n} s_i}{n}$.   □

The MCEM approach to evaluating $\mathbb{E}[\log(g)]$ uses what we call "crude" Monte Carlo, which will be described in Chapter 6, where we shall also mention ways of improving efficiency. In MCEM the fundamental monotonicity result for the EM algorithm no longer holds. Wei and Tanner (1990) suggest proceeding in an adaptive way, starting with small n and then increasing n as the iterations progress.

Celeux and Diebolt (1985) introduced the *stochastic* EM algorithm, denoted by SEM. In the SEM algorithm, the E-step of the EM algorithm is replaced by an S-step. In contrast to MCEM, the S-step involves a *single* simulation from their conditional distribution, in order to impute the missing observations. This produces a sequence of parameter estimates, which are a realisation of a Markov chain. The chain is ergodic (Ip, 1994, see Appendix A) and converges to its unique stationary distribution (Diebolt and Ip, 1996), which is centred on the maximum-likelihood estimate of the parameters. SEM possesses certain advantages compared with standard EM; it may for example be faster. A key issue is being able to determine when the Markov chain involved has converged. This is an issue to which we return in Section 7.3.4, as it is central to the use of Markov chain Monte Carlo methods used for modern Bayesian analysis. A number of illustrations, for selection models for longitudinal data with dropout, are provided by Gad Attay (1999). We shall discuss aspects of simulation in Chapter 6.

The history of the EM algorithm is given by McLachlan and Krishnan (1997, Section 1.8), where it is traced back to Newcomb (1886). Its modern expression is due to Dempster et al. (1977), who integrated and developed the previous work in the area. For more recent research, see Meng and van Dyk (1997).

The EM algorithm possess an attractive simplicity. Its importance derives from thte fact that it can be applied to many different problems. In the next section we give two further applications of its use, and additional examples are to be found in the exercises. The Bayesian formulation of these ideas is through the Data Augmentation algorithm of Chapter 7.

5.6.4 Two further examples

Example 5.9: Censored data

We know that censored data are commonly encountered, for example in measurements of survival times. Consider a random sample $(x_1, \ldots, x_m)$ from a pdf $f(x)$, and what we call *right* censored data, $(x_{m+1}, \ldots, x_{m+n})$. We have already encountered right-censored data in Section 2.2, where we modelled the fecundability data of Table 1.1, and in Exercise 3.15. The likelihood has the form:

$$L(\boldsymbol{\theta}; \mathbf{x}) = \prod_{i=1}^{m} f(x_i) \prod_{j=m+1}^{m+n} S(x_j) \tag{5.17}$$

where, as usual, $\boldsymbol{\theta}$ denotes the parameter vector, to be estimated, and where $S(x)$ is the survivor function,

$$S(x) = \int_{x}^{\infty} f(y)dy \ .$$

For each of the censored times x_j we can assign an exact but unobserved response time of y_j. In the notation of Section 5.6.3, this leads to

$$g(\mathbf{y}|\boldsymbol{\theta}) = \prod_{i=1}^{m} f(x_i) \prod_{j=m+1}^{m+n} f(y_j) \ ,$$

and

$$\log g(\mathbf{y}|\boldsymbol{\theta}) = \sum_{i=1}^{m} \log f(x_i) + \sum_{j=m+1}^{m+n} \log f(y_j) \ .$$

In order to take $\mathbb{E}[\log g(\mathbf{y}|\boldsymbol{\theta})|\mathbf{x}, \boldsymbol{\theta}^{(m)}]$, we need an expression for $\mathbb{E}[\log f(y_j)|x_j \leq y_j]$. As a simple illustration, let $f(y) = \lambda e^{-\lambda y}$ for $y \geq 0$, i.e., the exponential distribution $Ex(\lambda)$. In this case,

$$\log f(y) = \log \lambda - \lambda y \ ,$$

and so we need

$$\mathbb{E}[y|y \geq x] = \frac{\int_{x}^{\infty} y\lambda e^{-\lambda y}dy}{S(x)} \ ,$$

and as $S(x) = e^{-\lambda x}$, a little algebra following integration by parts shows that

$$\mathbb{E}[y|y \geq x] = x + \frac{1}{\lambda} \tag{5.18}$$

which ties in with what we would expect due to the lack-of-memory aspect of the exponential distribution and the Poisson process — whatever x is, the expected additional value, which is $\mathbb{E}[y - x|y \geq x]$, is always $1/\lambda$.

(a)
```
function y=censor(lambda)
%
%CENSOR calculates the negative log-likelihood for
%  data, including right-censoring, from an exponential distribution.
%  'lambda' is the exponential distribution parameter;
%  the uncensored and censored data are respectively in the
%  global vectors, 'uncen' and 'cen', established in the calling program.
%-----------------------------------------------------------------
global uncen cen
loglik=length(uncen)*log(lambda)-lambda*sum(uncen)-lambda*sum(cen);
y=-loglik;
```

(b)
```
% Program to produce the mle of the parameter, 'lambda'
% in the exponential distribution,
% using the EM algorithm;
% 'nit' iterations are taken
%-----------------------------------------------------------------
nit=10;
lambda=1;                      % this is the first E-step
uncens=[1 3 3 6 7 7 10 12 14 15 18 19 22 26 29 34 40];
                               % the uncensored data
cen=[28 48 49];                % this is the right-censored data
join=[uncens,cen];len=length(join);
for i=1:nit
   true=cen-(1/lambda)         % the expectation step
   join=[true,uncens];
   lambda=len/sum(join);       % the maximisation step
   disp(lambda)
end
```

Figure 5.2 *(a)* MATLAB *program to evaluate the negative log-likelihood for a set of data including right censoring, with an exponential model; (b)* MATLAB *program to obtain the maximum likelihood estimate of the exponential.*

Of course in the example of right-censored data from the exponential distribution we have an explicit maximum-likelihood estimate, which has already been derived as the solution to Exercise 3.15; see also Exercise 5.42. However, for illustration we give in Figure 5.2 MATLAB programs in the exponential distribution case, to evaluate (a) the log-likelihood from Equation (5.17), which is then readily maximised using *fminbnd*, as there is only one parameter, and (b) the maximum-likelihood estimate of λ using the EM algorithm based on Equation (5.18). The maximisation stage simply involves forming the mean of the augmented sample of data — see Exercise 5.15.

Example 5.10: Mixtures

It is often the case that data are sampled from a heterogeneous population which can be regarded as a mixture of a number of homogeneous populations. The pdf (or probability function) of a measurement x is then written as

$$f(x) = \sum_{j=1}^{k} \alpha_j f_j(x) \qquad (5.19)$$

where the $\{\alpha_j, \quad 1 \leq j \leq k\}$ are probabilities summing to unity, the jth sub-population is sampled with probability α_j and the appropriate pdf (or probability function) is then $f_j(x)$. Several books have been devoted to this important topic — see, for example, Everitt and Hand (1981), McLachlan and Basford (1988), and Titterington et al. (1985). Quite often the histogram for a random sample from $f(x)$ will reveal clear heterogenity and suggest a mixture model. However any pdf can be written in mixture form, a result which is exploited in the computer simulation of random variables — see Morgan (1984, p.107).

If we have a random sample $(x_1, \ldots, x_n)$, of size n from the pdf of Equation (5.19), then the likelihood is given by:

$$L(\boldsymbol{\theta}; \, \mathbf{x}) = \prod_{i=1}^{n} \left\{ \sum_{j=1}^{k} \alpha_j f_j(x_i | \boldsymbol{\psi}_j) \right\}. \qquad (5.20)$$

In Equation (5.20) we suppose that the jth component pdf (or probability function) has parameters $\boldsymbol{\psi}_j$. Thus the overall parameter vector $\boldsymbol{\theta}$ contains the parameters $\{\alpha_j, \ \boldsymbol{\psi}_j, \ 1 \leq j \leq k\}$. Maximising $L(\boldsymbol{\theta}; \mathbf{x})$ is difficult, due to the presence of the summations, which reflect the fact that what we are typically lacking is knowledge of which component of the mixture any particular sample value comes from. That is the missing information in this example. The EM algorithm for this situation operates as follows:

$$\text{Define} \quad \mathbf{y}_i = (x_i, \, \mathbf{z}_i'), \quad 1 \leq i \leq n,$$

where each $\mathbf{z}_i$ is an *indicator* column vector of length k; it has a 1 in the position corresponding to the component of the mixture that x_i comes from, and 0's elsewhere.

Thus,

$$g(\mathbf{y}|\boldsymbol{\theta}) = \prod_{i=1}^{n} \prod_{j=1}^{k} \alpha_j^{z_{ij}} f_j(x_i | \boldsymbol{\psi}_j)^{z_{ij}}. \qquad (5.21)$$

We see this as follows: suppose x_i is drawn from the rth component of the mixture, then $(\alpha_r f_r)$ enters the likelihood, and the terms $(\alpha_s f_s)$ for $s \neq r$ are raised to the power zero and so do not enter the likelihood. Hence we can now write

$$\log g(\mathbf{y}|\boldsymbol{\theta}) = \sum_{i=1}^{n} \mathbf{z}_i'\{\mathbf{v}(\boldsymbol{\alpha}) + \mathbf{u}_i(\boldsymbol{\psi})\},$$

where $\mathbf{v}(\boldsymbol{\alpha})$ is a column vector with jth component, $\log \alpha_j$, $1 \le i \le k$ and $\mathbf{u}_i(\boldsymbol{\psi})$ is a column vector with jth component $\log f_j(x_i|\boldsymbol{\psi}_j)$.

Thus at the E-step in the EM algorithm,

$$Q(\boldsymbol{\theta}, \boldsymbol{\theta}^{(m)}) = \sum_{i=1}^{n} \mathbf{w}_i(\boldsymbol{\theta}^{(m)})'\mathbf{v}(\boldsymbol{\alpha}) + \sum_{i=1}^{n} \mathbf{w}_i(\boldsymbol{\theta}^{(m)})'\mathbf{u}_i(\boldsymbol{\psi}),$$

where

$$\mathbf{w}_i(\boldsymbol{\theta}^{(m)}) = \mathbb{E}[\mathbf{z}_i|x_i, \; \boldsymbol{\theta}^{(m)}].$$

Thus, the jth element of $\mathbf{w}_i(\boldsymbol{\theta}^{(m)})$ is given by

$$w_{ij}\boldsymbol{\theta}^{(m)}) = \frac{\alpha_j^{(m)} f_j(x_i|\boldsymbol{\psi}_j^{(m)})}{\sum_{j=1}^{k} \alpha_j^{(m)} f_j(x_i|\boldsymbol{\psi}_j^{(m)})}.$$

In order to carry out the M-step of the EM algorithm we now simply have to maximise the expression $Q(\boldsymbol{\theta}, \boldsymbol{\theta}^m)$ with respect to $\boldsymbol{\theta}$. The case of two univariate normal pdfs is considered in Exercise 5.14, while Figure 5.3 provides MATLAB code for fitting a mixture of two Poisson distributions — see also Exercise 5.12.   □

5.7 Alternative methods of model-fitting

The emphasis so far in this book has been on the method of maximum-likelihood for fitting models to data. The reasons for this are the good properties enjoyed by the method and the fact that fast modern computers combined with efficient optimisation algorithms frequently allow us to apply the method with ease. However, there may be more obvious ways of matching models to data, and there are also methods which are more robust than maximum-likelihood. The list of alternative methods of model-fitting is a very long one. We shall in this section just mention a sample of the methods available. Others will be encountered later in the book, especially in Chapter 7, which presents the modern Bayesian procedures.

In the past, when maximising likelihoods was harder than it now is, simple methods of model fitting were attractive. A good example is provided by the negative binomial distribution, and Exercises 5.23 and 5.28 examine two simple ways of fitting that distribution to data. Although maximising likelihoods is now often routine, iterative procedures perform best if they are given good starting values, and these can be provided by simpler alternative methods. Thus it still remains useful to consider simpler alternatives to the method of maximum-likelihood. As we shall mention in Section 5.7.4, there are also cases when forming and maximising likelihoods are difficult, and then alternative

(a)

```
% Program to fit a two-component Poisson mixture
% using the EM algorithm. We display just 'nit' iterations.
% Calls the function DIVIDE.
% global variables are 'data' and 'k'.
%_____
global data k
nit=5000;
data=[162 267 271 185 111 61 27 8 3 1];
s=sum(data);
k=length(data);
ind=0:k-1;
pie=0.359;                          % first guess at pi, the probability
theta1=1.2;theta2=3.7;              % of the first population, and
data1=divide(theta1,theta2,pie)     % theta1 and theta2
data2=data-data1; disp('     it        pi      theta1      theta2')
for j=1:nit                         % the maximisation stage
  s1=sum(data1); s2=(s-s1);
  pie=s1/s;
  theta1=(ind*data1')/s1;theta2=(ind*data2')/s2;
  x=[j pie theta1 theta2];
  data1=divide(theta1,theta2,pie);
  data2=data-data1;
  disp(x)                           % the expectation stage
end
```

(b)

```
function y=divide(theta1,theta2,pie)
%
%DIVIDE conducts the data division for the EM algorithm approach
%  to fitting a mixture of two Poisson distributions.
%  Called by the program to fit the mixture by the EM algorithm.
%  global variables are 'data', and 'k'.
%
%_____
global data k
p1(1)=exp(-theta1); p2(1)=exp(-theta2);
for i=1:k-1
  p1(i+1)=p1(i)*theta1;             % note that we do not need the
  p2(i+1)=p2(i)*theta2;             % 'i' divisor here
  data1(i)=data(i)/(1+(1/pie-1)*p2(i)/p1(i));
end
data1(k)=data(k)/(1+(1/pie-1)*p2(k)/p1(k));
y=data1;
```

Figure 5.3 MATLAB *programs for fitting a mixture of two Poisson distributions to data describing the death notices for women aged* 80 *or over, from* The Times *for each day during the period, 1910–1912. From Hasselblad (1969).*

methods are needed. Monte-Carlo inference is one possible approach, based on simulation, and this will be described later, in Section 6.5.

5.7.1 Least-squares

Suppose a vector random variable $\mathbf{Y}$ of dimension d has expectation $\boldsymbol{\mu}(\boldsymbol{\theta})$, where $\boldsymbol{\theta}$ denotes the vector of parameters to be estimated. In its simplest form, the method of least-squares allows us to estimate $\boldsymbol{\theta}$ by minimising

$$S(\boldsymbol{\theta}) = (\mathbf{y} - \boldsymbol{\mu}(\boldsymbol{\theta}))'(\mathbf{y} - \boldsymbol{\mu}(\boldsymbol{\theta})) = \sum_{i=1}^{d} (y_i - \mu_i(\boldsymbol{\theta}))^2,$$

where $\mathbf{y}$ is a realisation of $\mathbf{Y}$, and we use y_i and $\mu_i(\boldsymbol{\theta})$ to denote the ith components of $\mathbf{y}$ and $\boldsymbol{\mu}(\boldsymbol{\theta})$ respectively.

This approach provides an estimate $\widehat{\boldsymbol{\theta}}$ without the need to specify anything other than the expectation of $\mathbf{Y}$. More generally, if $\sum$ denotes the variance-covariance matrix of $\mathbf{Y}$, then *weighted* least squares estimates result from minimising

$$S(\boldsymbol{\theta}) = (\mathbf{y} - \boldsymbol{\mu}(\boldsymbol{\theta}))' \sum\nolimits^{-1} (\mathbf{y} - \boldsymbol{\mu}(\boldsymbol{\theta})).$$

The simplest example of this occurs when the elements of $\mathbf{Y}$ are independent random variables with different variances, so that $\sum$ is then a diagonal matrix. If $\mathbf{Y}$ has a multivariate normal distribution then the maximum-likelihood estimate of $\boldsymbol{\theta}$ is the same as the least-squares estimate, and we are then able to make statements regarding the sampling distribution of the parameter estimator $\widehat{\boldsymbol{\theta}}$. See Exercise 5.22.

If the elements of $\boldsymbol{\mu}(\boldsymbol{\theta})$ are linear combinations of the elements of $\boldsymbol{\theta}$, say $\boldsymbol{\mu} = \mathbf{X}\boldsymbol{\theta}$ for some suitable matrix $\mathbf{X}$, and $\sum$ does not involve $\boldsymbol{\theta}$, then

$$S(\boldsymbol{\theta}) = (\mathbf{y} - \mathbf{X}\boldsymbol{\theta})' \sum\nolimits^{-1} (\mathbf{y} - \mathbf{X}\boldsymbol{\theta}),$$

and the weighted least-squares estimator of $\boldsymbol{\theta}$ is then obtained as the solution to the *normal* equations,

$$\mathbf{X}' \sum\nolimits^{-1} \mathbf{X}\widehat{\boldsymbol{\theta}} = \mathbf{X}' \sum\nolimits^{-1} \mathbf{y}. \tag{5.22}$$

Special numerical minimisation techniques have been devised for non-linear least-squares problems, such as the Gauss-Newton method (Nash and Walker-Smith, 1987, Chapter 11). In MATLAB one may use the function, *lsqnonlin*, to obtain the solution to a non-linear least-squares problem. For certain types of *aggregate* data, collected over time, likelihood formation can be virtually impossible. For aggregate data, the behaviour of individuals is unknown, and all that is recorded is the total numbers of individuals in various states. Thompson and Tapia (1990, p.217) estimated that in an example involving aggregate data, it took approximately 18 man-months to obtain the likelihood. Kalbfleisch et al. (1983) show how weighted least squares may be used to fit models to such data. An example is given in Morgan (1992, p.217), in which flies sprayed with insecticide were observed and the numbers flying were counted on several occasions.

5.7.2 Minimum chi-square

Goodness-of-fit statistics such as the Pearson X^2 provide a single measure of discrepancy between data values and expectations under a model, which is similar to the discrepancy minimised in the method of least-squares. Thus a further way of estimating parameters is to choose them to minimise a goodness-of-fit statistic. A modification for fitting logit models to quantal response data, as in Exercises 3.10 and 4.14, results in *minimum-logit-chi-square* estimates, which, in contrast to maximum-likelihood estimates, are explicit, and do not require iterative methods of model-fitting — see Exercise 5.21 and Morgan (1992, p.57).

5.7.3 The method of moments

Moments are measures such as mean, variance, skewness and kurtosis. Model-fitting by the method of moments is achieved by equating sample and population moments and solving the resulting equations for the parameters of the model. If the model has k parameters then k suitably chosen moments are used. In some cases, such as estimating the mean and variance of a normal distribution, when $k = 2$, the method-of-moments estimates are identical to the maximum-likelihood estimates. Large k requires the use of higher-order moments, which can translate into parameter estimators of poor precision. For an application to the negative binomial distribution, see Exercise 5.23.

5.7.4 *Empirical transforms

An extension of the method-of-moments is to match *all* population and sample moments, by equating empirical and theoretical moment generating functions. This idea can be traced to Parzen (1962), and has spawned a literature that is reviewed in the introduction to Yao and Morgan (1999). We can see that this approach can be attractively simple through the following example.

Example 5.11: Fitting the Cauchy two-parameter distribution

As in Example 3.1, we have a random sample, $(x_1, \ldots, x_n)$, from the two-parameter Cauchy pdf:

$$f(x) = \frac{\beta}{\pi\{\beta^2 + (x - \alpha)^2\}} \,,$$

and we wish to estimate the parameters α and β. In this example we cannot use the moment generating function as it does not exist; however, we may instead equate the real and imaginary parts of the empirical and theoretical characteristic functions. The *empirical* characteristic function is defined as:

$$\widehat{\phi}(s) = \frac{1}{n} \sum_{j=1}^{n} e^{isx_j},$$

in which $i^2 = -1$ and s is the variable of the characteristic function.

We obtain explicit estimators,

$$\widehat{\alpha}(s) = \frac{1}{s}\tan^{-1}\left\{\sum_{i=1}^{n}\cos(sx_j)\,/\,\sum_{j=1}^{n}\sin(sx_j)\right\}$$

$$\widehat{\beta}(s) = \frac{1}{|s|}\log\left\{n^{-1}\sum_{j=1}^{n}\cos(sx_j)/\cos(\widehat{\alpha}s)\right\}.$$

For further discussion, see Exercise 5.29.   □

In the above example we need to choose s, and there are various ways of doing this in general — see for example Laurence and Morgan (1987) and Yao and Morgan (1999).

The estimators of Example 5.10 are less efficient than maximum likelihood estimators. However transform-based estimators can be more robust than maximum-likelihood estimators (Besbeas and Morgan, 2004, 2008; Campbell, 1992, 1993). Additionally, they have potential for fitting complex stochastic models, as outlined by Ball (1995) and Yao and Morgan (1999). An application to fitting mixtures of distributions is considered in Exercise 5.25.

5.8 *Non-regular problems

In Chapter 4 we presented standard asymptotic theory, and assumed that we could interchange the orders of differentiation and integration, and suitably truncate Taylor Series expansions. Much of the time, the regularity conditions that underpin these assumptions hold. However it is also quite easy for these conditions to be violated (see, e.g., Cox and Hinkey, 1974, pp.112–113). There are various common reasons for this, and often the modifications of standard tests that need to be carried out when regularity conditions are violated are easily done. Cheng and Traylor (1995) define what is meant by a regular problem, and considers four examples resulting in non-regular situations. Two of these are the *indeterminate parameters* problem and the *constrained parameter* problem and we shall consider these here. In the indeterminate parameters problem, setting one parameter to zero results in the disappearance of all expressions containing another parameter (or other parameters). In the constrained parameter problem, the unknown true value of the parameter vector is on the boundary of the parameter space. A well-known example of the constrained parameter problem arises in the context of mixtures, one example of which is the model, which we have already encountered in Exercise 3.3, where the various terms are fully specified:

$$pr(X = x|n) = \gamma Q(x,n,\theta) + (1-\gamma)\binom{n}{x}\nu^x(1-\nu)^{n-x}, \qquad \text{for } 0 \leq x \leq n.$$

The mixing parameter γ is usually constrained to lie in the interval, $0 \leq \gamma \leq 1$. Thus testing the null hypothesis that $\gamma = 1$ using a likelihood-ratio test involves maximising the likelihood when the parameter vector is located on the boundary of the parameter space. Simulation is used in Example 6.9 to study the sampling distribution of the test statistic of the null hypothesis that $\gamma = 1$ using a likelihood-ratio test. An example of the indeterminate parameters problem is given below.

Example 5.12: Indeterminate parameters

Here we consider another discrete mixture problem which results in the probability function,

$$pr(X = j) = \alpha \binom{k}{j} p^j (1 - p)^{k-j} + (1 - \alpha) \binom{k}{j} 2^{-k}. \qquad (5.23)$$

Thus in this example, the random variable X is supposed to have a binomial $Bin(k, p)$ distribution with probability α, and the binomial $Bin(k, \frac{1}{2})$ distribution the rest of the time, with probability $1 - \alpha$. We can see that setting $\alpha = 0$ removes the parameter p from the specification of the probability function for X. The same effect results from setting $p = \frac{1}{2}$, which removes α from Equation (5.23). Chernoff and Lander (1995) showed that the re-parameterisation:

$$\theta_1 = \frac{\alpha}{2}(p - \frac{1}{2}),$$

$$\theta_2 = \alpha(p - \frac{1}{2})^2$$

simplifies testing the null hypothesis that either $p = \frac{1}{2}$ or $\alpha = 0$ to testing whether $\boldsymbol{\theta} = \mathbf{0}$, a point on the boundary of the parameter space.

This reparameterisation imposes constraints on $\boldsymbol{\theta}$. The simple case of $k = 2$ can be shown to result in an asymptotic distribution of the likelihood-ratio test statistic of the null hypothesis that $\boldsymbol{\theta} = \mathbf{0}$ which is a mixture of chi-square distributions, which we may denote by:

$$\frac{1}{2}\chi_1^2 + \lambda\chi_2^2 + \left(\frac{1}{2} - \lambda\right)\chi_0^2,$$

where $\lambda = \frac{1}{2\pi} \arctan(1/\sqrt{2}) = 0.098$, and χ_0^2 denotes a random variable which is zero with probability 1. $\qquad\square$

We can see from the last example that naive application of standard asymptotic theory to non-regular problems can be seriously misleading. For non-regular problems, early theoretical results were produced by Chernoff (1954) and Moran (1971). More recent discussion is to be found in Feng and McCulloch (1992) and Smith (1989). The paper by Self and Liang (1987) provides useful general principles for determining the appropriate asymptotic distribution for test statistics in a wide range of cases. For further discussion, see Exercises 5.26 and 5.27.

5.9 Discussion

The material of this chapter is wide-ranging, and it contains a large number of extremely important techniques for use in applied stochastic modelling. For instance, we have seen that it is essential to preface model fitting with a careful consideration of whether the model is appropriately parameterised; it is vital to standardise variables before attempting forms of regression analysis; it is important to check that regularity conditions are not violated.

The method of maximum-likelihood may always be used crudely, and without prior thought, but it may then fail. It may be beneficial to take care over the starting values for iterative methods; use of the EM algorithm may simplify computer programming, and ultimately the method of maximum-likelihood may prove to be impractical for particular, complex modelling situations. Following model fitting, the δ-method may provide desired estimates of standard error, and careful consideration should be given as to whether parts of the data may be exerting a particular influence on the conclusions of the analysis.

5.10 Exercises

5.1 Obtain the maximum-likelihood estimate of λ in Example 5.1.

5.2 In the MATLAB program of Exercise 4.1, experiment with an alternative procedure for ensuring that the parameters which are probabilities stay within range during the iterations leading to the maximum-likelihood estimate.

5.3 For the logit model and data of Exercise 3.10, reparameterise the model so that the probability of response to dose d_i is given by: $P(d_i) = 1/(1 + e^{\{a+b(d_i-\bar{d})\}})$, where $\bar{d}$ is the sample mean of the $\{d_i\}$. Use $fmax$ to estimate the correlation between $\hat{a}$ and $\hat{b}$ and discuss your findings. Draw contours of the log-likelihood surface.

5.4 Show that when the random variable X has a Weibull, Wei (ρ, κ) distribution, the random variable X^κ has an exponential distribution and give its mean value. What is the distribution of $(\rho X)^\kappa$?

5.5 Provide an example of nested optimisation.

5.6* If X is an exponential random variable with mean unity, $E[X \log X] = 1 - \gamma$, where γ is Euler's constant ($\gamma \approx 0.577215$). Use this result to derive an orthogonal parameterisation for the Weibull distribution.

5.7 Use the MATLAB function $fmax$, of Figure 4.1, to obtain the maximum-likelihood estimates of the parameters a and b in the logit model fitted to the data of Exercise 3.10. Use the δ-method and the estimated covariance matrix for the estimators $\hat{a}$ and $\hat{b}$ to estimate the variance of the maximum-likelihood estimator of the ED$_{50}$, given by: $\hat{\mu} = -\hat{a}/\hat{b}$.

Also fit the model using the (μ, b) parameterisation, and $fmax$, and compare the direct estimate of the variance of $\hat{\mu}$ with that obtained from using the δ-method.

5.8 Study the MATLAB program given below for fitting the animal survival model, $(\lambda, \phi_1, \phi_a)$, to a set of simulated data, using the EM algorithm.

```
% Program to carry out 'nit' iterations of an EM approach to
% likelihood maximisation, for a simple model for the survival of wild
% animals. The model is fitted to just a single cohort of
% data, of size 'n', resulting in  a 4-year study.
% There are two annual survival probabilities, 'phi_1' and 'phi_2'.
% The remaining probability is 'lam', which is the probability that
% a ring from a dead bird is found and reported. The data are generated
% externally, and in the global vector, 'data'.
% (see for example, Exercise 6.16).We estimate the numbers dead each
% year, by the vector 'd'. The number of birds still alive at the end
% of the study is estimated by the scalar,'a'
%-------------------------------------------------------------------
global data
sr=sum(data);
nit=50;
phi_1=0.422;phi_2=0.763;lam=0.192;   % initial estimates of the
                                     % parameters
n=1000;                              % this is the cohort size
m=n-sr;                              % the total unaccounted for at
                                     % the end of the study
d(1)=n*(1-phi_1)*(1-lam);            % estimate 'd'
d(2)=d(1)*phi_1*(1-phi_2)/(1-phi_1);
d(3)=d(2)*phi_2;d(4)=d(3)*phi_2;
a=n*(phi_1*phi_2^3);                 % estimate 'a'
for i=1:nit
                                     % now we form the mles.
  sd=sum(d);                         % This is the M-step.
  lam=sr/(sr+sd);
  phi_1=(sr-data(1)+sd-d(1)+a)/(a+sr+sd);
  phi_2=(3*a+d(3)+2*d(4)+data(3)+2*data(4))/(2*d(3)+3*d(4)+...
  3*a+data(2)+2*data(3)+3*data(4)+d(2));
                                     % Now we carry out the E-stage
  kap=1-lam*(1-phi_1*phi_2^3);
  d(1)=m*(1-phi_1)*(1-lam)/kap;
  d(2)=d(1)*phi_1*(1-phi_2)/(1-phi_1);
  d(3)=d(2)*phi_2;d(4)=d(3)*phi_2;
  a=m*phi_1*(phi_2^3)/kap;
  disp(data);disp(d);disp(a);
  disp(lam);disp(phi_1);disp(phi_2);
end
```

5.9 Check, using both direct optimisation and the EM algorithm, that $\widehat{\mu} = 0.627$ in the multinomial model with data: [125   18   20   34] and cell probabilities $\left[\frac{1}{2} + \frac{\pi}{4}, \quad \frac{1}{4}(1-\pi), \quad \frac{1}{4}(1-\pi), \quad \frac{\pi}{4}\right]$.

Obtain the quadratic equation satisfied by the maximum-likelihood estimate $\widehat{\pi}$. As an exercise in using the *fzero* command in MATLAB use *fzero* to check the value for $\widehat{\pi}$.

5.10 Abbott's formula for quantal dose-response data gives the probability of response at dose d as,

$$P(d) = \lambda + (1 - \lambda)F(d),$$

where λ is the probability of natural mortality, and $F(d)$ is a customary cdf, as used when there is no natural mortality. For this question, set $F(d) = 1/[1 + \exp\{a + b\log(d)\}]$.

The data below describe the mortality of aphids exposed to nicotine. Write down the likelihood based on Abbott's formula. Evaluate the probability of natural mortality, (a) directly, using *fminsearch*, and (b) using the EM algorithm.

% concentration	Number tested	Number dead
Control	45	3
0.0025	50	5
0.005	46	4
0.01	50	3
0.02	46	11
0.03	46	20
0.04	49	31
0.06	50	40
0.08	50	43
0.10	50	48
0.15	50	48
0.20	50	50

5.11 (Tanner, 1993, p.41.) Ten motorettes were tested at each of four temperatures: 150°, 170°, 190°, and 220°C. The failure times, in hours, are given below, with a '*' indicating a right-censored value.

150°	170°	190°	220°
8064*	1764	408	408
8064*	2772	408	408
8064*	3444	1344	504
8064*	3542	1344	504
8064*	3780	1440	504
8064*	4860	1680*	528*
8064*	5196	1680*	528*
8064*	5448*	1680*	528*
8064*	5448*	1680*	528*
8064*	5448*	1680*	528*

The data are to be described by the model:

$$t_i = \beta_0 + \beta_1 v_i + \sigma \epsilon_i \,,$$

where $\epsilon_i \sim N(0,1)$; $v_i = 1000/(\text{temperature} + 273.2)$ and $t_i = \log_{10}$ (ith failure time).

Show that the log-likelihood has the form:

$$l = \text{constant} - n \log \sigma - \sum_{i=1}^{m} (t_i - \beta_0 - \beta_1 v_i)^2/(2\sigma^2) - \sum_{i=m+1}^{n} (Z_i - \beta_0 - \beta_1 v_i)^2/(2\sigma^2),$$

where $\{Z_j\}$ are the unobserved failure times for the censored observations. Thus to perform the E step of the EM algorithm we need:

$$\mathbb{E}\left[Z_i \mid \beta_0,\ \beta_1,\ \sigma,\ Z_i > c_i\right] \text{ and}$$

$$\mathbb{E}\left[Z_i^2 \mid \beta_0,\ \beta_1,\ \sigma,\ Z_i > c_i\right].$$

Let $\mu_i = \beta_0 + \beta_1 v_i$.

Show that, in order, these expectation terms are:

$$\mu_i + \sigma\, h\left(\frac{c_i - \mu_i}{\sigma}\right),\quad \text{and}$$

$$\mu_i^2 + \sigma^2 + \sigma(c_i + \mu_i)\, h\left(\frac{c_i - \mu_i}{\sigma}\right),$$

where $h(x) = \phi(x)/\{1 - \Phi(x)\}$, and $\phi(x)$ and $\Phi(x)$ are respectively the pdf and cdf of the standard normal distribution. ($h(x)$ is called the *hazard* function.) Suppose that the variable indicating censoring of the ith observation is w_i ($w_i = 1$ for the non-censored case, and $w_i = 0$ for the censored case). At each E step, therefore, we have

$$y_{1i} = w_i t_i + (1 - w_i)\left\{\mu_i + \sigma\, h\left(\frac{c_i - \mu_i}{\sigma}\right)\right\}$$

$$y_{2i} = w_i t_i^2 + (1 - w_i)\left\{\mu_i^2 + \sigma^2 + \sigma(c_i + \mu_i)\, h\left(\frac{c_i - \mu_i}{\sigma}\right)\right\}.$$

Write down the equations to be solved at the M step, and show that the maximum-likelihood estimates are special cases of:

$$\hat{\beta} = \left(\sum_i \mathbf{v}_i \mathbf{v}_i'\right)^{-1}\left(\sum_i \mathbf{v}_i\, y_{1i}\right),$$

$$\hat{\sigma}^2 = \sum_i (y_{2i} - 2\mu_i\, y_{1i} + \mu_i^2)/n,$$

$$\text{when}\quad \mu_i = \boldsymbol{\beta}' \mathbf{v_i}.$$

5.12   Use the EM algorithm to fit a $\pi : (1-\pi)$ mixture of two Poisson distributions, $P_0(\theta_1)$ and $P_0(\theta_2)$, to the data:

No.	0	1	2	3	4	5	6	7	8	9
frequency	162	267	271	185	111	61	27	8	3	1

(Note that the point estimates and associated standard errors, obtained from direct optimisation, are given by:

$$\hat{\pi} = 0.359(0.043), \quad \hat{\theta}_1 = 1.25(0.02), \quad \hat{\theta}_2 = 2.66(0.08).)$$

5.13 (Morgan and Titterington, 1977.) The mover-stayer model has state transition probabilities of the form:

$$\begin{array}{rcll} q_{ij} & = & (1-s_i)p_j & (i \neq j = 1,2,\ldots,m), \\ q_{ii} & = & (1-s_i)p_i + s_i & (i = 1,2,\ldots,m), \end{array}$$

where $\{p_k\}$ is a probability distribution and

$$\begin{array}{l} 1 - s_i \geq 0\,, \\ (1-s_i)p_i + s_i \geq 0\,, \quad \text{for } i = 1,2,\ldots,m \ . \end{array}$$

Show that conditional probabilities of state change are, for $i \neq j$,

$$P_{ij} = p_j/(1-p_i)\ .$$

Consider how you would fit this model to data using the EM algorithm.

5.14 The log-likelihood for a model which is a mixture of two univariate normal pdfs has the form:

$$\ell(\boldsymbol{\psi}) = \sum_{i=1}^{n} \log\{\alpha\phi(x_i|\mu_1,\sigma_1) + (1-\alpha)\phi(x_i|\mu_2,\sigma_2)\},$$

using a self-explanatory notation. We wish to fit this model using the EM algorithm.

Let

$$n_j^{(m)} = \sum_{i=1}^{n} w_{ij}(\boldsymbol{\psi}^{(m)}) \quad \text{for } j = 1,2.$$

Show that

$$\alpha_j^{(m+1)} = n_j^{(m)}/n,$$

$$\mu_j^{(m+1)} = \frac{1}{n_j^{(m)}} \sum_{i=1}^{n} w_{ij}^{(m)} x_i,$$

$$\sigma_j^{2(m+1)} = \frac{1}{n_j^{(m)}} \sum_{i=1}^{n} w_{ij}^{(m)} (x_i - \mu_j^{(m+1)})^2,$$

$$\text{where} \quad w_{ij}^{(m)} = w_{ij}(\boldsymbol{\psi}^{(m)}).$$

5.15 Run the MATLAB program of Figure 5.2(b), and comment on the speed of convergence of the EM algorithm.

5.16 Adapt the MATLAB program of Figure 5.2(b) to fit the gamma, $\Gamma(2,\lambda)$, distribution using the EM algorithm, and compare the approach with that of Example 5.2. Why would you not consider using the EM algorithm for fitting the Weibull distribution?

5.17 In Example 5.10, we encountered the empirical characteristic function. Write down an expression for the empirical cumulative distribution function.

5.18 In the zero-truncated Poisson distribution of Example 5.8, if n_0 denotes the estimated missing number of zeros, verify that

$$\hat{\lambda} = \log\left(1 + \sum_{i=1}^{\infty} n_i/n_0\right) = \left(\sum_{i=1}^{\infty} i n_i\right) \bigg/ \left(\sum_{i=0}^{\infty} n_i\right).$$

5.19 One way of reparameterising to obtain uncorrelated parameter estimators is to use principal component analysis (Jolliffe, 2002). Apply principal component analysis to the correlation matrices of Example 4.1, by using the MATLAB command *eig*. Discuss the results and indicate whether you think this approach might be useful in general.

5.20 We have already seen that the annual survival rates of wild animals may be estimated by means of ring-recovery experiments, resulting in data of the kind displayed in Table 1.5, and using models such as that of Exercise 5.8. An alternative approach is to make observations of previously marked *live* animals resulting in *capture-recapture* experiments. In some cases capture-recapture and ring-recovery data may be combined; see, e.g., Lebreton et al. (1995), Catchpole et al. (1998a), and Example 6.4. In the classical Cormack-Jolly-Seber model for capture-recapture data, the probabilities of annual survival are allowed to vary from year to year, but do not depend on the age of the animal. The probability that a live marked animal is captured in any year is also allowed to vary from year to year. Write down the probability that an animal captured and marked in year i is next seen alive in year j. Explain why this model is parameter-redundant, and suggest how to proceed.

5.21 In quantal response data we observe r_i individuals responding out of n_i exposed to dose d_i, $1 \leq i \leq k$. The logit model specifies the probability of response to dose d_i as:

$$\begin{aligned}
P(d_i) &= \frac{1}{1 + e^{-(\alpha + \beta d_i)}}, \text{ so that} \\
\text{logit}\,(P(d_i)) &= \log\{P(d_i)/(1 - P(d_i))\} \\
&= \alpha + \beta d_i,
\end{aligned}$$

and so to judge whether the model may provide a good fit to data we may plot logit (r_i/n_i) vs d_i. (cf. Exercise 3.10.) Produce this plot for the age of menarche data of Exercise 4.14. Fit the model to the data by means of a linear regression of logit (r_i/n_i) on d_i. Note that you will need to make an adjustment to the data when $r_i = 0$ and $r_i = n_i$. The minimum-logit-chi-square method results when the linear regression is extended by minimising a weighted sum of squares in which the weight for the ith dose is the inverse of an approximation to $\mathrm{Var}(\log(R_i/(n_i - R_i)))$. Use the δ-method to show that

$$\mathrm{Var}(\log(R/(n - R))) \approx \{nP(d)(1 - P(d))\}^{-1}.$$

Consider how this weight may be estimated and then estimate α and β by the minimum-logit-chi-square method for the age-of-menarche data.

5.22* Ridout et al. (1999) develop the following model for the strawberry branching structure of Exercise 2.14. Let X denote the number of flowers at rank r, given the number of flowers, m, at the previous rank. The branches at the previous rank are called *parent* branches. If Y_i denotes the number of *offspring* branches resulting from the ith parent branch, then we can write:

$$X = Y_1 + Y_2 + \cdots + Y_m.$$

Assuming the parent branches produce no more than two offspring branches, then we may model the distribution of Y by the probability function:

$$pr(Y = 0) = (1 - \pi)^2 + \pi(1 - \pi)\rho$$
$$pr(Y = 1) = 2\pi(1 - \pi)(1 - \rho)$$
$$pr(Y = 2) = \pi^2 + \pi(1 - \pi)\rho.$$

(Although occasionally parent branches may produce more than two offspring branches, as we have seen from the data of Exercise 2.14, this is relatively rare. Thus good progress can be made by basing modelling on this assumption.) Interpret the parameters π and ρ. Show that $\mathbb{E}[X] = 2m\pi$. For any row of the data table of Exercise 2.14, write down a likelihood. Use an EM-type argument to show that the maximum-likelihood estimate of π results from equating the sample mean of X to its expectation.

5.23 A random variable X has the negative binomial distribution,

$$pr(X = k) = \binom{n + k - 1}{n - 1} \left(\frac{p}{q}\right)^k \left(1 - \frac{p}{q}\right)^n, \quad \text{for } k \geq 0,$$

in which $q = 1 + p$. Show that the expectation and variance of X are given by $\mathbb{E}[X] = np$, $\mathrm{Var}(X) = np(1 + p)$. Use these expressions to derive estimates of n and p using the method of moments.

5.24 Suppose that $\mathbf{Y}$ has a multivariate normal distribution with mean, $\mathbb{E}[\mathbf{Y}] = \mathbf{X}\boldsymbol{\theta}$ and $V(\mathbf{Y}) = \sigma^2\mathbf{I}$, where $\boldsymbol{\theta}$ is a vector of parameters to be estimated, and $\mathbf{I}$ is the identity matrix. Write down the likelihood and show that the maximum-likelihood estimate of $\boldsymbol{\theta}$ results from minimising with respect to $\boldsymbol{\theta}$ the sum-of-squares,

$$S(\boldsymbol{\theta}) = (\mathbf{y} - \mathbf{X}\boldsymbol{\theta})'(\mathbf{y} - \mathbf{X}\boldsymbol{\theta}),$$

where $\mathbf{y}$ is the observed value of $\mathbf{Y}$. Examine the case when the components of $\mathbf{Y}$ are correlated, and when $\mathbb{E}[\mathbf{Y}] = \boldsymbol{\mu}(\boldsymbol{\theta})$.

5.25* (Quandt and Ramsey, 1978.) The random sample $(y_1, \ldots, y_n)$ is observed for the random variable Y which has a pdf which is a mixture of two normal pdfs:

$$f(y) = \left[\lambda/\{\sigma_1(2\pi)^{\frac{1}{2}}\}\right] \exp\left[-\frac{1}{2}\{(y-\mu_1)/\sigma_1\}^2\right] + \left[(1-\lambda)/\{\sigma_2(2\pi)^{\frac{1}{2}}\}\right] \exp\left[-\frac{1}{2}\{(y-\mu_2)/\sigma_2\}^2\right].$$

Write down the moment generating function of Y. Construct an empirical moment generating function, and consider how you might use it to estimate the parameters of $f(y)$.

5.26 In enzyme kinetics, two alternative models relating y to x_1 and x_2 are specified as follows, where V, κ_1, κ_2 and κ_3 are parameters.

(i)

$$y = \frac{V x_1}{\kappa_1\left(1 + \frac{x_2}{\kappa_2}\right) + x_1\left(1 + \frac{x_2}{\kappa_3}\right)},$$

(ii)

$$y = \frac{V x_1}{\kappa_1\left(1 + \frac{x_2}{\kappa_2}\right) + x_1}.$$

Identify which non-regular problem results when we compare these two models.

5.27 In radioligand binding, two alternative models relating y to x are specified as follows, where α, κ, κ_1 and κ_2 are parameters.

(i) $y = \dfrac{p}{1 + 10^{x-\kappa_1}} + \dfrac{(1-p)}{1 + 10^{(\alpha x - \kappa_2)}}$,

(ii) $y = \dfrac{1}{1 + 10^{x-\kappa}}$.

Identify which non-regular problem results when we compare these two models.

5.28 For the negative binomial distribution of Exercise 5.23, write down pr $(X = 0)$, and hence derive a method of 'mean and zero-frequency' for estimating n and p. Derive a mean and zero-frequency method for estimating the parameters of the Poisson process with the two rates model of Section 2.4.2 (Griffiths, 1977).

5.29* The estimator, $\widehat{a}(s)$ of Example 5.10 is not unique. Consider how you would obtain a unique estimate of α. Discuss generally how you would select the variable, s, of the characteristic function.

5.30* (Catchpole and Morgan, 1998.) Compartment models are used to describe the movement of material through different compartments, which may for example be parts of an organism. An example is illustrated in Figure 5.4.

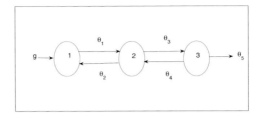

Figure 5.4 *An example of a three-compartment model. The* $\{\theta_i\}$ *are the parameters that determine the rate of transfer of material, while g is the arrival rate of material in the first compartment. At time t, the ith compartment contains an expected amount of* $x_i(t)$.

In general, the state vector $\mathbf{x}$ satisfies the differential equation,

$$\mathbf{x}' = \mathbf{A}\mathbf{x} + \mathbf{B}\mathbf{g},$$

where the components of the vector $\mathbf{g}(t)$ describe the flows into all of the compartments. In our example,

$$\mathbf{A} = \begin{bmatrix} -\theta_1 & \theta_2 & 0 \\ \theta_1 & -(\theta_2 + \theta_3) & \theta_4 \\ 0 & \theta_3 & -(\theta_4 + \theta_5) \end{bmatrix}$$

$$\mathbf{B}\mathbf{g} = [g(t)\ 0\ 0]'.$$

What is observed is $\mathbf{y}(t) = \mathbf{C}\mathbf{x}(t)$, for a matrix $\mathbf{C}$. For example, if we only observe the scalar quantity, $y(t) = x_3(t)$, then $y(t) = \mathbf{c}'\mathbf{x}(t)$ where $\mathbf{c}' = (0, 0, 1)$. Model-fitting is by the method of least squares. We shall consider the case, $y(t) = x_3(t)$. If $\tilde{y}$ and $\tilde{g}$ are the Laplace transforms of y and g respectively, show that if $\mathbf{x}(0) = \mathbf{0}$, then

$$\tilde{y} = \mathbf{c}'(s\mathbf{I} - \mathbf{A})^{-1}\mathbf{B}\tilde{g},$$

where s is the transform variable. If we write $\tilde{y} = Q(s)\tilde{g}$, find an expression for the *transfer function* $Q(s)$. Consider how to apply the methods of Sections 5.3.2 and 5.3.3 to this problem, to identify which parameters may be estimable.

5.31* Biologists are interested in the distribution of the cytoplasmic heritable determinant of the yeast, *Saccharomyces cerevisiae*, which exhibits prion-like properties. In an experiment, samples of yeast cells each have effectively, at time zero, an unknown number, m, of inanimate markers, which distribute themselves randomly over the two daughter cells, when cells divide at some unknown rate λ. What is recorded, in a number of replicated samples, is the percentage of cells containing markers. In each of four experiments, the replicate percentages are given below. Experiments A and B are essentially equivalent, but experiments C and D involve unknown time-shifts, which differ between the two experiments.

(i) Suggest a probability model for these data, and consider how you would fit the model using least-squares.

(ii) Consider how you might also make use of the following data on the growth of a yeast colony:

time (hrs)	1	2.17	3	4	5
OD	0.2	0.286	0.392	0.54	0.746

(OD is directly proportional to the number of yeast.) Data supplied by Simon Eaglestone, Lloyd Ruddock, and Mick Tuite from the Research School of Biosciences, University of Kent.

Experiment A

Time (hrs)	
8	100,100,100
9	100,100,100
10	99.4,100,100
11	97.8,98.9,98.6
12	96.6,94.0,97.5
13	87.6,89.1,86.4
14	81.0,78.1,76.6
15	67.4,67.5,66.4
16	66.0,66.1,64.2
17	52.4,53.2,52.3
18	52.9,44.7,44.5
19	36.1,37.8,43.2
20	37.5,33.3,19.6

Experiment B

Time(hrs)	
8	100,100,100
9	100,100,100
10	100,100,100
11	98.7,100,95.2
12	97.9,97.2,94.9
13	84.9,86.4,89.5
14	75.9,73.4,78.7
15	71.9,64.7,74.4
16	62.2,55.6,56.4
17	46.5,52.8,48.5
18	50.5,43.5,46.2
19	40.9,38.1,38.5
20	27.4,31.8,32.5

Experiment C

Time (hrs)	
18	100,100,100
20	100,100,100
22	100,100,100
24	100,100,100
26	100,100,100
28	100,100,100
30	100,100,99.3
32	98.3,99.2,99.1
34	98.9,96.5,95.6
36	93.6,92.4,91.7
38	85.0,88.0,88.5
40	73.8,81.2,86.3
42	67.2,73.3,71.3
44	70.0,66.7,65.4
46	58.2,60.0,64.6
48	54.4,45.9,50.3

Experiment D

Time (hrs)	
18	91.1,88.5,88.6
19	83.5,79.6,81.1
20	73.0,74.9,68.5
21	68.0,63.3,69.6
22	58.4,64.9,70.6
23	53.8,54.5,47.8
24	48.7,37.1,42.7
25	35.3,32.7,27.7
26	29.3,30.5
27	27.4,29.4,27.7
28	19.5,20.0,23.6
29	14.4,13.1,13.1
30	10.1,16.9,9.7
31	10.9,9.7,13.3
32	8.8,7.4,6.6
33	8.5,2.6,7.9
34	3.3,6.7,3.4

For more discussion, see Cole et al. (2004) and Morgan et al. (2003). Note also the relationship with the data of Table 1.15.

5.32 Experiment with GEM, applied to the data of Example 5.1, by modifying one step of a Newton-Raphson iteration, as suggested by Rai and Matthews (1993).

5.33 Experiment with SEM for the application of Example 5.1. Produce a sequence of estimates of π, and try to ascertain when the sequence has reached equilibrium.

5.34 Suppose that data $y_1, \ldots, y_n$ came from a multinomial distribution with probabilities $\{p_i\}$, where $\Sigma p_i = 1$ and $\Sigma y_i = m$. Show that the score vector is given by

$$\mathbf{U} = \mathbf{D}\mathbf{\Pi}^{-1}\mathbf{y},$$

where $\mathbf{\Pi}$ is a suitable diagonal matrix.

Note the correspondence between the expression for the expected information matrix, $\mathbf{J}$, given in Exercise 4.10, and the corresponding expected information matrix in Equation (5.9).

5.35 Discuss how the δ-method may be used

(a) for the transformation of Section 5.4, and

(b) in the re-parameterisation of Example 5.1.

5.36 In an experiment to estimate the annual survival probabilities of a particular species of bird, a single cohort of n birds ringed as nestlings is released, and the numbers r_1, r_2, r_3, r_4 of rings recovered from birds reported dead during the first year, second year, third year and fourth year of the study respectively are recorded. A possible probability model has three parameters: λ, the probability that the ring of a dead bird will be recovered soon after death; ϕ_1, the probability that any bird survives its first year of life; and ϕ_a, the probability of annual survival of any bird that successfully completes its first year of life. Write down and identify the likelihood. Identify information that may be regarded as missing, and then indicate how the EM algorithm would operate for this example.

5.37 In the re-parameterisation of Equation (5.1), the new parameter, θ, is defined as

$$\theta = (1 - w)(1 - e^{-\lambda}).$$

The maximum-likelihood estimator, $\hat{\theta}$, is given by $\hat{\theta} = 1-$ (proportion of litters with $Y = 0$). The δ-method for approximating the variance of $\hat{\theta}$ specifies that

$$\text{Var}(\hat{\theta}) \approx \mathbf{a}'\mathbf{S}\mathbf{a},$$

for suitable terms $\mathbf{a}$ and $\mathbf{S}$. Define $\mathbf{S}$, and evaluate $\mathbf{a}$ for the above example.

5.38 In the ABO blood group system there are four blood groups, A, B, AB and O, occurring with respective relative frequencies, $(p^2 + 2pr)$, $(q^2 + 2qr)$, $2pq$ and r^2, where p, q and r are probabilities and $p + q + r = 1$.

(i) Write down the likelihood, L, when the different groups are observed with frequencies n_A, n_B, n_{AB} and n_0 respectively, with $n = n_A + n_B + n_{AB} + n_0$.

(ii) Writing $r = 1 - p - q$, and treating L as a function of p and q, show that

$$\frac{\partial \log L}{\partial p} = \frac{2r}{p(p + 2r)} n_A - \frac{2}{(q + 2r)} n_B + \frac{1}{p} n_{AB} - \frac{2}{r} n_0.$$

(iii) Given that

$$-\frac{\partial^2 \log L}{\partial p \partial q} = \frac{2}{(p + 2r)^2} n_A + \frac{2}{(q + 2r)^2} n_B + \frac{2}{r^2} n_0,$$

show that

$$-\mathbb{E}\left[\frac{\partial^2 \log L}{\partial p \partial q}\right] = \frac{2n(4r + 3pq)}{(p + 2r)(q + 2r)}.$$

(iv) An iterative method for maximising the likelihood is called the *gene-counting* method. Here

$$\widehat{p}^{(i+1)} = \left(\frac{n_A + n_{AB}}{2n}\right) + \frac{\widehat{p}^{(i)}}{(\widehat{p}^{(i)} + 2\widehat{r}^{(i)})} \frac{n_A}{2n}$$

$$\widehat{q}^{(i+1)} = \left(\frac{n_B + n_{AB}}{2n}\right) + \frac{\widehat{q}^{(i)}}{(\widehat{q}^{(i)} + 2\widehat{r}^{(i)})} \frac{n_B}{2n},$$

where $\widehat{q}^{(m)}$ and $\widehat{p}^{(m)}$ are, respectively, the mth iterates of the maximum-likelihood estimates, $\widehat{q}$ and $\widehat{p}$. By separately writing each of the frequencies n_A and n_B as the sum of two quantities, one of which is missing, explain how gene-counting arises from applying the EM algorithm.

5.39 (Examination question, University of Nottingham.)

The EM algorithm is to be used as an alternative to the Newton-Raphson method for maximising the likelihood of Exercise 3.21 by imputing the values of the censored observations.

(i) Show that the E step of the algorithm involves the term

$$\mathbb{E}(T_i | T_i > t_i, \ \lambda^{(r)}) \quad \text{for} \quad m + 1 \le i \le m + n,$$

where $\lambda^{(r)}$ is the current estimate of λ.

(ii) Prove that

$$\mathbb{E}(T_i | T_i > t_i, \ \lambda^{(r)}) = \frac{2}{\lambda^{(r)}} + \frac{\lambda^{(r)} t_i^2}{1 + \lambda^{(r)} t_i}. \tag{*}$$

[You may assume that for $\theta > 0$

$$\int_s^\infty \frac{\theta^n y^{n-1} e^{-\theta y}}{(n-1)!} \ dy = \sum_{j=0}^{n-1} \frac{s^j \theta^j e^{-\theta s}}{j!}.]$$

(iii) Show how the M step is used to produce the next estimate of λ in the iteration, denoted by $\lambda^{(r+1)}$. Give an intuitive explanation of your result.

(iv) Suppose that you had been unable to derive the formula for the expectation in (*) above. How else could you have calculated $\mathbb{E}(T_i | T_i > t_i, \ \lambda^{(r)})$?

5.40   *Timed Species Counts* are used to estimate the abundance of wild birds. An observer spends one hour in an area of interest and records in which of 6 consecutive 10-minute periods (if any) a bird of a particular species is first heard. This experiment is repeated on 15 separate occasions, and results in data such as that illustrated below for birds in Kiwumulo in Uganda (data provided by Derek Pomeroy):

Species	n_0	n_1	n_2	n_3	n_4	n_5	n_6
Red-eyed dove, *Streptopelia semitorquata*	1	0	0	0	3	2	9
Common Bulbul, *Pycnonotus barbatus*	2	0	0	0	1	2	10
Trilling Cisticola *Cisticola woosnami*	2	0	0	1	1	4	7
Speckled Mousebird, *Colius striatus*	2	0	0	2	2	2	7

Here n_6 is the number of times a bird species is heard in the first interval, n_5 is the number of times it is heard in the second interval, etc. and n_0 is the number of times the bird species is not heard in the hour. A model for these data has the multinomial probabilities given below, where ν is the probability that the species is present, and $(1-p)$ is the probability that a bird species which is present is detected in each 10-minute interval.

n_6	n_5	n_4	n_3	n_2	n_1	n_0
$\nu(1-p)$	$\nu(1-p)p$	$\nu(1-p)p^2$	$\nu(1-p)p^3$	$\nu(1-p)p^4$	$\nu(1-p)p^5$	$\nu p^6 + (1-\nu)$

Explain how the EM algorithm would operate in this example, and demonstrate an advantage of the EM approach. Show how to provide estimates of the standard errors and correlation for the maximum-likelihood estimators of ν and p.

An alternative approach is to re-parameterise. Show that when the model is re-parameterised in terms of $\theta = \nu(1-p^6)$ and p then there exists an explicit maximum-likelihood estimator for θ. Interpret θ, and provide its estimates for the data above. Use this result to explain why the expected frequency corresponding to n_0 equals the observed frequency.

5.41   On each of T separate visits to N sites it is recorded whether or not a particular bird species is present or absent. An illustration of the data that may result from such a study is given below, for the American blue jay, *Cyanocitta cristata*. Here $T = 11$ and $N = 50$.

	Number of detections											
Species	0	1	2	3	4	5	6	7	8	9	10	11
Jay	17	9	11	6	5	2	0	0	0	0	0	0

For any site, let Y denote the number of visits on which a jay was detected. A particular model for detection gives the following expressions for the probability distribution of Y:

$$pr(Y = 0) = (1 - \psi) + \psi p^T$$

$$pr(Y = i) = \psi \binom{T}{i} p^{T-i}(1 - p)^i \quad \text{for} \quad 1 \leq i \leq T.$$

The data recorded are the numbers of sites with $0, 1, 2, \ldots, T$ detections, denoted $\{m_y, y = 0, 1, \ldots, T\}$. Here we have $m_0 = 17$, $m_1 = 9$, etc. Explain how the EM algorithm would operate in this example, and demonstrate an advantage of the EM approach.

An alternative procedure is to reparameterise. Show that when the model is reparameterised in terms of $\theta = \psi(1 - p^T)$ and p then there exists an explicit maximum-likelihood estimator for θ. Interpret θ and estimate it for the data above. If you have estimated θ and p, and also have expressions for their standard errors, explain how you would use the δ-method to obtain the standard error of ψ.

5.42 In a study of the failure times of transplanted kidneys, $m > 0$ individuals have failure times that are observed directly, leading to the data: $\{x_i\}$. A further $(n - m)$ individuals have failure times that are right-censored at time τ. Explain how you would model these data using an exponential distribution, resulting in the maximum-likelihood estimator,

$$\hat{\theta} = \frac{\sum_{i=1}^{m} x_i + (n - m)\tau}{m}.$$

In a study of how long it takes couples to conceive, it is observed that n_j couples take j conception cycles to conceive, for $j = 1, \ldots, r$, and that for n_{r+1} couples, all that is recorded is that they take longer than r cycles to conceive. Explain how you would model these data using a geometric random variable with probability p of successful conception per cycle, and write down the likelihood. It is shown in Equation (2.1) that the maximum-likelihood estimate of p is given by

$$\hat{p} = \frac{\sum_{j=1}^{r} n_j}{\sum_{j=1}^{r} j n_j + r n_{r+1}}.$$

Discuss the similarities and differences between the two models of this question, with particular reference to the expressions for the two maximum-likelihood estimators.

5.43 Continuation of Exercise 4.35. If p denotes the probability of an egg escaping infection, write down an expression for the variance of the maximum-likelihood estimator $\hat{p}$. Use the δ-method to approximate the variance of $\hat{\mu}$, where $p = e^{-\mu}$, and compare the result with that given in exercise 4.35.

5.44 The following data are a random sample from an exponential distribution, with all observations that are greater than 4 censored at 4, and denoted by a $*$.

2.11 4* 2.45 4* 2.30 2.39 2.25 2.06 4* 0.03 1.49 0.25 3.47 3.25 4* 2.76 2.00 0.99 3.13 3.67.

Write and run a MATLAB program to estimate the exponential parameter by means of SEM.

5.45   Write a MATLAB program to display the contours of the log-likelihood surface for the model and data of Example 5.1, in the parameterisation of Equation (5.2).

CHAPTER 6

Simulation Techniques

6.1 Introduction

Fast computers have revolutionised statistics. The ease with which we can now optimise functions numerically, as described in Chapter 3, provides one primary illustration. Simulation provides another. Our aim in simulation is to mimic the probability rules of models and we typically do this by obtaining realisations of random variables. Thus the histogram in Figure 6.1, for example, summarises the results of 200 simulations of a random variable with a particular distribution. To obtain such results without a computer would be tedious. Early simulations for use in statistics were obtained using mechanical means, but only for the simplest situations. As we shall see shortly, simulation using a computer is extremely valuable. It releases us from the straight-jacket of relying on asymptotic procedures. Furthermore, using fast simulation procedures, Bayesians can now effectively fit highly complex models to data, a discovery which has changed the face of statistics, and which is the subject of the next chapter. Later in this chapter we shall discuss the various uses of simulated random variables, but before that we shall describe the essentials of how they are generated.

What follows is just a sample from a wide area, covered in books such as Morgan (1984), Devroye (1986), Ripley (1987), and Dagpunar (1988). The MATLAB functions that we provide are written to be illustrative, rather than efficient. Efficiency in simulation is very important. However different applications have different requirements. In classical simulation work we often require a large number of random variables from a fixed distribution. In computational Bayes work, as we shall see in Chapter 7, the requirement is often quite different, when we regularly update the distribution that we simulate from. Thus an efficient simulation procedure for Bayesian work would be one with low 'start-up' costs, whereas for other applications we might well be prepared to pay start-up costs in order thereafter to obtain a cheap supply of simulated variables, all with the same distribution.

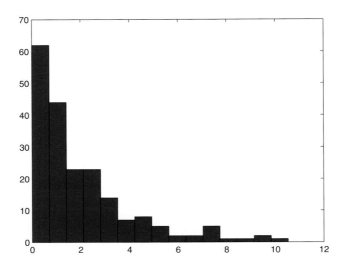

Figure 6.1 *A histogram summarising 200 simulations of a particular random variable.*

6.2 Simulating random variables

6.2.1 Uniform random variables

The building blocks of all simulations are realisations of a $U(0,1)$ random variable. Computers generate what are called *pseudo-random* variables, as a stream of approximately $U(0,1)$ variables which are obtained by iterating a deterministic formula starting from a *seed*. The seed is a number which initiates the sequence of pseudo-random variables resulting from the formula used. Changing the seed changes the sequence. Thus in MATLAB for example $rand(m,n)$ produces an $m \times n$ matrix with entries that we treat as independent $U(0,1)$ variates. This takes values from a particular deterministic formula called a *multiplicative congruential generator*, which produces the sequence of values of $u_{r+1}/(2^{31}-1)$, where $u_{r+1} = 7^7 u_r \; (mod(2^{31}-1))$, for $r \geq 0$, starting from the seed u_0. The seed in MATLAB is the same for any session, unless it is changed by means of the command, *rand* ('state', n), where n is a positive integer. Changing n changes the seed. In general, congruential random number generators produce a cycle of distinct numbers, which then repeats indefinitely. Good generators have long cycle lengths — see Morgan (1984, p.59) for discussion.

Pseudo-random numbers have passed rigorous tests and are usually treated as random. However in any particular application they have the potential for

```
function y=betasim(a,b,n)
%
% BETASIM simulates n independent realisations of a Be(a,b)
%    random variable.
%    Calls the function GAMMASIM.
%_____
x1=gammasim(a,n); x2=gammasim(b,n);
y=x1 ./(x1+x2);
```

Figure 6.2 MATLAB *function for simulating beta random variables distribution, making use of a supply of gamma random variables through the function gammasim, given in Figure 6.4.*

producing aberrant results and this should always be borne in mind. Partly for this reason, national lottery numbers are normally obtained by mechanical means.

6.2.2 Transformations

Suppose that U_1 and U_2 are two independent $U(0,1)$ random variables; then Box and Müller (1958) showed that

$$N_1 = (-2\log_e U_1)^{\frac{1}{2}} \cos(2\pi U_2)$$

$$N_2 = (-2\log_e U_1)^{\frac{1}{2}} \sin(2\pi U_2) \qquad (6.1)$$

are a pair of independent $N(0,1)$ random variables, a result which is readily verified using standard theory on the transformation of random variables.

This is just one example by which standard theory of the algebra of random variables finds use in simulation. One of several further examples presented by Morgan (1984, Chapter 4) is that

$$Y = X_1/(X_1 + X_2) \qquad \text{has the beta,}$$

$Be(a,b)$, distribution when X_1 and X_2 are independent $\Gamma(a,1)$ and $\Gamma(b,1)$ random variables with $a > 1$ and $b > 1$. Thus the function of Figure 6.2 allows us to generate beta random variables in MATLAB .

Binomial and Poisson random variables can be obtained using the functions of Figure 6.3. The binomial function makes direct use of the definition of the distribution, while the Poisson function utilises the mechanism of the Poisson Process (see Exercise 6.1). Both the transformation of Equations (6.1) and the function for simulating Poisson random variables in Figure 6.3 make use of the fact that when U is $U(0,1)$, $X = -\frac{1}{\lambda}\log_e U$ is a realisation of an $\text{Ex}(\lambda)$ exponential random variable with pdf

$$f(x) = \lambda e^{-\lambda x}, \qquad \text{for } x \geq 0. \qquad (6.2)$$

This is a particular example of the use of the *inversion* method, which is a general method for simulating continuous random variables with cdf $F(x)$ and

(a)
```
function y=binsim(m,p,n)
%
% BINSIM  simulates n independent realisations of a Bin(m,p) random
%    variable.
%_____
y=sum((rand(m,n)-p<0));      % uses the relational operator, '<' :
                             % matrix entry = 1 if and only if
                             % the corresponding  U(0,1) variate < p.
                             % Ref: Hunt et al (2006, p.87)
```

(b)
```
function y=poissim(lambda,n)
%
%POISSIM simulates n independent Po(lambda) random variables
%
%_____
y=zeros(n,1);
e1=exp(-lambda);
z=rand(n,1);
for i=1:n
  k=0;
  u=z(i);
  while u>e1;                % makes use of features of the
    u=u*rand;                % Poisson Process
    k=k+1;
  end
  y(i)=k;
end
```

Figure 6.3 MATLAB *functions for simulating (a) binomial and (b) Poisson random variables.*

with a well-defined inverse F^{-1}. If U is $U(0,1)$, then

$$Pr(F^{-1}(U) \leq x) = Pr(U \leq F(x)), \tag{6.3}$$

as the cdf is a strictly monotonic increasing function.

Thus $Pr(F^{-1}(U) \leq x) = F(x)$, as U is $U(0,1)$, and so if we set $X = F^{-1}(U)$ we obtain realisations of the random variable with cdf $F(x)$. In the exponential case of Equation (6.2), $F(x) = 1 - e^{-\lambda x}$, for $x \geq 0$, so that we obtain $X = -\frac{1}{\lambda} \log_e(1 - U)$, and as $(1 - U)$ is a $U(0,1)$ random variable, we see that in order to obtain a random variable with the desired exponential distribution, it is equivalent to set $X = -\frac{1}{\lambda} \log_e U$. The inversion method is convenient when it is easy to form F^{-1}, which is true, for example, of the Weibull and logistic distributions (Exercise 6.2) but is not true of the gamma and normal distributions. An alternative general approach which sees much use in computational Bayesian analysis employs the idea of *rejection*, which we shall now consider.

6.2.3 Rejection

In rejection we simulate a desired process which is complex by instead simulating an alternative process which is relatively simple and rejecting a suitable fraction of the results, thereby ensuring that the *accepted* fraction is exactly what we need, as realisations of the complex process. Thus for example we might simulate logistic random variables using the inversion method, which is straightforward, and then reject a suitable fraction of these so that the remaining sample has a normal distribution.

Example 6.1: Normal variates by rejection (von Neumann — see Kahn, 1956, p.36)

If X_1 and X_2 are independent, identically distributed Ex(1) exponential random variables then we reject the pair of values if $(X_1-1)^2 \geq 2X_2$. However if $(X_1-1)^2 < 2X_2$, then assigning a random sign to X_1 (positive or negative, each with probability $\frac{1}{2}$) simulates an $N(0,1)$ random variable.

We can prove this as follows. The joint pdf of X_1 and X_2 has the form

$$f_{X_1 X_2}(x_1, x_2) = e^{-(x_1+x_2)} \quad \text{for } x_1 \geq 0; \ x_2 \geq 0. \quad \text{Hence we can write}$$

$$Pr((X_1 - 1)^2 < 2X_2) = \int_0^\infty e^{-x_1} \int_{\frac{1}{2}(x_1-1)^2}^\infty e^{-x_2} dx_2 dx_1 ,$$

which is readily shown to be $\sqrt{\frac{\pi}{2e}}$. Hence conditional upon the event, $(X_1 - 1)^2 < 2X_2$, the resulting X_1 has the pdf

$$f_{X_1}(x_1) = \frac{e^{-x_1} \int_{\frac{1}{2}(x_1-1)^2}^\infty e^{-x_2} dx_2}{\sqrt{\pi/(2e)}}$$

$$= \sqrt{\frac{2}{\pi}} e^{-x_1^2/2} \quad \text{for } x_1 \geq 0.$$

This is the pdf of a half-normal distribution, and setting $W = -X_1$ with probability $\frac{1}{2}$ and $W = X_1$ with probability $\frac{1}{2}$ results in $W \sim N(0,1)$. See Exercise 6.3 for further discussion of this example. □

Example 6.2: Gamma variables by rejection

The random variable X has the $\Gamma(r,1)$ distribution with pdf,

$$f(x) = \frac{x^{r-1}e^{-x}}{\Gamma(r)} \quad \text{for } x \geq 0, \text{ and } r > 1.$$

We may simulate X by sampling from the exponential pdf, $h(x) = \frac{1}{r}e^{-x/r}$, resulting in the value $\tilde{x}$, say, and then checking to see if $kUh(\tilde{x}) < f(\tilde{x})$, when U is an independent $U(0,1)$ random variable, and k is a suitable specified

```
function y=gammasim(r,n)
%
% GAMMASIM simulates n independent realisations of a gamma(r,1) random
%    variable, using rejection.
%_____
for j=1:n              % inefficient MATLAB programming
  t=2;u=3;             % values to start the 'while' command
  while(t<u)
    e=-r*log(rand);
    t=((e/r)^(r-1))*exp((1-r)*(e/r-1));
    u=rand;
  end
  y(j)=e;
end
```

Figure 6.4 MATLAB *function for simulating* $\Gamma(r,1)$ *random variables, using the rejection method of Example 6.2.*

constant. If this condition is satisfied then we accept $\tilde{x}$ as a realisation of X. Otherwise we reject U and $\tilde{x}$ and start again. The constant $k > 1$ is chosen so that for all x, the inequality $kh(x) \geq f(x)$ is satisfied. Furthermore, the probability of rejection can be shown to be $1 - k^{-1}$, so that the smaller k is the better. See Exercise 6.19. In this example the optimum value of k is given by (see Exercise 6.4)

$$k = r^r e^{1-r}/\Gamma(r). \tag{6.4}$$

In this procedure we envelope the pdf $f(x)$ by the function, $kh(x)$. By simulating from $h(x)$ we generate points uniformly at random beneath $kh(x)$ and then, following rejection, below $f(x)$. The abscissae of these points are then realisations of X. For further discussion, see Morgan (1984, Section 5.3).

A MATLAB program to simulate gamma random variables by this approach is given in Figure 6.4. In order for the rejection method to operate as in this example, it is only necessary to specify the pdf we wish to simulate from up to a possibly unknown multiplicative constant term. □

Example 6.3: Ratio-of-uniforms method

If points are scattered at random over a disc centered at the origin and of unit radius (we call this the unit disc) then the Cartesian co-ordinates (V_1, V_2) of any such point allow us to carry out the Box-Müller method of Equation (6.1) without having to evaluate the trigonometric terms. We set

$$N_1 = V_2 \left(\frac{-2\log_e W}{W}\right)^{\frac{1}{2}},$$

$$N_2 = V_1 \left(\frac{-2\log_e W}{W}\right)^{\frac{1}{2}}, \tag{6.5}$$

where $W = V_1^2 + V_2^2$ – see Morgan (1984, p.80).

This approach is called the *Polar Marsaglia* method (Marsaglia and Bray, 1964), and produces points uniformly distributed over the unit disc by scattering points uniformly over a square with sides of length 2 enveloping the disc and then rejecting those points which lie outside the disc. We can see that in this case the rejection probability is $1 - \pi/4$.

The ratio of two independent $N(0,1)$ random variables has a Cauchy distribution, and so we can see from Equation (6.5) that the rejection approach of the Polar Marsaglia method provides a very simple way of simulating Cauchy random variables as a ratio of two independent uniform $U(-1,1)$ random variables which have not been rejected due to resulting in a point outside the unit disc. We shall now see how this result generalises. It is interesting that changing the area over which the points are uniformly scattered allows random variables with other distributions also to be simulated as the ratio of two independent uniform random variables (Kinderman and Monahan, 1977). The result, which is easily proved (see Exercise 6.5), is as follows (Ripley, 1987, p.66).

If $g(x)$ is any non-negative function which is proportional to a pdf $f(x)$, that is

$$g(x) = \kappa f(x) \quad \text{for some } \kappa,$$

then if the pair of variables, (V_1, V_2), is uniformly distributed over an area

$$A = \{(w, z) | 0 \le w \le \sqrt{g(z/w)}\}$$

then $X = V_1/V_2$ has the pdf $f(x)$. □

As with the rejection method of Example 6.2, the ratio-of-uniforms method does not require full specification of the target pdf that we wish to simulate from, that is we do not need to know κ above, but only the *shape* of $f(x)$, which we obtain via $g(x)$. This is a useful property of the method for Bayesian analysis, as we shall see in Chapter 7. Generalisations and improvements are described by Wakefield et al. (1991), especially designed for Bayesian work.

6.3 Integral estimation

6.3.1 Numerical analysis

As we shall see in Chapter 7, the key to Bayesian analysis lies in the derivation of marginal pdfs by integrating joint pdfs. The importance of integration will also be apparent from Chapter 8 when we integrate over random effects in generalised linear mixed models. Good robust numerical analysis procedures for evaluating integrals exist. MATLAB has two such functions for single integrals, viz; *quad*, which uses an adaptive recursive Simpson's rule, and *quad8*, which uses an adaptive recursive Newton-Cotes eight panel rule. We know that the normal cdf needs to be evaluated numerically. For example, the MATLAB command *quad* ('*norml*', -5, -0.5) approximates $\Phi(-0.5)$, by integrating the

```
function y=norml(y1,mu,sigma)
%
%NORML calculates the ordinates of a
% normal, N(mu, sigma^2), pdf, given a grid of 'x-values', in
% y1, and the value of the mean, 'mu' and the std. deviation
% 'sigma'.
%_____
y=(exp((-0.5*(y1-mu).^2)/sigma^2))/(sigma*sqrt(2*pi));
```

Figure 6.5 MATLAB *function specifying the standard normal pdf, for use with a numerical integration routine to obtain cdf values.*

standard normal pdf $\phi(x)$, from -5 to -0.5. It produces 0.3085, which is the correct value to 4 decimal places, when *npdf* is specified as in Figure 6.5.

In Bayesian analysis, integration is often high-dimensional, when Monte Carlo methods are found to be especially useful and outperform numerical procedures.

6.3.2 Crude Monte Carlo

In order to illustrate the basic ideas of Monte Carlo methods for integral estimation, we shall consider the integral, $I = \int_0^1 f(x)dx$. This integral may be regarded as

$$I = \mathbb{E}[f(U)], \quad \text{where } U \text{ is } U(0,1).$$

Thus we can estimate I by

$$\widehat{I_1} = \frac{\sum_{i=1}^n f(u_i)}{n}, \tag{6.6}$$

when $(u_1, u_2, \ldots, u_n)$ is a random sample from the $U(0,1)$ distribution. This very simple way of estimating I is called Crude Monte Carlo, and as the name suggests, it can readily be made more efficient. One way of doing this is to use importance sampling.

6.3.3 Importance sampling

We may also write $I = \int_0^1 \left\{ \frac{f(x)}{g(x)} \right\} g(x)dx$, for some function $g(x)$. If $g(x)$ is a pdf over $(0, 1)$, then we see that we can now write $I = \mathbb{E}\left[\frac{f(X)}{g(X)}\right]$, where $X \sim g(x)$, and so we can obtain a second estimate of I by $\widehat{I_2}$, given by:

$$\widehat{I_2} = \frac{1}{n} \sum_{i=1}^n \frac{f(x_i)}{g(x_i)} \tag{6.7}$$

where $(x_1, x_2, \ldots, x_n)$ is a random sample from $g(x)$. Importance sampling is often useful for the Bayesian methods of Chapter 7. For an example of the use of importance sampling, see Exercise 6.6.

6.4 Verification

Model fitting may not work the first time. This could, for instance, be due to problems with multiple optima if a deterministic search method is used to maximise a likelihood (see Section 3.4.1) or due to parameters such as probabilities straying out of range during a method of iterative search, possibly causing arguments of logarithms to become negative (see Section 5.4). Alternatively the problem may simply be the mundane one of an incorrect coding of the likelihood in a computer program. A simple verification check, of running the program on artificial data simulated from the model for a 'reasonable' set of parameter values, may quickly reveal basic errors. Much valuable scientific time can be wasted in attempting complex solutions, such as a reparameterisation for a problem with iterative maximum-likelihood estimation, when only a simple correction may be needed. When we analyse data that are simulated from a model, we *know* beforehand the values of the parameter estimates used in the simulation, and so we know what the model-fitting procedure should produce. If we cannot fit a model to simulated data there is little hope of fitting the model to real data. Fitting simulated data should be a standard procedure; when verification works correctly, it is very satisfying. Furthermore, the mechanism can then be repeated several times in order to provide estimates of error, as we shall see in Section 6.6. Sometimes verification reveals odd results. In such cases it is of vital importance to be able to repeat the simulation for the identical stream of pseudo-random numbers, and this is easily done as long as one knows the value of the seed used to initiate the string of pseudo-random numbers.

Example 6.4: An illustration of verification

Catchpole et al. (1993) were interested in fitting the model $(\lambda, \{\phi_{1i}\}, \phi_2, \phi_3, \phi_4, \phi_5, c)$ to avian recovery data augmented with recapture data: here the model fitted is the Freeman and Morgan (1992) model of Example 4.10, with constant reporting probability for dead birds. But in addition to the standard recovery experiment, the data are augmented by a single recapture study, with birds being recaptured alive or not, with probability c irrespective of age and ringing cohort. This approach was adopted in practice by Nicholas Aebischer in a survey of shags (*Phalacrocorax aristotelis*) on the Isle of May in Scotland. The result of a verification simulation study is presented in Table 6.1.

□

A verification study can in itself be revealing, as in Table 6.1, where the magnitude of c appears to have relatively little effect on the estimation of the survival probabilities, as long as $c > 0$.

Table 6.1 *A summary of a simulation study on the effect of the size of the recapture probability c. The model simulated had parameters:* $[\lambda, \{\phi_{1i}\}, \phi_2, \phi_3, \phi_4, \phi_5, c]$, *with* $\lambda = 0.1$, $\phi_{11} = 0.4$, $\phi_{12} = 0.5$, $\phi_{13} = 0.6$, $\phi_{14} = 0.5$, $\phi_{15} = 0.4$, $\phi_2 = 0.5$, $\phi_3 = 0.6$, $\phi_4 = 0.5$, $\phi_5 = 0.4$; *100 simulations were done using this parameter configuration, with 1000 birds ringed in each year for each of 5 years. For each simulation we have investigated the effect of varying c, as follows: (a) c = 0,   (b) c = 0.1,   (c) c = 0.5. Shown for each parameter is the frequency with which it was estimated in each interval. In all cases the frequency shown in* **bold** *corresponds to the interval containing the value of the parameter used in the simulations.*

			$\hat\phi_{11}$	$\hat\phi_{12}$	$\hat\phi_{13}$	$\hat\phi_{14}$	$\hat\phi_{15}$	$\hat\phi_2$	$\hat\phi_3$	$\hat\phi_4$	$\hat\phi_5$	$\hat\lambda$	c
(a)	0.-	0.05								4	51		
	0.05	0.15								2		**84**	
	0.15	0.25					2			3	1	3	
	0.25	0.35	22				21	4	2	16	3	2	
	0.35	0.45	**43**	17		25	**31**	21	14	18	**4**	1	
	0.45	0.55	18	**50**	24	**40**	23	**38**	23	**16**	4	1	
	0.55	0.65	4	16	**42**	15	6	7	**24**	3	1	1	
	0.65	0.75	1	4	19	7	4	10	7	8	6		
	0.75	0.85	2	2	2	2	1	4	12	12	6	1	
	0.85	0.95	10	10	11	11	11	5	7	5	11		
	0.95	1.00		1	2			8	11	13	13	7	
(b)	0.-	0.05									5		
	0.05	0.15									4	**100**	99
	0.15	0.25					2				16		1
	0.25	0.35	19			1	36			7	24		
	0.35	0.45	**67**	18		20	**27**	18	3	20	**13**		
	0.45	0.55	14	**67**	26	**68**	21	**73**	24	**39**	13		
	0.55	0.65		15	**63**	11	4	9	**57**	27	7		
	0.65	0.75			11				16	7	10		
	0.75	0.85									4		
	0.85	0.95											
	0.95	1.00									4		
(c)	0.-	0.05											
	0.05	0.15									2	**100**	
	0.05	0.25									10		
	0.25	0.35	22			1	23				26		1
	0.35	0.45	**67**	23		25	**60**	17		28	**30**		21
	0.45	0.55	11	**64**	20	**64**	15	**79**	25	**48**	22		**50**
	0.55	0.65		13	**71**	10	2	4	**62**	23	5		22
	0.65	0.75			9				13	1	3		4
	0.75	0.85									1		2
	0.85	0.95											
	0.95	1.00									1		

6.5 *Monte Carlo inference

Likelihood-based inference is very attractive, because of the known, good properties of the resulting procedures, as we have discussed in Chapter 4. However sometimes the likelihood is difficult to construct. In such cases simulation can be used to obtain an approximation to the likelihood surface. Consider, for example, the histogram in Figure 6.1. While we did not specify the mechanism which generated the values summarised there, it is clear that we are not simulating a normal random variable. For one thing, the variable in question is almost certainly positive, and in fact it seems likely that the variable is exponential, of expectation about 2, which is in fact the case. This kind of conclusion forms the basis for Monte Carlo inference, as we shall soon see.

Suppose we observe a random sample $\{y_i, 1 \leq i \leq n\}$ from a model with pdf $f(y; \boldsymbol{\theta})$, dependent on parameters $\boldsymbol{\theta}$. Our objective is to estimate $\boldsymbol{\theta}$, but let us suppose that $f( )$ is difficult to write down. Many examples of this arise in statistics – for example in the theory of queues (Gross and Harris, 1974). A simple illustration is if the random variable in question, Y, can be written as the convolution, $Y = X_1 + X_2$, where X_1 has a gamma distribution and X_2 is an independent normal random variable. It is easy to simulate a sample from Y, but the expression for its pdf is complicated by the fact that it involves the normal cdf, $\Phi(x)$. For Monte Carlo inference the ability to simulate Y in general is essential and not usually restrictive, even for quite complex models.

We need the log-likelihood function,

$$\ell(\boldsymbol{\theta}; \mathbf{y}) = \sum_{i=1}^{n} \log f(y_i; \boldsymbol{\theta}),$$

and in Monte Carlo inference we use:

$$\ell^*(\boldsymbol{\theta}; \mathbf{y}) = \sum_{i=1}^{n} \log \hat{f}(y_i; \boldsymbol{\theta}),$$

where we use simulation to estimate f by $\hat{f}$.

For any $\boldsymbol{\theta}$ we can produce an estimate of f, as a result of simulating from $f(y; \boldsymbol{\theta})$. This is analogous to using the histogram of Figure 6.1 as an aid to judging the form of the pdf that gave rise to that histogram. However before a crude estimate of f can be used, the result needs to be smoothed and made an explicit function of y. Diggle and Gratton (1984) accomplished this by using a kernel approach (see Appendix C), which results in the expression:

$$\hat{f}(y; \boldsymbol{\theta}) = \frac{1}{sh} \sum_{k=1}^{s} K\left(\frac{y - x_k}{h}\right), \qquad (6.8)$$

where $\{x_k, \ 1 \leq k \leq s\}$ is a random sample from $f(y; \boldsymbol{\theta})$ for fixed $\boldsymbol{\theta}$, for some sample size s, and $K( )$ is a kernel function. In this case, the form adopted for the kernel was:

$$K(u) = \begin{cases} 0.75(1 - u^2), & -1 \leq u \leq 1 \\ 0, & \text{otherwise.} \end{cases}$$

As usual with kernel methods, alternative forms may be used for $K(u)$, and h determines the smoothness of the approximation. Guidelines for choosing s, h and K, together with other aspects of this procedure, were provided by Diggle and Gratton (1984), who employed a simplex optimisation method (see Section 3.3.3) and called the approach Monte Carlo inference. Armed with the approximation of Equation (6.8) we can now form

$$\ell^*(\boldsymbol{\theta}; \mathbf{y}) = \sum_{i=1}^{n} \log \widehat{f}(\mathbf{y}_i; \boldsymbol{\theta}).$$

Simulated annealing, discussed in Section 3.4, might be used to good effect here in order to avoid the local optima which might well arise as a result of the likelihood being constructed from simulated data.

The mechanics of Monte Carlo inference are readily programmed, as we can see from the illustration of Example 6.5.

Example 6.5: Using Monte Carlo inference

We have a random sample of size 50 from a $N(\theta, 1)$ distribution and we wish to obtain the maximum-likelihood estimate of θ. We take $s = 200$ and $\theta = 1$ in the illustration of Figure 6.6.

The results are given in Figure 6.7, from using a normal kernel – see Appendix C. A coarse grid for θ results in a jagged approximation to the likelihood curve, which is known in this simple example and is therefore shown for comparison. Also shown is a three-point moving average, which smooths out the variation in the approximate likelihood $\widehat{f}(y; \theta)$. We shall discuss alternative ways of smoothing in Section 8.4.

The essential message of Monte Carlo inference is that if you can simulate from a model which is not parameter redundant, then you can fit it. This is evidently an extremely useful and powerful conclusion, which in many cases will justify the labour involved in carrying out Monte Carlo inference.

6.6 Estimating sampling distributions

For a given model, selected for a data set, estimates of aspects of interest, such as confidence regions for parameters, readily follow if we can obtain a supply of replicate data sets. In such a case the model could be fitted to each data set in turn, each time resulting in an alternative estimate of the parameters. The variation in the alternative estimates provides a measure of the spread of the parameter estimators. Typically, however, replicate data sets are not available, and traditionally statisticians have instead used the asymptotic criteria of Chapter 4 in order to approximate to sampling distributions.

However, once a model has been fitted to data, and parameter estimates have been obtained, then we can simulate new data sets which share the same basic characteristics as the original data set, such as size, and thereby obtain as many replicate data sets as we require. The model may then be fitted to each of the replicate data sets in turn, and each time this results in parameter

```
% Program to illustrate Monte-Carlo inference.
% We have a random sample from  a normal distribution with unit variance
% and we plot the log-likelihood as a function of the mean, without using
% the form of the pdf.
%_____
% n is the sample size taken
% s is the sample size for the simulations at each theta
% k1 determines the kernel smoothing.
% Calls the function KERNEL.
%_____
n=50;
s=200;
k1=1;
y=randn(n,1);
y=y+ones(size(y));         % y is the basic N(1,1) sample;
y2=y .^2;
y3=sum(y2);
                           % we plot the log-likelihood, which is
                           % known in this example, over the
                           % range, 0-2, for comparison
theta=0.1:0.01:2;
s2=ones(size(theta));
knownlogl=((-n*log(2*pi))/2)*s2...
          -s2*(y3/2)+sum(y)*theta-(n/2)*(theta.^2);
plot(theta,knownlogl);

hold;                      % now we move on to the kernel stage

k=0;
for thet=0.1:0.05:2
  k=k+1;
  tt(k)=thet;
                           % now for each value of theta we simulate
                           % a sample of size s. This goes into x.
  x=randn(s,1);
  x=x+thet;
  mclogl(k)=0;
  for i=1:n
    mclogl(k)=mclogl(k)+log(kernel(y(i),x,k1));
  end
end
plot(tt,mclogl,'*')
                           % finally we form and plot the moving
                           % average

movav=mclogl(2:k-1)+mclogl(1:k-2)+mclogl(3:k);
plot(tt(2:k-1),movav/3,'--')
```

Figure 6.6 *A* MATLAB *program for Monte Carlo inference. Note that the command* randn *simulates variates from* $N(0,1)$*, and the function* kernel *is specified in Appendix C.*

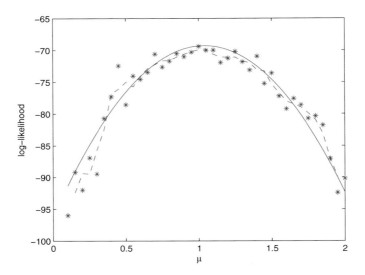

Figure 6.7 *The result of running the* MATLAB *program of Figure 6.6, with a normal kernel. The solid line denotes the known log-likelihood. In addition to the approximate likelihood, indicated by asterisks, we also show its three-point moving average (Chatfield, 1984, p.41).*

estimates. If there are k replicate data sets then, by this means, we obtain a sample of size k for each of the model parameters, and this allows us to draw histograms of the samples for parameters of interest and to calculate sample correlations between the samples for different model parameters. This is easily done, as we can see from the illustration of Example 6.6. In this example the only matching that needs to be made to the original data is to ensure that we have the same number of couples in each data set as in the original data set. If we were basing the simulations on several multinomial distributions, as is the case in the models we have considered for data on the survival of wild animals, then it would be appropriate to match the cohort numbers of animals marked each year, when these numbers are available.

Example 6.6: The parametric bootstrap

In Figure 6.8 we give a MATLAB program for simulating from the beta-geometric model of Section 2.3. We are matching the simulation to the smokers data of Table 1.1 and using $\mu = 0 \cdot 275$, $\theta = 0 \cdot 091$, corresponding to the maximum-likelihood estimates of these parameters for that data set. In Figures 6.9 and 6.10 we summarise the results from 100 simulations by means of kernel density estimates. Also shown in Figures 6.9 and 6.10 are the approximating asymptotic normal sampling distributions, which result from Equation

Table 6.2 *A comparison of the sample standard errors and correlation for $\widehat{\theta}$ and $\widehat{\mu}$ in the beta-geometric model for the smokers data of Table 1.1. We use the format of Table 4.1, and take sample means for the point estimates.*

	Using asymptotic normality (Eqn. 4.2)		Parametric bootstrap		Bootstrap	
$\widehat{\mu}$	0.275	0.036	0.278	0·032	0.277	0.037
$\widehat{\theta}$	0.091	0.739 0.060	0.096	0.672 0.058	0.090	0.711 0.069

(4.2). What we are generating here by simulation are finite-sample approximations to the sampling distributions of $\hat{\mu}$ and $\hat{\theta}$. Clearly in this case the asymptotic and parametric bootstrap approximations agree well. In the standard beta-geometric distribution, $\theta \geq 0$, and this constraint has been disregarded in Figures 6.9 and 6.13.

In Table 6.2 we compare the sample standard errors and correlations for $\widehat{\theta}$ and $\widehat{\mu}$, obtained in this way, using separately the bootstrap, to be described in the next section, and the parametric bootstrap, with those resulting from the classical asymptotic normal approximation. For both the bootstrap results we also show the sample means of $\widehat{\mu}$ and $\widehat{\theta}$, which provide a verification of the procedure.   □

A defect in this very attractive computational procedure is that our inferences are based on a *single* value of the model parameters, viz the maximum-likelihood estimate, which is itself just an estimate. We can investigate the robustness of our conclusions by conducting a *sensitivity* analysis, whereby we repeat the entire procedure for a small number of alternative values for the model parameters; for instance, we could decide on the alternative values to be used by considering the classical asymptotic distribution for the estimators (see Exercise 6.11). This approach can be extended by obtaining parameter values to be used in simulations by sampling from the asymptotic distribution. As we shall see in the next chapter, a Bayesian analysis provides a natural framework for carrying out this kind of procedure.

The approach of this section is called the *parametric bootstrap*, and we shall now consider how the bootstrap approach works.

6.7 Bootstrap

The aim of the bootstrap is also to obtain replicates samples without any new real (as opposed to simulated) data. It neatly side-steps the issue of which model and which parameter values to use in a simulation by *simulating directly from the original data*, with no mention of any model, and so in this sense it is *non-parametric*. In comparison with the parametric bootstrap, Young (1994) regarded the non-parametric bootstrap as 'where the fun lies.'

If we denote a real random sample by $\{x_i,\ 1 \leq i \leq n\}$, then the kth

```
% Program to fit the beta-geometric model to fecundability data,
% and then to fit, k times, data simulated from the fitted model,
% and produce histograms of -the maximum log-likelihood values,
% and of the parameter estimates.
%-----------------------------------------------------------------
% k is the number of simulations
% data(i) is the number that wait i cycles;
% data(13) is the number that wait > 12 cycles.
% Calls the function BEGEO, which produces the
% beta-geometric negative log-likelihood.
% Calls the function MULTSIM, which simulates a
% multinomial distribution, and FMINSEARCH.
% 'data' are global, and set in the driving program.
%-----------------------------------------------------------------
global data                              % the same data are used in
                                         % 'begeo'
data=[29 16 17 4 3 9 4 5 1 1 1 3 7];     % the 'smokers' data
k=100; n=sum(data);
x0=[.5,.05];                             % start for the
x=fminsearch('begeo',x0);                % maximum-likelihood estimation
y=begeo(x);                              % 'x' contains the mle
disp(y);disp(x);                         % 'y' is the -log-lik at the mle

p=zeros(size(data));                     % now we set up the multinomial
                                         % and simulate from it.
p(1)=x(1);                               % this is mu
s=p(1);
for i=2:12                               % iterative specification of
  p(i)=p(i-1)*(1-sum(x)/(1+(i-1)*x(2))); % probabilities
  s=s+p(i);
end
p(13)=1-s;                               % probability of right-censoring
x1=zeros(1,k); x2=zeros(1,k);
for i=1:k
  data=multsim(n,p);                     % simulates new data set;
  y=fminsearch('begeo',x0);              % starts from x0
  x1(i)=y(1);x2(i)=y(2);x3(i)=begeo(y);  % in order, this stores, mu,
end                                      % theta, and - the log-
                                         % likelihood at its maximum


figure; hist(x1); xlabel ('mu')          % we now plot the three
figure; hist(x2); xlabel ('theta')       % histograms
figure; hist(x3); xlabel ('-log-likelihood')
```

Figure 6.8 MATLAB *program to simulate k replicate data sets from a fitted beta-geometric distribution. The model is then fitted to each replicate data set, resulting in the approximations to the sampling distributions of the estimators of the model parameters shown in Figures 6.9 and 6.10.*

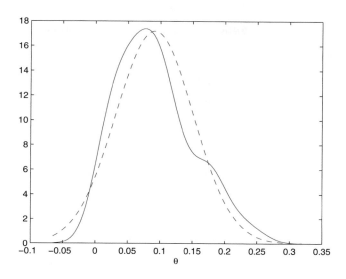

Figure 6.9 *Resulting from the program of Figure 6.8, the kernel density estimate of the histogram for $\widehat{\theta}$ (solid curve) and the asymptotic normal approximation of Equation (4.2) (dashed curve). A normal kernel is used.*

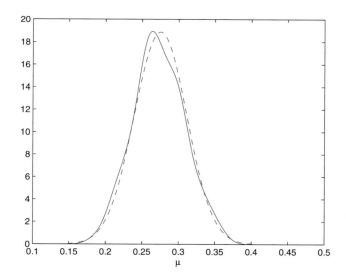

Figure 6.10 *Resulting from the program of Figure 6.8, the kernel density estimate of the histogram for $\widehat{\mu}$ (solid curve) and the asymptotic normal approximation of Equation (4.2) (dashed curve). A normal kernel is used.*

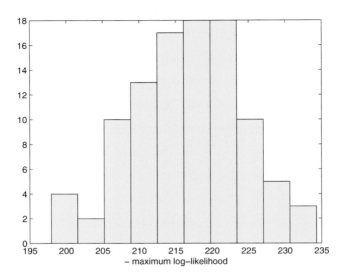

Figure 6.11 *Resulting from the program of Figure 6.8, the histogram for minus the maximum log-likelihood values.*

bootstrap sample will also be of size n, and consist of, and only of, elements of the original sample. In any bootstrap sample, at any point in that sample, the element x_i appears with probability $1/n$. Clearly bootstrap samples will probably contain replicates of the elements of the original sample.

Consider a random variable X, with cdf $F(x)$, which is unknown. In the bootstrap we approximate the distribution of X by the discrete distribution: $p_i = pr(X = x_i) = \frac{1}{n}$, $1 \leq i \leq n$. As $n \to \infty$, this distribution will tend to $F(x)$, so that in this sense the bootstrap procedure has an asymptotic justification.

The bootstrap approximation to the cdf is the empirical distribution function already encountered in Exercise 5.17:

$$\widehat{F}(x) = \frac{1}{n} \left\{ \text{number of } x_i \leq x \right\}.$$

The difference between the parametric and non-parametric bootstraps is readily appreciated by the difference between the MATLAB programs of Figures 6.8 and 6.12.

Example 6.6 continued

We now consider how to estimate the sampling distribution of the parameter estimators of the beta-geometric model fitted to the smokers data of Table 1.1, using the bootstrap. The MATLAB program is easily obtained by a simple modification of the program of Figure 6.8. It is given in Figure 6.12, and an

```
% Program to fit the beta-geometric model to fecundability data,
% and then to fit, k times, data obtained from the bootstrap,
% and produce histograms of -the maximum log-likelihood values,
% and of the parameter estimates.
%------------------------------------------------------------------
% k is the number of simulations
% data(i) is the number that wait i cycles;
% data(13) is the number that wait > 12 cycles.
% Calls the function BEGEO, which produces the
% beta-geometric negative log-likelihood.
% Calls the function MULTSIM, which simulates a
% multinomial distribution, and FMINSEARCH.
% 'data' are global, and are set in the driving program
%------------------------------------------------------------------
global data                         % the same data are used in
                                    % 'begeo'
data=[29 16 17 4 3 9 4 5 1 1 1 3 7]; % the 'smokers' data
k=100; n=sum(data);

                                    % now we set up the multinomial
p=data/n;                           % and simulate from it.

for i=1:k
   data=multsim(n,p);               % simulates new data set;
   y=fminsearch('begeo',[.5,.05]);  % starts from [.5,.05]
   x1(i)=y(1);x2(i)=y(2);
   x3(i)=begeo(y);                  % in order, this stores, mu,
end                                 % theta, and - the log-
                                    % likelihood at its maximum

figure; hist(x1); xlabel ('mu')     % we now plot the three
figure; hist(x2); xlabel ('theta')  % histograms
figure; hist(x3); xlabel ('-log-likelihood')
```

Figure 6.12 MATLAB *program to simulate k replicate data sets for the fecundability problem of Example 1.1, using the bootstrap. The beta-geometric model is then fitted to each replicate data set, resulting in approximations to the sampling distributions of the estimators of the model parameters.*

illustration of the resulting sampling distributions is provided by Figure 6.13. We can see from Table 6.2 the good agreement between the asymptotic normal approximation of Equation (4.2) and the bootstrap. The parametric bootstrap replicate data sets are in this case less variable than the bootstrap ones, and correspondingly the estimates of standard error are smaller. In this example, based on the "smoker" data set, which is described well by the beta-geometric model, there is good agreement between the bootstrap and the parametric bootstrap results. For further discussion see Exercise 6.26.   □

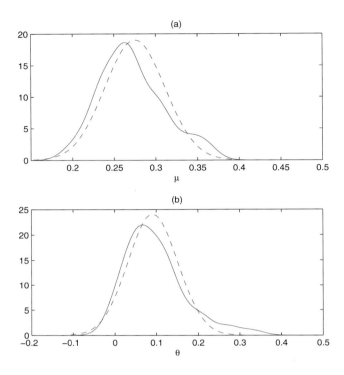

Figure 6.13 *The results from the program of Figure 6.12: the bootstrap estimates of the sampling distributions, with the number of sample replicates $k = 100$; (a) is for $\hat{\mu}$ and (b) is for $\hat{\theta}$. We show a kernel smoothing of the histogram (solid curves), as well as the asymptotic normal approximation from Equation (4.2) (dashed curves).*

Bootstrap samples provide approximations to the sampling distributions of parameters. Once we have such approximations we can obtain approximate confidence intervals by using the *percentile* method, in which we simply select parameter cut-off points, say θ_L, θ_U for a parameter θ, with a certain percentage, such as 5%, of simulated values lying outside the resulting interval.

The approximation of the cdf by the empirical cdf is one which in some cases may be improved. For example, rather than simulate from the original sample, we might smooth it first, so that in the *smoothed* bootstrap we simulate from:

$$\hat{f}(t) = \frac{1}{nh}\sum_{i=1}^{n} K\left(\frac{t - x_i}{h}\right),$$

where we have used a kernel approach to provide the smoothing (see Appendix

C). Thus each element of a smoothed bootstrap sample has the form

$$Y_i = x_{I_i} + \epsilon_i, \quad 1 \leq i \leq n, \tag{6.9}$$

where I_i is $U(1, 2, \ldots, n)$, i.e., $pr(I_i = k) = 1/n$, $\quad 1 \leq k \leq n$, and the $\{\epsilon_i\}$ are independent, identically distributed random variables with probability density function,

$$f(t) = \frac{1}{h} K\left(\frac{t}{h}\right).$$

An alternative to the approach of Equation (6.9), suggested by Kendall and Kendall (1980), is to set

$$Y_i = x_i + \epsilon_i, \quad 1 \leq i \leq n, \tag{6.10}$$

for $\{\epsilon_i\}$ with a suitable distribution. In this case we randomly distort members of the original sample, rather than sample from it. Young (1990) found in some examples that the best smoothed bootstrap outperformed the method of Equation (6.10), which was, in turn, better than the unsmoothed bootstrap. When smoothing is an improvement is considered by Silverman and Young (1987). The discussion of Hinkley (1988) suggests that general conclusions may prove elusive.

The basic bootstrap approximation to a sampling distribution is described by Thompson and Tapia (1990, p.156) as 'profoundly discontinuous,' and they give a dartboard illustration of how a bootstrap approach can provide a poor outcome: a sample of dart locations following a number of throws of a dart can easily exclude the bull's-eye. A bootstrap procedure would 'tell us the [bull's-eye] is a safe place.' A standard illustration of bootstrap *failure* is when a random sample of size n is taken from $U(0, \theta)$ and it is desired to estimate θ. The maximum-likelihood estimator, $\hat{\theta}$, of θ is the largest member of the sample (Exercise 2.17). In each bootstrap sample there is probability $1 - \left(1 - \frac{1}{n}\right)^n$ that the sample contains $\hat{\theta}$ as its largest element. For large n, $1 - \left(1 - \frac{1}{n}\right)^n \approx 1 - e^{-1} = 0.632$. The bootstrap estimate of the sampling distribution is discrete, rather than continuous, because of the atypical nature of the maximum-likelihood estimate in this case.

Bias in the bootstrap may be reduced by using balanced simulation. For example, we may want to ensure that the set of bootstrap samples contains a *representative* set of possibilities, and the set may also be reduced to exclude unnecessary replicates. Davison et al. (1986) suggested that bias may be reduced by using *balanced* simulation, in which s copies of the original sample, $(x_1, \ldots, x_n)$, are placed together to form a single string of length ns. This string is then randomly permuted, and the s bootstrap samples are then read off as successive blocks of length n in the permuted string.

The bootstrap was introduced by Efron (1979) and since then has developed into a technique of richness and complexity — see Efron and Tibshirani (1993) and Davison and Hinkley (1997). Its current limitations, especially for complex sampling situations, are described by Young (1994). Used with care and good sense it is an attractive and powerful tool for statistics.

6.8 Monte Carlo testing

In order to test how well a model fits a set of data, we may compute a statistic such as the Pearson X^2, or the deviance, which is described in Chapter 8, and refer it to a suitable standard. In Section 4.7 the standard we used was based on asymptotic considerations. However, as an alternative, consider the histogram of values of the log-likelihood in Figure 6.11. The empirical value of $218,772$, for the beta-geometric model fitted to the real data, fits well into the histogram — it is not extreme. This, therefore, demonstrates a key use of the bootstrap, in hypothesis testing. When we encounter sparse data then this may suggest that routine applications of asymptotic criteria may be inappropriate in that instance — see, for example, Table 1.5 — and now for testing goodness of fit we have an obvious and simple alternative.

Example 6.7: A model for flour beetle data

The following quotation is taken from Pack and Morgan (1990), in connection with fitting model (4.5) to the flour beetle data of Table 1.4:

> Although the model provides a satisfactory fit to the data, it does not model the endpoint mortalities very closely. The mortality counts in most of the cells are zero and this may have some effect on the test for goodness of fit. 100 sets of data were simulated using the maximum-likelihood estimates of the parameters, and the same group size. The range of deviances was 75.76 to 129.92, with a mean of 90.12 and a sample variance of 150.75. The fitted model (deviance $= 92.03$ on 95 degrees of freedom) is therefore clearly consistent with the simulated data.

□

The Monte Carlo test was introduced by Barnard (1963). In the last example only 100 simulations were used. Frequently not many simulations are needed before a clear picture emerges. Guidance is provided by Hope (1968) and Marriott (1979). Garren et al. (2000) provide an example of how the Monte Carlo test may lack power. In that case a better performance resulted from using the deviance, rather than the maximum log-likelihood.

Example 6.8: An illustration of a significant Monte Carlo test (Catchpole, pers.com.)

The histogram of Figure 6.14 demonstrates the poor fit of the model ($\{\lambda_i\}$, $\{\phi_{1i}\}, \phi_2, \phi_a$) to the data of Table 1.5, when the $\{\lambda_i\}$ and $\{\phi_{1i}\}$ were regressed on a covariate denoting winter conditions. In this model the reporting probabilities, $\{\lambda_i\}$, are allowed to vary from year to year, and also the survival probability of birds in their second year of life, ϕ_2, is estimated separately from those of first-year birds and birds older than 2 years, which have annual survival probability ϕ_a.   □

The Monte Carlo approach extends naturally to model selection and comparison, including also the case of non-nested models — see Williams (1970). The situation is slightly more complex here in that we can simulate from more than one model, and also consider a likelihood ratio, as well as the likelihoods

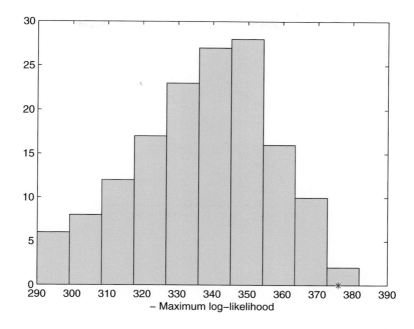

Figure 6.14 *An illustration of a Monte Carlo test indicating a poor fit of a model to data. The observed value of the minus maximum log-likelihood is denoted by a '*'.*

from each model. The procedure is illustrated by the following two examples.

Example 6.9: A Monte Carlo procedure to examine the sampling distribution of a likelihood-ratio test statistic

In Exercise 3.3 we were interested in testing whether $\gamma = 1$ in the mixture $\gamma p_1 + (1 - \gamma)p_2$. This is a constrained parameter non-regular problem — see Section 5.8. In Figure 6.15 we present three $Q - Q$ plots (see Gerson, 1975) of the likelihood-ratio test statistic of this hypothesis, resulting from 100 simulations matched to the original data. These suggest that a χ_2^2 distribution provides the appropriate cut-off for obtaining likelihood-ratio test confidence intervals for γ.   □

Example 6.10: A Monte-Carlo comparison between two non-nested alternatives

Consider again the litter mortality data of Exercise 3.3. The following analysis is taken from Pack (1986), who was comparing two alternative models for the data, a beta-binomial model (which we may think of as an infinite mixture of binomials), and a mixture of just two binomial distributions.

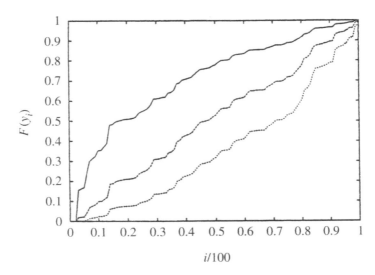

Figure 6.15 *(Brooks et al., 1997)   A $Q - Q$ plot of the likelihood-ratio test statistic of the hypothesis that a mixing parameter $\gamma = 1$ in a discrete mixture model, based on 100 matched simulations from the beta-binomial model (which arises when $\gamma = 1$) fitted to the litter mortality data of Exercise 3.3. Let y_i denote the ith ordered value of the likelihood-ratio test statistic. We plot $F(y_i)$ vs $i/100$, where $F(.)$ denotes the appropriate cumulative distribution functions: $- - -$ the order statistics from χ_1^2; $— — —$ the order statistics from χ_2^2; $———$ the order statistics from χ_3^2.*

The maximised log-likelihood value for the fitted mixture model is ranked 50th out of 100, when compared with the 99 maximised log-likelihood values from fitting the beta-binomial to simulated beta-binomial data. For the fitted beta-binomial model the corresponding rank is 45. However the difference in empirical maximum log-likelihood values (beta-binomial minus mixture values), of -4.33, is ranked only 4th largest out of 100, when compared with the 99 such differences when both the models are fitted to the beta-binomial simulated data. The simulated values range from -17.11 to 2.38, so the likelihood difference is somewhat smaller than we would expect if the data really were beta-binomial. If the roles of the two models are reversed then the maximised log-likelihoods are both consistent with a mixture model, being ranked 46th (beta-binomial) and 35th (mixture); in this case the empirical difference is ranked 39th. The evidence therefore favours the mixture model for these data.

□

6.9 Discussion

The value of mechanical simulation has been appreciated for many years (e.g., Student, 1908). However modern simulation techniques have irreversibly changed the face of statistics. This is a direct consequence of the speed of modern computers, which are now essential for most statistical modelling. Suppose that Rip van Winkle had been a statistician, had fallen asleep in, say, 1960 and then woken up to read this book. While he would undoubtedly have marveled at developments such as the EM algorithm and generalised linear models (discussed in Chapter 8), it is the simulation-based procedures which would have surprised him most. Statisticians have available to them now a far more flexible and richer set of tools than they could previously have dreamed of, and nowhere is this more evident than in the area of Bayesian statistics, which is the topic of the next chapter.

6.10 Exercises

6.1   Verify that the function of Figure 6.3(b) that simulates Poisson random variables does so through the mechanism of the Poisson Process, described in Appendix A.

6.2   Derive algorithms for simulating Weibull and logistic random variables by the inversion method. Provide MATLAB functions for these algorithms.

6.3   Show that the rejection algorithm for normal random variables, given in Example 6.1, is a consequence of enveloping the half-normal pdf with a multiple of an exponential $Ex(1)$ pdf. The procedure of using an exponential envelope in this way is outlined in Example 6.2 for simulating gamma random variables.

6.4   Verify the optimum value of k, the expansion factor for the enveloping function in Equation (6.4).

6.5*   Prove that the ratio-of-uniforms method specified in Example 6.3 works as stated. If the constant κ could be chosen, how would it be selected?

6.6   (Morgan, 1984, p.169.) Explain how you would use importance sampling to estimate the standard normal cdf, making use of the logistic pdf,

$$f(y) = \frac{\pi \exp(-\pi y/\sqrt{3})}{\sqrt{3}\{1 + \exp(-\pi y/\sqrt{3})\}^2}, \qquad -\infty < y < \infty,$$

which corresponds to a random variable with mean 0 and variance 1.

6.7*   Use Monte Carlo inference to estimate the parameter in an exponential $Ex(\lambda)$ distribution. Experiment with alternative kernels.

6.8   Use the bootstrap to simulate replicate data sets to that of Table 1.5. Use these replicates to estimate the standard errors of the parameters in the model $(\phi_1, \phi_a, \lambda)$, fitted to these data.

6.9   Check the goodness-of-fit of the model of Exercise 6.8 to the data of Table 1.5, using a Monte Carlo test.

6.10   Use a Monte Carlo test to compare the beta-binomial model with a mixture of two binomial distributions for the ant-lion data of Example 1.10.

6.11 For the data sets of Table 1.1, conduct a sensitivity analysis of the parametric bootstrap estimate of error for the parameters θ and μ in the beta-geometric fecundability model, by (a) changing both θ and μ by an amount equal to estimated asymptotic standard errors, and (b) obtaining values of θ and μ by sampling from the asymptotically normal joint pdf of the maximum likelihood estimators, $(\widehat{\theta}, \widehat{\mu})$, given in Equation (4.2).

6.12 Check that the 'naive' density estimate of Equation (C.1) in Appendix C is different from the traditional histogram.

6.13 For the geyser data below, (a) experiment in MATLAB with different histogram representations (try: help hist); (b) experiment with different kernels, in forming a kernel density estimate.

Durations (in minutes) of 107 eruptions of Old Faithful geyser (Weisberg, 1980).

4.37	3.87	4.00	4.03	3.50	4.08	2.25
4.70	1.73	4.93	1.73	4.62	3.43	4.25
1.68	3.92	3.68	3.10	4.03	1.77	4.08
1.75	3.20	1.85	4.62	1.97	4.50	3.92
4.35	2.33	3.83	1.88	4.60	1.80	4.73
1.77	4.57	1.85	3.52	4.00	3.70	3.72
4.25	3.58	3.80	3.77	3.75	2.50	4.50
4.10	3.70	3.80	3.43	4.00	2.27	4.40
4.05	4.25	3.33	2.00	4.33	2.93	4.58
1.90	3.58	3.73	3.73	1.82	4.63	3.50
4.00	3.67	1.67	4.60	1.67	4.00	1.80
4.42	1.90	4.63	2.93	3.50	1.97	4.28
1.83	4.13	1.83	4.65	4.20	3.93	4.33
1.83	4.53	2.03	4.18	4.43	4.07	4.13
3.95	4.10	2.72	4.58	1.90	4.50	1.95
4.83	4.12					

What is unsatisfactory about representing these data by means of histograms and kernel density estimates?

6.14 Suppose we were not supplied with the MATLAB routine *randn* for simulating normal random variables. A simple way to simulate approximately standard normal random variables is to make use of the Central Limit Theorem of probability theory, and set $Y = \sum_{i=1}^{12} U_i - 6$, where the $\{U_i\}$ are independent, identically distributed $U(0, 1)$ random variables. For example, in MATLAB one might set

$$Y = \text{sum (rand(12, 1)) - 6}.$$

Use this approximation for the remainder of this question.

(i) Decide on the parameters of a mixture of two normal pdfs.

(ii) Simulate a sample of size 50 from this mixture.

(iii) Select three possible smoothing parameters and for each of these, using a normal kernel, construct and plot the kernel density estimate of the underlying mixture pdf.

6.15 As an exercise in verification, simulate 200 samples from a zero-inflated Poisson distribution, using the function *zipsim* of Exercise 4.9. Fit the model to each of the simulated data sets, and construct histograms of the $\widehat{\lambda}$ estimates and of the $\widehat{w}$ estimates. Smooth the histograms using kernel probability density estimates. Use the kernel probability density estimates in order to construct 95% confidence intervals for λ and for w.

6.16 As an exercise in verification, simulate appropriate multinomial data using the simulation program, *ecosim*, below, and then obtain the maximum likelihood estimates of the parameters using the function *ecolik*.

(a)
```
% Program to simulate multinomial data corresponding to
% a simple bird survival model, for 4 years of recovery.
% There are three probabilites : the annual survival probabilities,
% 'phi_1' and 'phi_2', and the reporting probability, 'lam'
% 'n' denotes the number of birds in the cohort.
% The 'data' vector is global, as it may be used by the function
% ECOLIK, which forms the corresponding negative log-likelihood,
% to be minimised by fminsearch, for example.
% Calls MULTSIM
%_____
global data
n=1000;
phi_1=0.2;phi_2=0.5; lam=0.6;      % sets the model probabilities.
p(1)=lam*(1-phi_1);                % The vector 'p' contains the
p(2)=lam*phi_1*(1-phi_2);          % cell probabilities, apart from
p(3)=p(2)*phi_2;                   % the 'left-over' cell.
p(4)=p(3)*phi_2;
data=multsim(n,p);                 % can cope with probabilities
                                   % which do not sum to unity
```

(b)
```
function y=ecolik(x)
%
%ECOLIK constructs the negative log-likelihood corresponding to
%  a simple model for the annual survival of wild animals.
%  The data are simulated elsewhere, and introduced via the global
%  vector 'data'. The components of the vector 'x' are the model
%  probabilities.
%_____
global data
loglik=sum(data(1:4))*log(x(1))+(data(2)+data(3)+data(4))*...
    log(x(2)*(1-x(3)))+data(1)*log(1-x(2))+(data(3)+2*data(4))*...
    log(x(3))+data(5)*log(1-x(1)*(1-x(2)*x(3)^3));
y=-loglik;
```

6.17   For the geyser data of Exercise 6.13, construct 10 bootstrap samples, and provide a kernel probability density estimate for each. Comment on the poor results.

6.18   Simulate data from an exponential distribution, *Ex(2)*, and construct a standard kernel density estimate from the data. Comment on the poor result.

6.19   Show that the probability of rejection in the general rejection method of Example 6.2 is $1 - k^{-1}$.

6.20   In the general rejection method of Example 6.2, why does it suffice only to specify the pdf up to a multiplicative constant term?

6.21   Suppose we want to simulate from the bivariate normal distribution $N_2(\boldsymbol{\mu}, \boldsymbol{\Sigma})$, where $\boldsymbol{\mu} = (1, 2)'$ and $\boldsymbol{\Sigma} = \begin{pmatrix} 1 & 0.9 \\ 0.9 & 1 \end{pmatrix}$.

The Cholesky approach is to obtain a matrix $\mathbf{A}$ such that $\mathbf{A}'\mathbf{A} = \boldsymbol{\Sigma}$, and then set $\mathbf{X} = \boldsymbol{\mu} + \mathbf{A}'\mathbf{Y}$, where $\mathbf{Y} \sim N_2(\mathbf{0}, \mathbf{I})$. Find $\mathbf{A}$ in this case, and hence simulate from $N_2(\boldsymbol{\mu}, \boldsymbol{\Sigma})$.

6.22   Suggest verification procedures to check that the simulation functions of this book work correctly.

6.23   Suggest a way to simulate data from a beta-binomial distribution.

6.24   Experiment with alternative smoothing for the Monte Carlo inference program in Figure 6.6. Also try using an alternative kernel.

6.25   Modify the program of Figure 6.6 so that it becomes a function producing the Monte Carlo estimate of the log-likelihood at a single value of θ. Use *fminbnd* to try to obtain a Monte Carlo maximum-likelihood estimate of θ. Compare the performance of *fminbnd* with simulated annealing.

6.26   Compare bootstrap and parametric bootstrap results for the beta-geometric model fitted to a data set other than the 'smokers' data set from Table 1.1.

6.27   The rejection method of Example 6.2 simulates $\Gamma(r, 1)$ random variables. Explain how to simulate $\Gamma(r, \lambda)$ random variables.

6.28   Outline how the rejection method for simulating random variables works; use as an illustration an enveloping pdf which is exponential of mean unity and a target distribution (the one we wish to simulate from) which is half-normal, with pdf

$$f(x) = \sqrt{\frac{2}{\pi}}\, e^{-\frac{1}{2}x^2}, \quad \text{for } x \geq 0.$$

(i) Show that the optimal expansion factor for the exponential envelope is $k = \sqrt{\frac{2e}{\pi}}$.

(ii) Show that the method is equivalent to the following rule:

   Generate X_1 and X_2 as *iid* exponential random variables of mean unity. Reject the pair if

$$(X_1 - 1)^2 > 2X_2.$$

Otherwise accept X_1.

6.29 A $\Gamma(n,1)$ random variable has the pdf

$$f(x) = \frac{x^{n-1}e^{-x}}{\Gamma(n)} \quad \text{for} \quad x \geq 0, \quad n > 1.$$

Describe how to simulate such random variables, using the rejection method with envelope

$$g(x) = \frac{ke^{-x/n}}{n}, \quad \text{for} \quad x \geq 0$$

and suitable $k > 1$.

6.30* Edit and run the bootstrap program of Figure 6.12, so that the smoothed bootstrap is used.

6.31 One way to choose the smoothing parameter, h, in kernel density estimation is to minimise the *AMISE* criterion, given by :

$$AMISE = \left(\frac{R(K)}{nh}\right) + \left(\frac{1}{4}\right)\eta^4 h^4 R(f'').$$

Here $f(x)$ denotes the pdf being estimated, and K denotes the kernel being used (itself a pdf with mean zero and variance η^2) and $R(g)$ is the smoothness measure :

$$R(g) = \int g^2(x)dx.$$

Derive the value of h which minimises the *AMISE* , as a function of $R(f'')$.

Base the procedure on the case of a normal form for $f(x)$, with variance σ^2, for which $R(f'') = 3/(8\sqrt{\pi}\sigma^5)$ and assume also a $N(0,1)$ form for K, so that $\eta = 1$. Show that in this case the *AMISE* is minimised by setting $h = \left(\frac{4}{3}\right)^{1/5}\sigma n^{-1/5}$.

6.32 The random variable X has an extreme-value distribution, with cumulative distribution function:

$$P(X \leq x) = 1 - \left(\frac{k}{x}\right)^a, \quad \text{with parameters } a \text{ and } k, \text{ for } a > 0, \ x \geq k > 0.$$

Explain how to simulate X using the inversion method.

6.33 The random variable X has a logistic distribution, with pdf given by

$$f(x) = \frac{e^{-x}}{(1+e^{-x})^2} \quad \text{for} - \infty \leq x \leq \infty.$$

(i) Use the inversion method to construct an algorithm to simulate X.

(ii) If U_1, U_2 are independent $U(0,1)$ random variables, let $X_+ = -\log_e U_1$ if $U_2 < (1+U_1)^{-2}$. Show that this results from a rejection method for simulating X, based on an exponential envelope. Explain how you would modify X_+ to simulate X.

6.34 Experiment with the MATLAB program of Figure 6.6. For example, examine the effect of increasing s, and consider whether deterministic and/or stochastic optimisation procedures might deliver an optimum, after the script file is re-written as a function file.

CHAPTER 7

Bayesian Methods and MCMC

7.1 Basic Bayes

We can suppose that the 18th century Presbyterian minister, Thomas Bayes, died in complete ignorance of the possibility that his name would be attached to much statistical research in the late 20th century. Similarly, undergraduates who learn Bayes' Theorem, that for events A and B,

$$Pr(A|B) \propto Pr(B|A)Pr(A), \tag{7.1}$$

do not suspect its importance for statistical inference. The Bayesian approach to statistical modelling uses probability as a way of quantifying the *beliefs* of an observer about the parameters of a model, given the data that have been observed. Thus this approach is quite different from the classical approach which we have focussed on so far in this book, with its emphasis on the formation of a likelihood. In the classical approach it is assumed that the model has true fixed, but unknown, parameter values which are estimated by choosing those values which maximise the likelihood function associated with the data, or by other methods such as least squares. The Bayesian approach is to choose a *prior* distribution, which reflects the observer's beliefs about what values the model parameters might take, and then update these beliefs on the basis of the data observed, resulting in the *posterior* distribution. In Bayesian inference, the rule of (7.1) tells us how the prior beliefs about a parameter $\boldsymbol{\theta}$ are modified by data to produce the posterior set of beliefs:

$$\pi(\boldsymbol{\theta}|\mathbf{x}) \propto f(\mathbf{x}|\boldsymbol{\theta})\pi(\boldsymbol{\theta}). \tag{7.2}$$

Here $\pi(\boldsymbol{\theta})$ is the prior distribution, $\pi(\boldsymbol{\theta}|\mathbf{x})$ is the posterior distribution, and $f(\mathbf{x}|\boldsymbol{\theta})$ is the likelihood. Note that when we have maximised likelihoods then we have written the likelihood as a function of the model parameters, given the data, as in Chapter 2, for example. Thus for Bayesian inference the likelihood is still vitally important, but it is used in a quite different way than in classical inference. Prior distributions may be constructed after discussions with experts who have detailed knowledge of the relevant area of investigation (see for example Kadane and Wolfson, 1998, O'Hagan, 1998, and O'Hagan et al., 2006) or may be taken as flat if all values of a parameter appear to be equally likely before any data are collected.

As we have seen, the classical approach to statistical inference is to maximise the likelihood $f(\mathbf{x}|\boldsymbol{\theta})$ as a function of $\boldsymbol{\theta}$. A statistician following the Bayesian paradigm of expression (7.2) will obtain the entire joint posterior distribution of $\boldsymbol{\theta}$ without any labour such as numerical optimisation. However we often want more than just a joint distribution. For example, in a non-Bayesian context, if we consider the beta-geometric model for fecundability data, discussed in particular in Chapters 2 and 3, we want to make statements about the mean fecundability, μ, of women in each of the two (smoking and non-smoking) groups. In general, for subsets of the parameter vector $\boldsymbol{\theta}$ of special interest it is necessary to obtain the appropriate marginal distributions, by means of the usual integration of the joint distribution. This is where the labour for Bayesian inference is to be found. The normalisation constant in expression (7.2) is the integral, $\int f(\mathbf{x}|\boldsymbol{\theta})\pi(\boldsymbol{\theta})d\boldsymbol{\theta}$. Thus in general terms, to perform Bayesian inference, we need to evaluate

$$I = \frac{\displaystyle\int u(\boldsymbol{\theta})f(\mathbf{x}|\boldsymbol{\theta})\pi(\boldsymbol{\theta})d\boldsymbol{\theta}}{\displaystyle\int f(x|\boldsymbol{\theta})\pi(\boldsymbol{\theta})d\boldsymbol{\theta}},$$

for suitable functions $u(\boldsymbol{\theta})$. For instance, if we wanted to summarise the marginal distribution of the first component of $\boldsymbol{\theta}$, θ_1, by its expectation then we would take $u(\boldsymbol{\theta}) = \theta_1$.

The first exercises in Bayesian inference were simple academic ones, and examples of this nature are given in the next section. Application of the approach to real problems needed extensive numerical integration — see for example Racine et al. (1986). The modern Bayesian breakthrough is the result of Markov chain Monte Carlo (MCMC) methodology, which we describe later in this chapter. It allows us to undertake complex Bayesian analysis without using classical numerical analysis to form integrals. As a result we can now naturally formulate Bayesian approaches for even very complicated situations.

7.2 Three academic examples

(i) Suppose $X \sim N(\theta, \sigma^2)$, where σ^2 is known, and that we observe the random sample, $\mathbf{x} = (x_1, \ldots, x_n)$. If we take as the prior distribution, $\pi(\theta) = N(\mu_0, \sigma_0^2)$, for suitable known values of μ_0 and σ_0 then after an exercise in 'completing the square' (see Exercise 7.1) the posterior distribution is also normal, and of the form $N(\mu_n, \sigma_n^2)$, where

$$\mu_n = \left(\frac{\dfrac{n\bar{x}}{\sigma^2} + \dfrac{\mu_0}{\sigma_0^2}}{\dfrac{n}{\sigma^2} + \dfrac{1}{\sigma_0^2}} \right),$$

$$\sigma_n^{-2} = n\sigma^{-2} + \sigma_0^{-2}. \tag{7.3}$$

We can see how Bayes' Theorem sensibly combines the prior and likelihood information, with the emphasis that is placed on the data increasing as n increases. The same is true in the next two examples.

(ii) Suppose a binomial experiment of n independent trials each with probability θ of success results in r successes. If we adopt a beta prior density for θ, with pdf given by:

$$\pi(\theta) = \frac{\theta^{a-1}(1-\theta)^{b-1}}{B(a,b)}, \quad 0 \leq \theta \leq 1,$$

where $B(a,b)$ is the beta function, then the posterior distribution is also beta, with the pdf:

$$\pi(\theta|r) = \frac{\theta^{a+r-1}(1-\theta)^{b+n-r-1}}{B(a+r, \; b+n-r)}.$$

See Exercise 7.2.

(iii)* The multivariate generalisation of example (i) above is when $\mathbf{x} \sim N_p(\boldsymbol{\mu}, \boldsymbol{\Sigma})$. We may take as the prior distribution the multivariate normal form:

$$\pi(\boldsymbol{\mu}, \boldsymbol{\Sigma}) \propto |\boldsymbol{\Sigma}|^{-(\frac{m+1}{2})} \exp\left[-\frac{1}{2}\{\mathrm{tr}(\boldsymbol{\Sigma}^{-1}\mathbf{G} + (\boldsymbol{\mu}-\boldsymbol{\phi})'\boldsymbol{\Sigma}^{-1}(\boldsymbol{\mu}-\boldsymbol{\phi})\}\right],$$

where $\boldsymbol{\phi}, \mathbf{G}$ and m (which needs to be taken greater than $2p-1$) are parameters which specify the prior. The posterior distribution is then also multivariate normal, with respective posterior mean and variance,

$$(\boldsymbol{\phi} + n\bar{\mathbf{x}})/(n+1),$$

$$\{\mathbf{A} + \mathbf{G} + n(\bar{\mathbf{x}} - \boldsymbol{\phi})(\bar{\mathbf{x}} - \boldsymbol{\phi})'/(n+1)\}/(n+m-2p-2),$$

where $\bar{\mathbf{x}}$ is the multivariate sample mean and

$$\mathbf{A} = \sum_{i=1}^{n} (\mathbf{x}_i - \bar{\mathbf{x}})(\mathbf{x}_i - \bar{\mathbf{x}})'.$$

$\square$

In all three of the above examples the distributional form of the posterior was the same as that of the prior, because the form selected for the prior was said to be *conjugate* to the distribution sampled to produce the likelihood. A conjugate prior distribution results in a posterior distribution which has the same distributional form as the prior distribution. When modern computational methods are used for Bayesian inference, the restriction of the prior to a conjugate form can still be important, as it can facilitate the construction of conditional distributions used in Gibbs sampling, to be described in the next section.

7.3 The Gibbs sampler

We shall now consider the Gibbs sampler, which is one of the modern computational Bayesian methods for obtaining samples from *marginal* distributions, by simulating random variables sequentially from particular *conditional* distributions.

First, let us note that we can write a joint pdf $f(x,y)$ as the product of a conditional distribution and a marginal distribution, for example:

$$f(x,y) = f(x|y)f(y).$$

Thus if we know the functional form for the joint pdf $f(x,y)$, then we can obtain the form for the conditional pdf, $f(x|y)$, simply by treating y as a constant in $f(x,y)$. The marginal density for y, $f(y)$, becomes absorbed in the normalisation constant for the conditional pdf, $f(x|y)$. This argument holds for quite general joint pdfs, and is clearly demonstrated in Example 7.1, below.

7.3.1 Illustrations

Example 7.1:   A trivariate distribution (Casella and George, 1992)

Consider the trivariate joint distribution given below apart from a normalisation constant:

$$f(x,y,n) \propto \binom{n}{x} y^{x+\alpha-1}(1-y)^{n-x+\beta-1}\frac{e^{-\lambda}\lambda^n}{n!}, \qquad (7.4)$$

$$0 \le x \le n,$$

$$0 \le y \le 1, \quad n \ge 1,$$

where λ, α and β are parameters. There are aspects of this distribution which are already familiar to us. For example, we can see by inspection that the three conditional distributions for x, y, and n, conditional on the other two variables, are:

$$\mathrm{pr}(n|x,y) = \frac{e^{-(1-y)\lambda}\{(1-y)\lambda\}^{n-x}}{(n-x)!}, \quad n \ge x,$$

so that conditionally, the random variable $N = W + x$, where $W \sim P_o((1-y)\lambda)$,

$$\mathrm{pr}(x|y,n) \sim \mathrm{Bin}(n,y),$$

$$f(y|x,n) \sim \mathrm{Be}(x+\alpha, \ n-x+\beta). \qquad (7.5)$$

Also $f(y|n) \sim \mathrm{Be}(\alpha,\beta)$, so that $\mathrm{pr}(x|n)$ is beta-binomial.   □

The Gibbs sampler is a simulation device for obtaining realisations of the trivariate random variable (X,Y,N) with the joint distribution of (7.4) by simulating from the known conditional forms of (7.5). Trivially, we can see that obtaining realisations from (X,Y,N) implies that we automatically have realisations from all of the three marginal distributions. For X, for instance, we simply focus on the values for X, and ignore those for Y and N. Although we do not in general have explicit expressions for the marginal distributions, we can readily approximate these, as well as summary statistics such as the mean and variance, using standard statistical methods. We can see therefore that this approach, which is of quite general application, provides estimates of desired features of marginal distributions without the need to integrate a joint distribution. Evidently it is a boon for Bayesian inference which readily produces a joint posterior distribution by means of Equation (7.2). The fundamental simplicity of methods such as Gibbs sampling is the explanation for the explosion in Bayesian methodology in the late 20th century. We shall now provide an illustration of the Gibbs sampler at work.

Example 7.1 continued

For simplicity we fix n in the expression of (7.4) and suppress the dependence on n. Thus we focus on the joint distribution given below apart from a normalisation constant:

$$f(x, y) \propto \binom{n}{x} y^{x+\alpha-1}(1-y)^{n-x+\beta-1},$$
$$0 \le x \le n,$$
$$0 \le y \le 1.$$

The conditional distributions we shall sample from are therefore

$$\mathrm{pr}(x|y) \sim \mathrm{Bin}(n, y),$$

$$f(y|x) \sim \mathrm{Be}(x+\alpha, \ n-x+\beta). \tag{7.6}$$

When we have just two random variables, knowledge of the joint distribution and the two conditional distributions means that we also know the forms for the marginal distributions: for instance, in this example, $\mathrm{pr}(x) = f(x, y)/f(y|x)$, resulting in the beta-binomial form for $\mathrm{pr}(x)$,

$$\mathrm{pr}(x) = \binom{n}{x} \frac{\Gamma(\alpha+\beta)\Gamma(x+\alpha)\Gamma(n-x+\beta)}{\Gamma(\alpha)\Gamma(\beta)\Gamma(\alpha+\beta+n)}, \tag{7.7}$$

$$0 \le x \le n,$$

while similarly we can show that $f(y) \sim \mathrm{Be}(\alpha, \beta)$ — see Exercise 7.3. However we shall use the example to illustrate Gibbs sampling.

In common with methods for the numerical maximisation of a likelihood surface, the Gibbs sampler is iterative. However the similarity ends there. Gibbs sampling for this example starts with an initial value $y^{(0)}$ and then obtains the sequence: $y^{(0)}, x^{(0)}, y^{(1)}, x^{(1)}, y^{(2)}, x^{(2)}, \ldots$ by simulating:

$$x^{(j)} \sim \mathrm{pr}(x|Y^{(j)} = y^{(j)}), \quad j \ge 0, \text{ followed by}$$
$$y^{(j+1)} \sim f(y|X^{(j)} = x^{(j)}), \quad j \ge 0,$$

and then simulating $x^{(j+1)}, y^{(j+2)}$, etc. In time the sequence settles into an equilibrium state, when the x-values are realisations of the random variable with the beta-binomial distribution of Equation (7.7) and the y-values are realisations of a $\mathrm{Be}(\alpha, \beta)$ random variable.

The histogram of Figure 7.1 illustrates the result of this approach, from using the MATLAB program in Figure 7.2. The marginal distribution for Y in this illustration is $\mathrm{Be}(1,2)$, with pdf proportional to $(1-y)$. The extreme

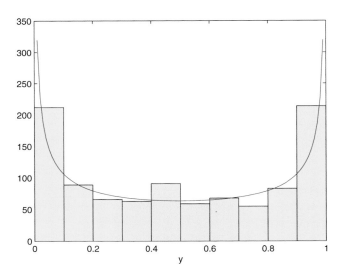

Figure 7.1 *Histogram of the values $y^{(j)}$ obtained by running the Gibbs sampling program of Figure 7.2, with $k = 200$ and $m = 100$. Also shown is a scaled version of the beta $Be(0.5, 0.5)$ pdf.*

simplicity of Gibbs sampling is readily apparent from Figure 7.2, which makes use of functions established in Chapter 6 for simulating from the conditional distributions of (7.6). A complication of the MATLAB program is that it samples the sequence after m cycles, and does this k times. Why this is sometimes done is discussed later in Section 7.3.4.

We can see that a more general Gibbs sampling program for more than two random variables would not be much more complicated than the one presented here, as long as the conditional distributions are easy to simulate from. However, we have here ignored a number of major issues to which we return later, such as how long it takes for equilibrium to be reached, and the fact that successive values, for instance in the sequence $\{y^{(j)}, j \geq 1\}$, are not independent. $\square$

7.3.2 General specification

In general, we have a set of random variables, $\{X_1, \ldots, X_k\}$, and our aim is to be able to simulate values from the multivariate distribution of these variables, when we know the forms of all of the univariate conditional distributions of the type: $f(x_i | x_1 \ldots, x_{i-1}, x_{i+1}, \ldots x_k)$.

We start with an initial value, $\{x_1^{(0)}, \ldots, x_k^{(0)}\}$, and we then simulate the

```
% Program to perform Gibbs sampling for Example 7.1, and fixed n.
% 'm': denotes the number of cycles before sampling
% 'k': denotes the number of repeats to be used
% 'n','alpha' and 'beta' are the known parameters of the
% bivariate distribution.
% Calls the functions BETASIM and BINSIM.
%------------------------------------------------------------------
m=50; k=100;
n=20;alpha=0.5;beta=0.5;
for j=1:k
  x=n*rand;                        % start for x
  for i=2:m
    y=betasim(alpha+x,n-x+beta,1);
    x=binsim(n,y,1);
  end
  y1(j)=y;                         % we record the values
  x1(j)=x;                         % after m cycles
end
figure; hist(y1); xlabel('y')
figure; hist(x1); xlabel('x')
```

Figure 7.2 *A* MATLAB *program to perform Gibbs sampling for the model of Example 7.1, for fixed n. This results in the histogram of Figure 7.1, corresponding to the beta pdf, Be(0.5, 0.5).*

values:

$$x_1^{(1)} \sim f(x_1|x_2^{(0)},\ldots,x_k^{(0)}),$$

$$x_2^{(1)} \sim f(x_2|x_1^{(1)},x_3^{(0)},\ldots,x_k^{(0)}),$$

$$x_3^{(1)} \sim f(x_3|x_1^{(1)},x_2^{(1)},x_4^{(0)},\ldots,x_k^{(0)})$$

$$\vdots$$

$$x_k^{(1)} \sim f(x_k|x_1^{(1)},\ldots,x_{k-1}^{(1)}).$$

We call this set of simulations a *cycle*.

Thus each variable is visited in the natural order, and a single cycle requires k simulations. After i cycles of the iterations we have the vector of simulated values, $(x_1^{(i)},\ldots,x_k^{(i)})$.

The theorem which provides the foundation for Gibbs sampling is that as i increases, under quite mild conditions which we discuss later, the joint distribution of the $\{x_j^{(i)}\}$ converges in distribution to the required joint distribution, that is as $i \to \infty$,

$$f(x_1^{(i)},\ldots,x_k^{(i)}) \longrightarrow f(x_1,\ldots,x_k)$$

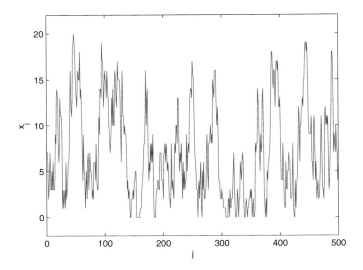

Figure 7.3 *A plot of the progress of the iterations resulting from running the program of Figure 7.2. Here we plot $x^{(j)}$ versus the iteration number, j. This is called a trace plot.*

in distribution (Geman and Geman, 1984).

Brooks (1998) makes the important observation that Gibbs sampling requires single realisations of random variables, many times over, with the specification of the details of the distribution changing from one simulation to the next. As already stated in Chapter 6, this is in contrast to standard classical simulations involving many realisations of a particular random variable from a fixed distribution, and means that for Gibbs sampling one would prefer simulation methods with low start-up costs. An efficient general-purpose random number generator would, in comparison, be expected to have high start-up costs.

7.3.3 Convergence

We see from the *trace plot* of Figure 7.3 the sequential progress of the simulations for Example 7.1. Rather than provide here a general proof that the Gibbs sampler converges to the distribution of interest, we shall now follow Casella and George (1992) and indicate why the procedure converges in the special case of a 2×2 table: here we have binary random variables X and Y

with the joint distribution:

$$
\left.
\begin{aligned}
pr(X = 0, Y = 0) &= p_1 \\
pr(X = 1, Y = 0) &= p_2 \\
pr(X = 0, Y = 1) &= p_3 \\
pr(X = 1, Y = 1) &= p_4
\end{aligned}
\right\}
\qquad
\begin{aligned}
\sum_{i=1}^{4} p_i &= 1, \\
p_i &\geq 0, \quad \text{for all } i.
\end{aligned}
$$

In this example, when we simulate from the conditional distributions using the Gibbs sampler, successive realisations result from a Markov chain. Markov chains are important stochastic processes. They have been widely studied and their properties are well known — see Cox and Miller (1965, p.76) and Appendix A. The transitions from X to Y are determined by the matrix,

$$
\mathbf{Q}_x =
\begin{bmatrix}
\dfrac{p_1}{p_1 + p_3} & , & \dfrac{p_3}{p_1 + p_3} \\[2ex]
\dfrac{p_2}{p_2 + p_4} & , & \dfrac{p_4}{p_2 + p_4}
\end{bmatrix},
$$

where the states are ordered with '0' coming before '1,' while those from Y to X are determined by the matrix

$$
\mathbf{Q}_y =
\begin{bmatrix}
\dfrac{p_1}{p_1 + p_2} & , & \dfrac{p_2}{p_1 + p_2} \\[2ex]
\dfrac{p_3}{p_3 + p_4} & , & \dfrac{p_4}{p_3 + p_4}
\end{bmatrix}.
$$

For example,

$$
\begin{aligned}
pr(Y = 1 | X = 0) &= pr(Y = 1, X = 0)/pr(X = 0) \\
&= p_3/(p_1 + p_3),
\end{aligned}
$$

and so on.

If we just focus on the transitions for X, then

$$
pr(X^{(i)} = j | X^{(i-1)} = k) =
$$

$$
\sum_h pr(X^{(i)} = j | Y^{(i)} = h) pr(Y^{(i)} = h | X^{(i-1)} = k).
$$

Thus the sequence of values taken by X also form a Markov Chain, with transition matrix,

$$
\mathbf{Q} = \mathbf{Q}_x \, \mathbf{Q}_y.
$$

Under standard conditions (Cox and Miller, 1965, p.101 and Appendix A), the distribution of the sequence of values taken by X converges to what is called an *equilibrium distribution* $\boldsymbol{\mu}$ which satisfies the equation,

$$
\boldsymbol{\mu}\mathbf{Q} = \boldsymbol{\mu}. \tag{7.8}
$$

We note finally that the marginal distribution of X has the form,

$$
\begin{aligned}
s_0 &= pr(X = 0) = (p_1 + p_3), \\
s_1 &= pr(X = 1) = (p_2 + p_4),
\end{aligned}
$$

and we can readily verify that $\boldsymbol{\mu} = \mathbf{s}$, as $\mathbf{s} = (s_0,\ s_1)$ satisfies the equation, $\mathbf{sQ} = \mathbf{s}$ — see Exercise 7.4.

Thus in this case, operation of the Gibbs sampler results in a sequence of values for X which converges to an equilibrium distribution which is the marginal distribution of X.

7.3.4 Implementation issues

We know from the discussion in Chapter 3 on iterative methods of optimisation that it is in general difficult to judge when an iterative method has converged. The same is true for the Gibbs sampler; the same issues arise as in the simulation of queueing systems in operational research with regard to judging when the system has settled down to equilibrium (see, e.g., Morgan, 1984, p.199). We saw in Section 5.6.3 that the same problem arises with stochastic EM. In particular, in Gibbs sampling one has to choose between using, on the one hand, a single long run of the sampler, and then making observations late in the sequence, by which time one would hope to have reached equilibrium, and on the other hand, forming a sample by taking the last value of each of a number of shorter runs, each with a different starting value. The advantage of the latter approach is that it is likely to provide results from a wider exploration of the sample space. However it may be wasteful in that for each run a 'burn-in' period of values needs to be discarded before the sequence is judged to be in equilibrium. This is an issue which has been resolved in the MATLAB program of Figure 7.2 by sampling the sequence after m cycles, and then repeating this procedure many times. For a single run, the discarding of a burn-in set of values would occur just once. A detailed review of the issues involved is provided by Brooks (1998). However this is an area where there are no completely general principles, and general reliable theoretical results for detecting convergence are elusive — see Brooks and Roberts (1998) for a review. We shall now consider a complex example, including a large number of parameters.

Example 7.2: A model for the survival of blue-winged teal, *Anas discors*

Brooks et al. (2000a) consider a model for recovery data of blue-winged teal, presented in Brownie et al. (1985). The data are shown in Exercise 7.5. In this model the probability of any bird surviving its first year of life is ϕ_1, and for birds of age at least 1 the probability of annual survival is taken as ϕ_a. The reporting probabilities of dead birds are assumed to be time-dependent, written as λ_i for birds found and reported dead in year i. All the parameters were given independent $U(0,1)$ prior distributions, the likelihood was formed in the usual way, as a product of multinomial distributions, and Gibbs sampling was used to obtain samples from the posterior distribution for the model parameters. Alternative prior distributions are considered later for this example, but independence is still assumed. A useful practical tool

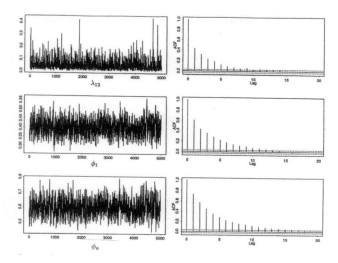

Figure 7.4 *Raw trace plots (left) and autocorrelations plots (right) for parameters* λ_{12}, ϕ_1 *and* ϕ_a, *in the model of Example 7.2. Autocorrelations that lie outside the dotted bands are significantly different from zero at the 5% level. From Brooks et al. (2000a); reproduced with permission from* Statistical Science.

was found to be the sample autocorrelation function (acf) of trace plots from runs of the sampler used. As we have stated already, successive terms in a sequence resulting from using the Gibbs sampler are not independent. The acf gives the product-moment correlations between successive terms in a series, between terms one step apart, and so on, and is routinely used in the analysis of time-series data (see Chatfield, 1984, p.23).

For models of this nature, a single run of the Gibbs sampler was used, with a burn-in period of 1000, and sampling from a sequence of length 10,000. The ratio-of-uniforms method was used to sample from non-standard posterior conditional distributions. As the procedures appeared to explore the full extent of the sample space with ease, the choice of starting value did not seem to be important, and the starting value was taken to be the mean of the prior distribution for each parameter. The use of more complex convergence diagnostics did not seem necessary for these models. We show in Figure 7.4 the trace plots for 5000 simulations, by way of illustration, as well as the autocorrelation plots, for parameters λ_{12}, ϕ_1, and ϕ_a. The autocorrelations suggest that the Gibbs sampling will provide a good coverage of the sample space, which justifies the use of a sequence of length 10,000, above. Formal checks for convergence of MCMC iterations are important. The method of Brooks and Gelman (1998) is employed by the computer package WinBUGS, described in Appendix B.   □

7.3.5 Goodness-of-fit

Suppose a model with parameters $\boldsymbol{\theta}$ has been fitted to data $\mathbf{x}$. A measure of the goodness of fit of the model may be written as $D(\mathbf{x}; \boldsymbol{\theta})$. For example, $D(\mathbf{x}; \boldsymbol{\theta})$ could be the Pearson X^2 value, or it could be the deviance. A classical approach is to evaluate $D(\mathbf{x}; \boldsymbol{\theta})$ at the maximum-likelihood estimate, $\hat{\boldsymbol{\theta}}$, to produce a single value of goodness of fit.

The significance of $D(\mathbf{x}; \hat{\boldsymbol{\theta}})$ might be gauged by reference to a suitable benchmark, which may, for example, be the chi-square distribution with the appropriate degrees of freedom. A Monte Carlo test may also be carried out, as discussed in Section 6.8, in which case a number of samples, $\{\mathbf{x}_i, 1 \le i \le m\}$, are simulated from the model when $\boldsymbol{\theta} = \hat{\boldsymbol{\theta}}$. The observed goodness-of-fit measure, $D(\mathbf{x}; \hat{\boldsymbol{\theta}})$, is then compared with the range of goodness-of-fit measures, $D(\mathbf{x}_i, \hat{\boldsymbol{\theta}}_i)$, $1 \le i \le m$, where $\hat{\boldsymbol{\theta}}_i$ is the maximum-likelihood estimate of $\boldsymbol{\theta}$ corresponding to the ith simulated sample, $\mathbf{x}_i$. A Bayesian approach to goodness of fit is also based on simulation, but does not involve any additional fits of a model to data. It proceeds as follows.

We can readily obtain a sample of parameter values $\boldsymbol{\theta}_i$, $1 \le i \le n$, from the posterior distribution of $\boldsymbol{\theta}$, through taking n appropriate values from the Gibbs sampler in equilibrium. For each $\boldsymbol{\theta}_i$ we calculate $D(\mathbf{x}; \boldsymbol{\theta}_i)$ and we also simulate a set of data, $\mathbf{x}_i$, of same size as $\mathbf{x}$, from the model when $\boldsymbol{\theta} = \boldsymbol{\theta}_i$, and form $D(\mathbf{x}_i; \boldsymbol{\theta}_i)$. A graphical impression of goodness of fit is then obtained from plotting $D(\mathbf{x}_i; \boldsymbol{\theta}_i)$ vs $D(\mathbf{x}; \boldsymbol{\theta}_i)$ for all i. We call such plots *discrepancy plots*. If the model describes the data $\mathbf{x}$ well, we would expect similar values for $D(\mathbf{x}_i; \boldsymbol{\theta}_i)$ and $D(\mathbf{x}; \boldsymbol{\theta}_i)$. We define a *Bayesian p-value* as the proportion of times that $D(\mathbf{x}_i; \boldsymbol{\theta}_i) > D(\mathbf{x}; \boldsymbol{\theta}_i)$. We note that there is total flexibility with regard to choice of $D(\mathbf{x}; \boldsymbol{\theta})$, which may be any appropriate measure of goodness-of-fit. We note also that changing the prior can change the Bayesian p-value. There are several alternative ways to define a Bayesian p-value. For discussion of a number of possibilities, see Bayarri and Berger (1998).

Example 7.2 continued

For the teal data, we show in Figure 7.5 the goodness-of-fit plots, for the model already considered, with parameters, $(\phi_1, \phi_a, \{\lambda_i\})$, and for the alternative model in which the reporting probability λ is constant but there is full age dependency of the survival probabilities. The first of the two models is seen to provide a better description of the data. The corresponding Bayesian p-values are (a) 0.50 and (b) 0.27.   □

The Bayesian p-value provides a simple summary of the information in discrepancy plots. It does not, however, take account of the spread of the points in those plots. We may note that discrepancy plots can also be obtained by simulating data from $\boldsymbol{\theta}_i$ sampled from the asymptotic normal distribution for the maximum-likelihood estimator, given by Equation (4.2).

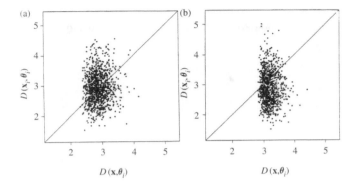

Figure 7.5 *Discrepancy plots for recovery data of dead teal. (a) Here the parameter set is* $(\phi_1, \phi_a, \{\lambda_i\})$. *(b) Here the parameter set is* $(\phi_1, \phi_2, \phi_3, \ldots, \phi_{12}, \lambda)$. *From Brooks et al. (2000a); reproduced with permission from* Statistical Science.

7.3.6 Prior sensitivity

The result of a Bayesian analysis is a blend of prior information and the information in the data. It is important to assess how sensitive conclusions are with regard to the choice of the prior distribution. While this can be done formally — see West, 1993, for example — a simple graphical procedure is to compare marginal histograms resulting from the sampler with the corresponding prior densities, and to observe the changes as the prior takes different forms. The ergodic theorem (Karlin and Taylor, 1975, Section 9.5) tells us that it is valid to consider histograms, means, variances, etc., of the output of the Gibbs sampler. The correlations between successive values resulting from the Gibbs sampler do not invalidate this. However, large autocorrelations would imply that the sampler would require much longer runs than if autocorrelations were low, in order to fully explore the sample space.

Example 7.2 concluded

The graphical comparisons of Figure 7.6 show, for the three parameters considered there, that posterior conclusions are largely unaffected by the choice of prior, with the exception of when a Be(4, 1) prior is adopted for the reporting probabilities. Reporting probabilities of dead birds in the wild are typically quite small. The Be(4, 1) prior is therefore unrealistic, and the results for the posterior distribution are inappropriate. For more discussion and explanation, see Brooks et al. (2000a). This example demonstrates that the choice of prior distribution may be important, and should always be examined through a sensitivity analysis of this kind.   □

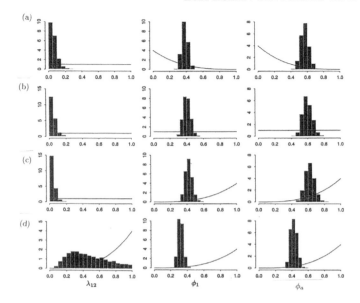

Figure 7.6 *A study of prior sensitivity for the model with parameters* $(\phi_1, \phi_a, \{\lambda_i\})$
for the teal data of Example 7.2. The histograms for the marginal distributions of the
parameters concerned are the result of Gibbs sampling. For the two ϕ*-parameters,*
the prior distributions, shown in the figure, are: (a) Be(1, 4), (b) uniform, $U(0,1)$,
and for both (c) and (d) Be(4, 1). A uniform $U(0,1)$ *prior is adopted for all of*
the reporting parameters, $\{\lambda_i\}$, *in (a), (b) and (c); a Be(4,1) prior is taken for the*
reporting parameters, $\{\lambda_i\}$, *in (d). In all cases, 10,000 observations are sampled from*
the posterior distributions, after the initial 1000 iterations are discarded. Reproduced
with permission from Statistical Science.

7.3.7 Efficient marginal distributions

Suppose, for simplicity, that we have a posterior sample (x_j, y_j), $1 \leq j \leq m$
from a Gibbs sampling procedure applied to a bivariate random variable,
(X, Y), and that we are interested in the marginal distribution of X. We may
produce a histogram of the sample values, as has been done in Figure 7.6,
or smooth this using kernel density estimation — see Appendix C. However
in order to conduct the Gibbs sampling, we have made use of the conditional
distribution, $f(x|y)$, which is therefore known. Hence a more efficient estimate
of the marginal distribution of X is given by

$$\hat{f}(x) = \frac{1}{m} \sum_{j=1}^{m} f(x|y_j). \tag{7.9}$$

This estimate is called the Rao-Blackwell estimate (Gelfand and Smith, 1990).
When the conditional distribution is only known up to a multiplicative

constant then the Gibbs sampler can still be used, for example by using the ratio-of-uniforms approach described in Example 6.3 for simulating from the conditional distribution. In this case, however, we cannot simply use the Rao-Blackwell estimate of Equation (7.9).

7.4 The Metropolis-Hastings algorithm

7.4.1 How it works

We now consider a more general way of simulating from a desired distribution, which makes use of the idea of rejection, which has already been encountered in Section 6.2.3, in the context of simulating random variables. We have a random variable $\mathbf{X}$, with distribution $\pi(\mathbf{x})$, and we construct a series of random variables $\mathbf{X}^{(i)}$, which take values $\mathbf{x}^{(i)}$ in such a way that as $i \to \infty$, the distribution of $\mathbf{X}^{(i)}$ tends to $\pi(\mathbf{x})$. We proceed as follows:

We have a transition probability function $q(\mathbf{a}, \mathbf{b})$ which provides possible changes in $\mathbf{X}^{(i)}$, for any value of i. Thus if $\mathbf{X}^{(i)} = \mathbf{b}$, then with probability $q(\mathbf{b}, \mathbf{a})$, the value $\mathbf{a}$ is a contender for $\mathbf{X}^{(i+1)}$ and indeed with subsequent probability $\alpha(\mathbf{b}, \mathbf{a})$, for some specified *acceptance* function $\alpha(\mathbf{b}, \mathbf{a})$, we accept $\mathbf{a}$, so that $\mathbf{X}^{(i+1)} = \mathbf{a} = \mathbf{x}^{(i+1)}$. If, however, we reject $\mathbf{a}$, which occurs with probability $1 - \alpha(\mathbf{b}, \mathbf{a})$, then $\mathbf{X}^{(i+1)} = \mathbf{b} = \mathbf{x}^{(i+1)}$, that is the value of $\mathbf{X}$ remains unchanged. Thus

$$pr(\mathbf{X}^{(i+1)} = \mathbf{a} | \mathbf{X}^{(i)} = b) = q(\mathbf{b}, \mathbf{a})\alpha(\mathbf{b}, \mathbf{a}), \quad \text{for } \mathbf{a} \neq \mathbf{b}$$
$$\text{and} \quad pr(\mathbf{X}^{(i+1)} = \mathbf{b} | \mathbf{X}^{(i)} = \mathbf{b}) = 1 - \sum_{\mathbf{i} \neq \mathbf{b}} q(\mathbf{b}, \mathbf{i})\alpha(\mathbf{b}, \mathbf{i}).$$

We note that there are obvious similarities between this procedure and simulated annealing, encountered in Chapter 3. We shall comment on this link further below.

Remarkably, as long as the acceptance probability function α is suitably chosen, and this can be done in a variety of ways, then the resulting sequence $\mathbf{x}^{(i)}$ converges as $i \to \infty$ to a series of values from $\pi(\mathbf{x})$. This is dependent upon the transition probability function q satisfying certain standard conditions, which allow a surprising degree of flexibility in how q is chosen. For given forms for $\pi(\ )$ and $q(\ ,\ )$, the standard choice for α is:

$$\alpha(\mathbf{b}, \mathbf{a}) = \begin{cases} \min\left\{\dfrac{\pi(\mathbf{a})q(\mathbf{a}, \mathbf{b})}{\pi(\mathbf{b})q(\mathbf{b}, \mathbf{a})}, 1\right\}, & \text{if } \pi(\mathbf{b})q(\mathbf{b}, \mathbf{a}) > 0 \\ 1, & \text{if } \pi(\mathbf{b})q(\mathbf{b}, \mathbf{a}) = 0. \end{cases} \quad (7.10)$$

We know that if $\pi(\mathbf{x})$ is a posterior distribution from a Bayesian analysis, it is often very convenient to specify its form without evaluating the scaling factor which ensures that it integrates to unity — see expression (7.2). We therefore appreciate the immediate attraction of Equation (7.10) for Bayesian work, which is that as it only depends upon the *ratio* of $\pi(\mathbf{a})/\pi(\mathbf{b})$ then any unspecified scaling factor cancels, and therefore does not need to be evaluated.

*7.4.2 *Why it works*

In Markov chain theory we usually start with a transition matrix $\mathbf{P} = \{p_{ij}\}$ and when $\mathbf{P}$ is irreducible and aperiodic, we seek the unique equilibrium distribution $\boldsymbol{\pi}$, satisfying:

$$\boldsymbol{\pi}' = \boldsymbol{\pi}'\mathbf{P} \qquad (7.11)$$

— see Appendix A. In the work which follows we consider the converse problem: we start with a distribution $\boldsymbol{\pi}$, and we want to construct a transition matrix $\mathbf{P}$ to satisfy Equation (7.11). Once that is done, then by simulating from the Markov chain with transition matrix $\mathbf{P}$, when the Markov chain settles down to equilibrium then the simulated values are from $\boldsymbol{\pi}$, as desired. The key to constructing $\mathbf{P}$ lies in the idea of the detailed balance equations of stochastic reversibility (see, for example, Kelly, 1979, Section 1.2 and Morgan, 1976): a Markov chain with transition matrix $\mathbf{P}$ and equilibrium distribution $\boldsymbol{\pi}$ said to be *reversible* if we can find a vector $\boldsymbol{\gamma}$ such that

$$\gamma_i p_{ij} = \gamma_j p_{ji} \quad \text{for all } i \neq j. \qquad (7.12)$$

This provides a 'balancing' of movement between the ith and jth states in equilibrium. Now for a reversible Markov chain, with vector $\boldsymbol{\gamma}$ satisfying Equation (7.12), the jth element of $\boldsymbol{\gamma}'\mathbf{P}$ is given by

$$
\begin{aligned}
(\boldsymbol{\gamma}'\mathbf{P})_j &= \sum_i \gamma_i p_{ij}, \\
&= \sum_i \gamma_j p_{ji}, \qquad \text{because of Equation (7.12)}, \\
&= \gamma_j \sum_i p_{ji} \;=\; \gamma_j,
\end{aligned}
$$

since $\mathbf{P}$ is a *stochastic* matrix with unit row sums (see Appendix A). Thus the direct result of stochastic reversibility is that

$$\boldsymbol{\gamma}'\mathbf{P} = \boldsymbol{\gamma}',$$

and hence we know that $\boldsymbol{\gamma} = \boldsymbol{\pi}$, as the equilibrium distribution is unique (see Appendix A). Hence in order to construct $\mathbf{P}$, we just need to ensure that the resulting Markov chain is reversible.

Suppose now that $\mathbf{Q} = \{q_{ij}\}$ is a *symmetric* transition matrix and that we construct a realisation of a Markov chain in which the states selected by $\mathbf{Q}$ are accepted with probabilities $\{\alpha_{ij}\}$. By this we mean that if the ith state is followed by the jth state, with probability q_{ij}, then the jth state is accepted with probability α_{ij}. Also, if a candidate new state suggested by $\mathbf{Q}$ is not accepted then we suppose that the system remains in the current state. The resulting sequence of states, generated by the transition probabilities of $\mathbf{Q}$ and accepted according to the above rules using probabilities $\{\alpha_{ij}\}$, is also a Markov chain, with transition matrix,

$$
\begin{aligned}
p_{ij} &= q_{ij}\alpha_{ij}, \qquad \text{for } i \neq j, \\
p_{ii} &= 1 - \sum_{j \neq i} p_{ij}.
\end{aligned}
$$

Suppose we take

$$\alpha_{ij} = \left(\frac{\pi_j}{\pi_i + \pi_j} \right), \tag{7.13}$$

as suggested by Barker (1965), where the $\{\pi_i\}$ form the probability distribution of interest to us, then

$$
\begin{aligned}
\pi_i p_{ij} &= \frac{\pi_i q_{ij} \pi_j}{(\pi_i + \pi_j)}, \quad \text{for } i \neq j, \\
&= q_{ji} \alpha_{ji} \pi_j, \quad \text{as } \mathbf{Q} \text{ is symmetric, and } \alpha_{ji} = \pi_i/(\pi_i + \pi_j), \\
&= p_{ji} \pi_j,
\end{aligned}
$$

which, from Equation (7.12), means, that constructed in this way, $\mathbf{P}$ is the transition matrix of a reversible Markov chain. Thus, when simulated, this rejection procedure, based on $\mathbf{Q}$ and on the $\{\alpha_{ij}\}$ specified by Equation (7.13), will give rise to the sequence we desire.

The acceptance function suggested by Metropolis et al. (1953), also for symmetric $\mathbf{Q}$, has

$$\alpha_{ij} = \begin{cases} 1, & \text{if } \pi_j/\pi_i \geq 1, \\ \dfrac{\pi_j}{\pi_i}, & \text{if } \pi_j/\pi_i \leq 1, \end{cases} \tag{7.14}$$

which is also easily shown to result in a reversible chain – see Exercise 7.7. The likelihood ratio, π_j/π_i, enters very sensibly into the acceptance function. The acceptance probabilities of Equations (7.13) and (7.14) were generalised by Hastings (1970) and the resulting form for α has already been given in Equation (7.10). In this case we do not require $\mathbf{Q}$ to be symmetric. So far $\mathbf{Q}$ has been allowed to be quite general. We can see that the acceptance probability of Equation (7.14) is precisely how simulated annealing proceeds (see Section 3.4.2).

We have here dealt only with the discrete case. However the theory generalises readily to the continuous case — see for example Chib and Greenberg (1995), Tanner (1993, p.137), and Exercises 7.8 and 7.9.

What is remarkable in this way of simulating reversible chains with the desired target equilibrium distribution is that there are several choices of $\{\alpha_{ij}\}$ and that $\mathbf{Q}$ can be quite general, as long as it produces a transition matrix $\mathbf{P}$ which is irreducible and aperiodic (see Exercise 7.11). Peskun (1973) has shown that the form of $\{\alpha_{ij}\}$ specified in Equation (7.10) minimises the probability of rejection. We shall now consider various ways of choosing $\mathbf{Q}$.

7.4.3 *Choice of transition matrix

We start our discussion with a particular important form for $\mathbf{Q}$. We shall write the vector $\mathbf{z}_{(i)}$ for the vector $\mathbf{z}$ apart from its ith entry, i.e.,

$$\mathbf{z}_{(i)} = (z_1, z_2, \ldots z_{i-1}, z_{i+1}, \ldots).$$

Let the ith component of vector $\mathbf{y}$ be y_i, for some i.

Suppose we focus on x_i, the ith element of $\mathbf{x}$, and choose $\mathbf{Q}$ so that we update the value of x_i according to the rule:

$$q(\mathbf{x}, \mathbf{y}) = \begin{cases} \nu(y_i | \mathbf{x}_{(i)}) & \text{when } \mathbf{y}_{(i)} = \mathbf{x}_{(i)}, \\ 0, & \text{otherwise,} \end{cases} \qquad (7.15)$$

where ν specifies the distribution of the ith value of $\mathbf{x}$, conditional upon specifying the values of $\mathbf{x}$ apart from x_i, that is $\mathbf{x}_{(i)}$.

In this case, the key selection probability of Equation (7.10), which we shall now call β, becomes:

$$\beta = \frac{\pi(\mathbf{y}) q(\mathbf{y}, \mathbf{x})}{\pi(\mathbf{x}) q(\mathbf{x}, \mathbf{y})} = \frac{\pi(\mathbf{y})/\nu(y_i | \mathbf{x}_{(i)})}{\pi(\mathbf{x})/\nu(x_i | \mathbf{y}_{(i)})}$$

$$= \frac{\pi(\mathbf{y})/\nu(y_i | \mathbf{y}_{(i)})}{\pi(\mathbf{x})/\nu(x_i | \mathbf{x}_{(i)})}, \quad \text{from Equation (7.15)}.$$

But, for example, $\pi(\mathbf{y}) = \nu(y_i) | \mathbf{y}_{(i)}) \eta(\mathbf{y}_{(i)})$, by the definition of conditional probability, where $\eta(\cdot)$ specifies the distribution of $\mathbf{y}_i$. Hence,

$$\beta = \frac{\eta(\mathbf{y}_{(i)})}{\eta(\mathbf{x}_{(i)})} = 1, \quad \text{as } \mathbf{y}_{(i)} = \mathbf{x}_{(i)}.$$

Thus this procedure simply updates the ith component of the vector $\mathbf{x}$ according to the conditional distribution of Equation (7.15), always maintaining the other elements of the vector $\mathbf{x}$. By sequentially changing $\mathbf{Q}$, so that i changes in the above, a feature which is commonly done, we can see that the Metropolis-Hastings procedure produces Gibbs sampling. The same convergence issues apply to the Metropolis-Hastings procedure in general as to the Gibbs sampler.

General ways of choosing $\mathbf{Q}$ are discussed by Chib and Greenberg (1995). Three examples are:

(i) to mimic the acceptance-rejection approach of Section 6.2.3 for identically distributed random variables (Tierney, 1994).

(ii) set $q(\mathbf{x}, \mathbf{y}) = g(\mathbf{y})$, for some distribution $g(\cdot)$. Here potential transitions are made independently of $\mathbf{x}$, but $\mathbf{x}$ still influences the selection of candidate transitions through the acceptance function α.

(iii) set $q(\mathbf{x}, \mathbf{y}) = g(\mathbf{y} - \mathbf{x})$ for some distribution $g(\cdot)$. This approach produces a candidate observation

$$\mathbf{y} = \mathbf{x} + \mathbf{e},$$

where $\mathbf{e}$ has the distribution $g(\cdot)$. This is termed random-walk Metropolis, and if $g(\cdot)$ is symmetric, which is usually the case, e.g., by taking $g(\cdot)$ as the pdf of the multivariate normal distribution, then Equation (7.10) reduces to the simple Metropolis algorithm of Equation (7.14).

As we might expect, the choice of $\mathbf{Q}$ has an important bearing on efficiency — see, e.g., Roberts et al. (1994), and Exercise 7.18. In practice it is necessary to tune Metropolis-Hastings methods, for instance with regard to appropriately selecting the variance of a proposal distribution, with an aim to

(a)

```
% Program for a Metropolis-Hastings analysis for Example 7.1, with fixed
% n. We sample after m cycles, and repeat k times.
% n, alpha and beta are the known parameters of the bivariate distribution
% and are global variables.
% Calls the function BIV.
%_____
global n alpha beta
m=50;k=50;
n=21; alpha=3;beta=2;
for j=1:k
  b=[n/3,1/3];
  for i=2:m
    a=[fix((n+1)*rand),rand];        % rand is equivalent to rand(1,1)
    alphacrit=min(1,biv(a)/biv(b));
    if alphacrit>rand;
      b=a;
      else
      end
    end
  x(j)=b(1);y(j)=b(2);
end
figure;hist(x); xlabel ('x')
figure;hist(y); xlabel ('y')
```

(b)

```
function z=biv(arg)
%
% BIV calculates the joint pdf, for the Metropolis selection probability
% of Equation (7.14). The three parameters, 'n','alpha' and 'beta', are
% global variables.
%_____
global n alpha beta
x=arg(1);y=arg(2);
z=(y^(x+alpha-1))*((1-y)^(n-x+beta-1));
z=z/(factorial(n-x)*factorial(x));        % MATLAB function for factorial
```

Figure 7.7 MATLAB *programs to carry out a Metropolis-Hastings analysis for Example 7.1.*

obtaining rejection probabilities which are greater than about 30%.

Example 7.1 concluded

We show in Figure 7.7 MATLAB programs for the Metropolis algorithm at work for this example, using the transition matrix of (b) above and a bivariate uniform distribution for $g(\cdot)$. Results are displayed in Figure 7.8.   □

An attraction of the Metropolis-Hastings method is how easily it is applied

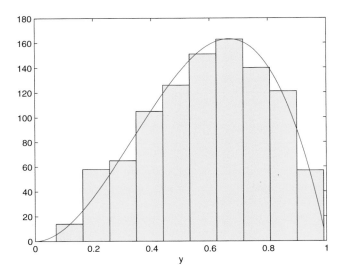

Figure 7.8 *Histogram of the y-values from running the program of Figure 7.7, with m = k = 200. This corresponds well to the known marginal Be(3,2) pdf for the random variable Y, and illustrated in the figure.*

to different problems. We can see this from the application of Example 7.3.

Example 7.3: Use of Metropolis-Hastings to fit the beta-geometric model

The program of Figure 7.9 provides a Metropolis-Hastings analysis for a beta-geometric model. We can see that the program is very similar to that of Figure 7.7(a). Note that we can make use of the begeo function of Figure 4.4; however we need to take account of the fact that the begeo function delivers a negative log-likelihood, and what we need is a positive likelihood. The result of running the program of Figure 7.9 is given in Table 7.1.

□

7.5 A hybrid approach

In the model of Example 7.1, the conditional distributions are of standard form. They are straightforward to simulate and Gibbs sampling proceeds easily, as we have seen. By contrast, the conditional distributions for the models of Example 7.2 are non-standard; Brooks et al. (2000a) show how they may usefully be written as proportional to beta distributions but with constants of proportionality which are functions of the model parameters. In such a case one might still use Gibbs sampling, and a method of random-variable simulation such as the ratio-of-uniforms method or the method of adaptive rejection

```
% Program for a Metropolis-Hastings analysis for the beta-geometric
% model of Section 2.2. We sample after m cycles, and repeat k times
% Calls the function BEGEO.
%_____
global data
data=[29 16 17 4 3 9 4 5 1 1 1 3 7];        % the smokers data
m=500;k=100; for j=1:k
  b=[rand,rand];                            % starting values
  for i=2:m
    a=[0.2+0.4*rand,rand];                  % proposed values
    alpha=min(1,exp(begeo(b))/exp(begeo(a)));
    if alpha>rand;
      b=a;
    else
    end
  end
mu(j)=b(1);theta(j)=b(2);
end
```

Figure 7.9 *Metropolis-Hastings for fitting the beta-geometric model to the smokers data of Table 1.1. Simulated values from the joint posterior distribution for μ, θ are given in the respective vectors.*

Table 7.1 *Estimates of (μ, θ) when the beta-geometric model of Section 2.2 is fitted to the data of Table 1.1, using the Metropolis-Hastings method and independent uniform prior distributions. Shown are estimates, together with estimated standard deviations and the estimated correlation between the parameters. These values have been obtained from the simulations resulting from the program of Figure 7.9. The values compare well with the maximum-likelihood estimates and the asymptotic estimates of standard error and correlation, obtained from expression (4.2), given previously in Table 4.1(i).*

Smokers data

Parameter	Estimate	Standard deviations/correlation	
		μ	θ
μ	0.286	0.038	
θ	0.122	0.716	0.066

sampling, when conditional distributions are log-concave — see Gilks and Wild (1992). Alternatively, this is when one might prefer to use the Metropolis-Hastings method, since we have seen that the acceptance probability given in Equation (7.10) does not require knowledge of proportionality constants in the form of the target distribution. In some cases certain of the conditional distributions in Gibbs sampling may be standard while others would not be. It is then possible to use a hybrid approach, updating the non-standard conditional distributions by using a univariate Metropolis-Hastings method — see Tierney (1994). This is known as Metropolis-within-Gibbs and is the approach adopted by the computer package BUGS, a popular user-friendly computer program for much MCMC work — see Spiegelhalter et al. (1996a; 1996b). The Windows version of BUGS (WinBUGS) is described in Appendix B. BUGS automatically decides how to deal with non-standard conditional distributions. Brooks (1998) and Brooks et al. (2000a) discuss the advantages and disadvantages of the different methods.

7.6 The data augmentation algorithm

This algorithm is the Bayes version of the EM algorithm, and provides a simple application of Gibbs sampling. Suppose we have parameters $\boldsymbol{\theta}$ with prior distribution given by $\pi(\boldsymbol{\theta})$. Suppose also that we observe incomplete data $\mathbf{y}$, and that the missing data are in the vector $\mathbf{z}$. We assume that we know the missing data conditional distribution, $f(\mathbf{z}|\mathbf{y}, \boldsymbol{\theta})$. We require the posterior distribution $\pi(\boldsymbol{\theta}|\mathbf{y})$ and we obtain it iteratively as follows:

At the kth iterate, we have the distribution $\pi^{(k)}(\boldsymbol{\theta}|\mathbf{y})$, and we simulate a random sample of size m from this estimate of the posterior distribution, to give the set of values: $(\boldsymbol{\theta}_1^{(k)}, \ldots, \boldsymbol{\theta}_m^{(k)})$. As we know $f(\mathbf{z}|\mathbf{y}, \boldsymbol{\theta})$, then we can simulate the sample: $(\mathbf{z}_1^{(k)}, \ldots, \mathbf{z}_m^{(k)})$, obtaining $\mathbf{z}_i^{(k)}$ from $f(\mathbf{z}|\mathbf{y}, \boldsymbol{\theta}_i^{(k)})$. This is analogous to the expectation step of the EM algorithm. What corresponds to the maximisation step is the formation of:

$$\pi^{(k+1)}(\boldsymbol{\theta}|\mathbf{y}) = \frac{1}{m} \sum_{i=1}^{m} \pi(\boldsymbol{\theta}|\mathbf{z}_i^{(k)}, \mathbf{y}),$$

which makes use of the form of the conditional distribution of $\boldsymbol{\theta}$, given the observed and imputed data, $\pi(\boldsymbol{\theta}|\mathbf{z}, \mathbf{y})$. If $m = 1$, then we can see that the system reduces to a Gibbs sampler, with alternate simulation from $f(\mathbf{z}|\mathbf{y}, \boldsymbol{\theta})$ and $\pi(\boldsymbol{\theta}|\mathbf{z}, \mathbf{y})$. In situations where it is difficult to simulate from $\pi^{(k+1)}(\boldsymbol{\theta}|\mathbf{y})$, importance sampling (see Section 6.3.3) can be used. An illustration involving censored regression data is given by Wei and Tanner (1990).

7.7 Model probabilities

For any problem involving data, we may need to choose between a number of alternative statistical models. We have seen in Sections 4.5 and 4.6 how this can be done using classical hypothesis tests, and the difficulties associated with this are discussed in Section 4.8. The Bayesian framework is attractively simple in contrast. Suppose we have a family of alternative models, $M_1, \ldots, M_k$, and that before data $\mathbf{x}$ are analysed, they have prior probabilities $P(M_i), 1 \leq i \leq k$ of being the models to use. We might, for example, set $P(M_i) = \frac{1}{k}$ for all values of i, if we have no a priori preference between the models.

Under model M_i, we suppose that we have parameters $\boldsymbol{\theta}_i$, with associated prior probability $\pi(\boldsymbol{\theta}_i)$ and likelihood $L_i(\boldsymbol{\theta}_i; \mathbf{x})$. Note that the standard normalisation constant from the basic Bayesian updating formula of expression (7.2) is given by

$$c_i(\mathbf{x}) = \int L_i(\boldsymbol{\theta}_i; \mathbf{x})\pi(\boldsymbol{\theta}_i)d\boldsymbol{\theta}_i.$$

If we write $P(M_i|\mathbf{x})$ for the posterior model probabilities, then, by Bayes' theorem,

$$P(M_i|\mathbf{x}) = \frac{c_i(\mathbf{x})P(M_i)}{\psi(\mathbf{x})},$$

where $\psi(\mathbf{x}) = \sum_{i=1}^{k} c_i(\mathbf{x})P(M_i)$.

In order to choose between any two models M_1 and M_2, we may consider the ratio

$$\frac{P(M_1|\mathbf{x})}{P(M_2|\mathbf{x})} = \frac{P(M_1)c_1(\mathbf{x})}{P(M_2)c_2(\mathbf{x})},$$

which conveniently no longer involves $\psi(\mathbf{x})$ – see Kass and Raftery (1995). If $P(M_1) = P(M_2)$ then this ratio is called the *Bayes factor*. If we are only choosing between the pair of competing models, M_1 and M_2, so that $P(M_2|\mathbf{x}) = 1 - P(M_1|\mathbf{x})$, then we can see that

$$P(M_i|\mathbf{x}) = \frac{P(M_i)c_i(\mathbf{x})}{P(M_1)c_1(\mathbf{x}) + P(M_2)c_2(\mathbf{x})}, \quad \text{for} \quad i = 1, 2.$$

More generally, when the number of models involved in the comparison is k, then

$$P(M_i|\mathbf{x}) = \frac{P(M_i)c_i(\mathbf{x})}{\sum_{j=1}^{k} P(M_j)c_j(\mathbf{x})} \quad \text{for} \quad 1 \le i \le k. \tag{7.16}$$

There is a very attractive simplicity to Equation (7.16), especially in comparison with the classical techniques of model selection. However, the normalisation constants $c_i(\mathbf{x})$ are usually analytically intractable, and evidently also depend on the prior probabilities $\pi(\boldsymbol{\theta}_i)$. In principle, we can use crude Monte Carlo to estimate $c_i(\mathbf{x})$, by means of the equation:

$$\hat{c}_i(\mathbf{x}) = \frac{1}{n}\sum_{j=1}^{n} L_i(\boldsymbol{\theta}_{i,j}|\mathbf{x}),$$

where $(\boldsymbol{\theta}_{i,1}, \ldots, \boldsymbol{\theta}_{i,n})$ is a random sample of size n simulated from $\pi(\boldsymbol{\theta}_i)$. See Section 6.3.2, and Gammerman (1997, p.195), for alternatives. When forming the likelihood, care must be taken not to encounter numerical problems.

Example 7.4: Geometric and beta geometric models

Consider the application of these two models to the data of Table 1.1. For the geometric model, the integral to form $c(\mathbf{x})$ is simply an appropriate beta function. However it is necessary to use Monte Carlo integration when the

Table 7.2 *Capture-recapture data for European dippers, banded in 1981–1986. Data taken from Lebreton et al. (1992).*

Year of release	Number released	Year of recapture (1981+)					
		1	2	3	4	5	6
1981	22	11	2	0	0	0	0
1982	60		24	1	0	0	0
1983	78			32	2	0	0
1984	80				45	1	2
1985	88					51	0
1986	98						52

model is beta geometric. Discussion of this example is continued in Exercise 7.32. See Price and Seaman (2006) for Bayesian analyses of fecundability data with digit preference. □

In practice, precise calculation of the $\{c_i(\mathbf{x})\}$ can sometimes require excessively large values of n. If this is the case, then a convenient alternative approach is provided by the reversible jump MCMC proposed by Green (1995), and which we discuss in Section 7.9.

The effect that $\pi(\boldsymbol{\theta})$ can have on posterior model probabilities can be appreciated from Example 7.5. We can see there that the calculation of model probabilities depends on two separate prior distributions.

Example 7.5: Survival of European dippers, *Cinclus cinclus*
The data of Table 7.2 describe the live recaptures of marked European dippers in eastern France. A Bayesian analysis of these data is provided by Brooks et al. (2000a). Models are described by a pair of letters, A/B; the first letter, A, indicates the assumption made in the model regarding the annual survival probability — either varying over time (T), constant (C), or two constants (C2) corresponding to different survival caused by a flood in 1983; the second letter, B, indicates the assumption made in the model regarding the probability of recapture. Here we shall only consider the case of a constant (C) probability of recapture and the case of time-varying probability of recapture (T).

When there is time-dependence of survival probability, we write ϕ_i to denote the probability of survival of the ith year; when there is time-dependence of recapture probability, we write p_i to denote the appropriate probability. The C2/C model has constant recapture probability, but two survival rates, to allow for the probability of survival being affected by a flood in 1983, viz.,

$$\phi = \begin{cases} \phi_f, & \text{for } i = 2,3 \\ \phi_n, & \text{for } i = 1,4,5,6. \end{cases}$$

Table 7.3 *The p-values and posterior model probabilities associated with different model and prior combinations for the dipper data. Prior 1 is a uniform prior on all model parameters. Prior 2 puts independent Be(1,9) distributions on the ϕ parameters and a uniform prior on the p parameters. Prior 3 is a Be(1,9) for all model parameters.*

	p-values			P($M|x$)		
Model	Prior 1	Prior 2	Prior 3	Prior 1	Prior 2	Prior 3
T/T	0.086	0.049	0.000	0.000	0.003	0.000
T/C	0.068	0.078	0.000	0.000	0.003	0.000
C/C	0.069	0.056	0.023	0.205	0.951	1.000
C2/C	0.125	0.153	0.050	0.795	0.004	0.000

The T/T model is the Cormack-Jolly-Seber model already described in Exercise 5.20. We know that this model is parameter-redundant. We are able to estimate the product $(p_7\phi_6)$ by maximum-likelihood, but not the components of that product.

Details of the MCMC analysis are given by Brooks et al. (2000a). The results are given in Tables 7.2 and 7.3. The posterior model probabilities are based on simulated samples of size $n = 100,000$ from the prior distribution, $\pi(\boldsymbol{\theta})$. The Bayesian p-values indicate that the C2/C model provides the best fit to the data, which agrees with the classical analysis of Lebreton et al. (1992). The posterior model probabilities for Prior 1 are clear-cut, giving greatest weight to the C2/C model, but some weight also to the C/C model. The priors for the survival probabilities in Prior 2 and Prior 3 have mean 0.1. In these cases the priors are in conflict with the data. This is less of a problem for the C/C model, as it has the smallest number of parameters. This explains why the C/C model is preferred under Prior 2 and Prior 3.

□

7.8 Model averaging

We can use posterior model probabilities to average over different models. Thus if we consider the survival probability of dippers from Example 7.3, and focus on annual survival in non-flood years, we obtain the point estimate $0.599 = (0.561 \times 0.205) + (0.609 \times 0.795)$, as the values 0.561 and 0.609 are the point estimates of annual survival in non-flood years under the C/C and C2/C models respectively. Model averaging conveniently avoids issues of model-selection bias, discussed in Section 4.8. It may also be accomplished using information criteria; see Buckland et al. (1997) and Anderson and Burnham (1998). Model averaging takes place over a previously selected set of models. It therefore requires a sensible selection of the family of models to be matched to the data. Additionally, as we have seen from Table 7.2, prior

Table 7.4 *Posterior means and standard deviations under independent uniform priors, for the dipper data.*

Model	T/T Mean	T/T SD	T/C Mean	T/C SD	C/C Mean	C/C SD	C2/C Mean	C2/C SD
ϕ	–	–	–	–	0.561	0.025	–	–
ϕ_1	0.723	0.133	0.622	0.107	–	–	–	–
ϕ_2	0.450	0.071	0.458	0.065	–	–	–	–
ϕ_3	0.482	0.061	0.481	0.058	–	–	–	–
ϕ_4	0.626	0.060	0.623	0.056	–	–	–	–
ϕ_5	0.603	0.058	0.608	0.054	–	–	–	–
ϕ_6	–	–	0.587	0.057	–	–	–	–
ϕ_f	–	–	–	–	–	–	0.472	0.043
ϕ_n	–	–	–	–	–	–	0.609	0.031
p	–	–	0.893	0.030	0.896	0.029	0.892	0.030
p_2	0.666	0.134	–	–	–	–	–	–
p_3	0.867	0.082	–	–	–	–	–	–
p_4	0.879	0.064	–	–	–	–	–	–
p_5	0.875	0.058	–	–	–	–	–	–
p_6	0.904	0.052	–	–	–	–	–	–

distribution can have an important effect, and so the rôle of prior distributions needs careful thought.

Suppose that we write $\pi(\boldsymbol{\theta}|M_i, \mathbf{x})$ for the posterior distribution of $\boldsymbol{\theta}$ under model M_i. We can then form the *averaged* posterior distribution, $\bar{\pi}(\boldsymbol{\theta}|\mathbf{x})$, from:

$$\bar{\pi}(\boldsymbol{\theta}|\mathbf{x}) = \sum_i P(M_i|\mathbf{x})\pi(\boldsymbol{\theta}|M_i, \mathbf{x})$$

(see Carlin and Louis, 1996, Equation (2.28)). It then follows that

$$\mathbb{E}[\boldsymbol{\theta}|\mathbf{x}] = \sum_i P(M_i|\mathbf{x})\mathbb{E}[\boldsymbol{\theta}|M_i, \mathbf{x}] \quad \text{and}$$

$$\mathbb{E}[\boldsymbol{\theta}\boldsymbol{\theta}'|\mathbf{x}] = \sum_i P(M_i|\mathbf{x})\mathbb{E}[\boldsymbol{\theta}\boldsymbol{\theta}'|M_i, \mathbf{x}],$$

from which we can estimate the model-averaged posterior means and standard deviations of the parameters. Discussion of this example is continued in Exercise 7.14. and its solution, where we see how in this application precision is reduced following model averaging. This is due to the incorporation of model uncertainty.

If individual models make different predictions, then it is important to know that, so that model-averaging should not be done routinely, and without considering the results of component models. An illustration of this is provided by King et al. (2008).

7.9 Reversible jump MCMC: RJMCMC

The Metropolis-Hastings method can be extended to produce a single Markov chain that moves between models as well as over the parameter spaces of individual models.

Within each iteration of the Markov chain, the algorithm involves two steps:

(i) For the current model, update the model parameters using the Metropolis-Hastings algorithm.

(ii) Update the model, conditional on the current parameter values using a RJMCMC algorithm.

The posterior model probabilities are then estimated as the proportion of time that the constructed Markov chain is in any given model.

The second, reversible jump, step of the algorithm involves two components, viz:

(i) Proposing to move to a different model with some given parameter values;

(ii) Accepting this proposed move with some probability.

The simplest case, which we now consider, is when there are just two nested alternative models. For each step of the MCMC algorithm we firstly update each parameter, conditional on the model, using the Metropolis-Hastings algorithm; then, we update the model, conditional on the current parameter values of the Markov chain, using the reversible jump algorithm. We only describe the updating of the model here and consider each step of the reversible jump procedure in turn:

Suppose that at iteration j, the Markov chain is in state $(\boldsymbol{\theta}, m)_j$. We propose to move to model $m' = m_2$, with parameter values $\boldsymbol{\theta}'$. If this involves additional parameters, $\mathbf{u}$, say, then we simulate them from some suitably selected proposal distribution q.

We accept the proposed model move to state $(\boldsymbol{\theta}', m')$ and set $(\boldsymbol{\theta}, m)_{j+1} = (\boldsymbol{\theta}', m')$ with probability $\min(1, A)$, where

$$A = \frac{\pi(\boldsymbol{\theta}', m'|\text{data})\mathbb{P}(m|m')}{\pi(\boldsymbol{\theta}, m|\text{data})\mathbb{P}(m'|m)q(\mathbf{u})} \left| \frac{\partial(\boldsymbol{\theta})}{\partial(\boldsymbol{\theta}')} \right|,$$

where $\mathbb{P}(m|m')$ denotes the probability of proposing to move to the model m, given that the current state of the chain is m', and vice versa. In the case of just two alternative models, this probability is always equal to unity. The final expression denotes an appropriate Jacobian. Otherwise, if the proposed move is rejected, then we set $(\boldsymbol{\theta}, m)_{j+1} = (\boldsymbol{\theta}, m)_j$.

For the reverse move, where we propose to move from state $(\boldsymbol{\theta}', m')$, to state $(\boldsymbol{\theta}, m)$, involving reducing the number of parameters, we set the additional parameters equal to zero. This move is then accepted with probability $\min(1, A^{-1})$, where A is given above, and $\mathbf{u}$ corresponds to the current values of the additional parameters.

A simple illustration of this is provided in the MATLAB program of Figure

7.10, where we compare geometric and beta-geometric models for the smokers data of Table 1.1.

In this case the Jacobian is given by

$$
\left| \frac{\partial(\mu, \theta)}{\partial(\mu, u)} \right| = \left| \begin{array}{cc} \frac{\partial \mu}{\partial \mu} & \frac{\partial \theta}{\partial \mu} \\ \frac{\partial \mu}{\partial u} & \frac{\partial \theta}{\partial u} \end{array} \right|
$$

$$
= \left| \begin{array}{cc} 1 & 0 \\ 0 & 1 \end{array} \right| = 1.
$$

The situation is more complex when models are non-nested and/or there are more than two alternative models. However, the approach is very similar to that outlined above. See, for example, King et al. (2009) for illustrations. For instance, if there are more than two models, then one moves through model space at random, according to the probabilities $\mathbb{P}$. Typically, moves between models would be to similar models. In the case of MCMC for a single model, efficiency depends upon the choice of the proposal distribution, and the same issue arises in RJMCMC, with regard to how one selects $\mathbb{P}$. In practice, in comparison with MCMC, RJMCMC methods can have high rejection rates, and it can be difficult to determine when the iterations have converged. The classical counterpart to RJMCMC is provided by trans-dimensional simulated annealing (Brooks et al., 2003).

7.10 Discussion

In this chapter we have met the basic methods of modern, computational Bayesian modelling and analysis. The examples considered have been straightforward; however we can now appreciate the fundamental appeal of the MCMC approach. If models can be accommodated by the WinBUGS package then the analysis is readily carried out. If individual computer programs need to be written then we have seen that the basic structure can be straightforward. Of course there are always associated complex issues, such as judging when convergence to an equilibrium distribution has been achieved. This particular problem is circumvented by the use of Markov chain coupling, resulting in *exact samplers* — see Propp and Wilson (1996). We have seen in Chapter 3 that classical maximum-likelihood methods usually need likelihoods to be maximised using numerical, iterative methods. Here too it can be difficult to judge when convergence has been achieved.

A more complex application is given by Brooks (1998), involving a non-linear regression model with a single change-point to describe data on the shape of corbelled tombs at sites throughout the Mediterranean (Buck et al., 1993). In this case the likelihood surface has discontinuities, but the Bayesian analysis is straightforward. Albert and Chib (1993) provide various applications, including logit modelling. Besag et al. (1995) provide applications to spatial statistics. The use of RJMCMC to overcome difficult issues in the fitting of mixture models to data is described by Richardson and Green (1997).

```
%
% RJMCMC program, to discriminate between geometric and beta-geometric models
% for data on times to conception; smokers data from Table 1.1. Note that
% we use likelihoods, and not log-likelihoods, so that we have to exponentiate
% functions that produce log-likelihoods.
% Calls GEO and BEGEO
%-------------------------------------------------------------------------
global data
data=[29 16 17 4 3 9 4 5 1 1 1 3 7];
n=5000; nbg=0;
ng=0; i=0; alpharj=0;
%
% We start in the geometric model
%
p(1)=0.5;k=0;
while i < n
    while alpharj < rand
        ng=ng+1; i=i+1; k=k+1;
                % We use MH to decide on a new value for p
        pnew=rand;
        alpha=min(exp(-geo(pnew))/exp(-geo(p(k))),1);
        if rand<alpha;
            p(k+1)=pnew;
        else
            p(k+1)=p(k);
        end
                % We now do the RJ step to compare models.
                % We take q(u) to be U(0,1).
        x(1)=p(k+1); x(2)=rand;
        alpharj=min(exp(-begeo(x))/exp(-geo(p(k+1))),1);
    end
    alpharj=0;
    while alpharj < rand
        nbg=nbg+1; i=i+1; k=k+1;
                % We use MH to decide on a new value for mu and theta
        xnew=rand(1,2);
        alpha=min(exp(-begeo(xnew))/exp(-begeo(x)),1);
        if rand<alpha;
            x=xnew; p(k+1)=x(1);
        else
            p(k+1)=x(1);
        end
                % We now do the RJ step to compare models
        x(2)=0;
        alpharj=min(exp(-geo(p(k+1)))/exp(-begeo(x)),1);
    end
    alpharj=0;
end
```

Figure 7.10 *A simple illustration of RJMCMC. Here we compare the geometric and beta-geometric models for the smokers data on times to conception, from Table 1.1. The variables* **ng** *and* **nbg** *record the times spent in the geometric and beta-geometric models, respectively.*

It is often interesting to observe a high degree of agreement between classical and Bayesian methods of model-fitting, which are based on such different paradigms. Each approach has its strengths and weaknesses, and modern statistical modelling is likely to contain elements of both the classical and Bayesian approaches.

7.11 Exercises

7.1 Verify the expression for the normal posterior distribution, given by Equations (7.3).

7.2 Verify the expression for the beta posterior distribution in Example (ii) of Section 7.2.

7.3 Verify in Example 7.1 that for fixed n, $f(y) \sim Be(\alpha, \beta)$.

7.4 Verify the expression given for the equilibrium distribution specified in Equation (7.8).

7.5 Shown below are the reported recoveries of dead blue-winged teal marked as young from 1961 to 1973. The data are taken from Brownie et al. (1985).

Year of ringing	Number ringed	\multicolumn{12}{c}{Year of recovery (1961+)}
		1	2	3	4	5	6	7	8	9	10	11	12
1961	910	6	2	1	1	0	2	1	0	0	0	0	0
1962	1157		11	5	6	1	1	1	1	0	0	0	1
1963	1394			19	4	4	4	0	0	1	1	0	0
1964	3554				65	25	8	4	2	4	4	1	0
1965	4849					65	17	2	1	6	2	3	1
1966	2555						52	9	8	3	4	2	1
1967	305							3	1	0	1	0	0

Consider what might be a set of appropriate models, $\{M_i\}$, to describe these data. What prior probabilities might you assign to these models? How might you construct prior distributions for the model parameters?

7.6 Use the MATLAB program of Figure 7.2 to construct a sequence of values from the Be(1,2) pdf. Use the fact that $f(y|x) \sim Be(x + 1, n - x + 2)$, to construct the Rao-Blackwell estimate of $f(y)$, for the case $n = 20$.

7.7 Verify that the Metropolis acceptance function of Equation (7.14) results in a reversible Markov chain. Verify that the Hastings acceptance function of Equation (7.10) also results in a reversible Markov chain.

7.8 (Chib and Greenberg, 1995.) We saw in Exercise 6.21, how to simulate from a particular bivariate normal pdf. Consider how the Metropolis-Hastings method can be extended to deal with continuous random variables, using this bivariate distribution as an illustration. Suggest a suitable transition probability function and construct a sample from the pdf.

7.9 (Tanner, 1993, p.138.) Consider again the genetics model of Example 5.7. Assume a uniform prior for the single model parameter. Suggest a suitable transition probability function, and simulate a random sample from the posterior distribution for the model parameter using the Metropolis-Hastings method as in Exercise 7.8.

7.10 Show that the Hastings acceptance probability of Equation (7.10) includes the Barker acceptance probability of Equation (7.13).

7.11 Verify that when the transition matrix $\mathbf{P}$ is constructed from the transition $\mathbf{Q}$ in the Metropolis-Hastings method, then $\mathbf{P}$ is irreducible if and only if $\mathbf{Q}$ is.

7.12 Experiment with running the MATLAB programs of Figure 7.7 for different values of m and k.

7.13 Consider how you would use data augmentation for the problem of Example 5.1, for the censored survival data problem in Figure 5.2(b), and for fitting mixtures of distributions, as in Example 5.9.

7.14 In Table 7.2, use the posterior model probabilities for Prior 1 to calculate the model-averaged variance of ϕ_n.

7.15 In the MATLAB program of Figure 7.7 to carry out a Metropolis-Hastings analysis for Example 7.1, discuss why you think potential transitions were made from setting $q(\mathbf{x}, \mathbf{y}) = g(\mathbf{y})$, rather than an alternative form for $q(\mathbf{x}, \mathbf{y})$.

7.16 When a prior distribution is completely flat, what is the relationship between the likelihood and the posterior distribution?

7.17 Why is the output from the Metropolis-Hastings method a Markov chain? How can the autocorrelation function for such output reveal appreciable correlations between values several steps apart?

7.18 (Link et al., 2002.) A Metropolis-Hastings method rejects candidate values in order to obtain $N(0, 1)$ random variables. Two sequences of candidate values are illustrated in Figure 7.11. The lighter line results from adding $N(0, 100)$ deviates to the current values; the darker line results from adding $N(0, 0.01)$ deviates to current values. The rejection probabilities are, respectively, 83% and 4%, while the first-order autocorrelations of accepted values are, respectively, 0.86 and 0.97. Discuss the results.

7.19 Shown below are the posterior means and standard deviations (SD) under independent $U(0, 1)$ priors, for survival probabilities of four models fitted to the teal data set of Exercise 7.5, from a sample of 10,000 observations from the posterior distribution, together with the Bayesian p-values and posterior model probabilities $P(M|\mathbf{x})$ associated with each model. Why do we refer to standard deviations, rather than standard errors? Compare and contrast the p-values and the values of $P(M|\mathbf{x})$. What is notable about the means and standard deviations for ϕ_{12}?

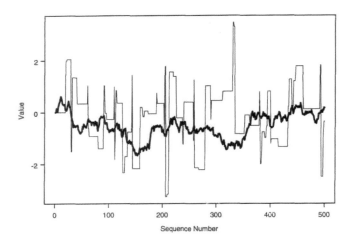

Figure 7.11 *Plots of two sets of candidate values for Metropolis-Hastings; reproduced with kind permission from Link et al. (2002).*

	Model							
	1		2		3		4	
Parameter	Mean	SD	Mean	SD	Mean	SD	Mean	SD
$\phi_{1,1}$			0.531	0.126	0.525	0.248		
$\phi_{1,2}$			0.588	0.089	0.531	0.114		
$\phi_{1,3}$			0.431	0.081	0.321	0.083		
$\phi_{1,4}$			0.428	0.047	0.484	0.070		
$\phi_{1,5}$			0.396	0.047	0.259	0.055		
$\phi_{1,6}$			0.352	0.052	0.264	0.077		
$\phi_{1,7}$			0.459	0.179	0.158	0.108		
ϕ_1	0.420	0.026					0.403	0.039
ϕ_2	0.604	0.040	0.602	0.038	0.445	0.071		
ϕ_3	0.694	0.047	0.690	0.049	0.538	0.087		
ϕ_4	0.640	0.062	0.636	0.059	0.459	0.094		
ϕ_5	0.690	0.073	0.682	0.070	0.578	0.100		
ϕ_6	0.624	0.093	0.616	0.091	0.545	0.108		
ϕ_7	0.410	0.133	0.397	0.130	0.372	0.134		
ϕ_8	0.494	0.207	0.474	0.206	0.468	0.202		
ϕ_9	0.729	0.217	0.722	0.219	0.749	0.205		
ϕ_{10}	0.688	0.232	0.695	0.230	0.691	0.233		
ϕ_{11}	0.348	0.240	0.352	0.240	0.355	0.240		
ϕ_{12}	0.493	0.285	0.511	0.285	0.504	0.288		
ϕ_A							0.598	0.053
p-value	0.27		0.28		0.54		0.50	
P($M|x$)	0.009		0.054		0.000		0.937	

7.20   The data of Exercise 3.3 describe the number of dead fetuses in litters of mice. This is one of six data sets considered by Brooks et al. (1997). Brooks (2001) has analysed these data from a Bayesian perspective and considered five alternative models for each data set, viz., a beta-binomial model, a mixture of two binomials, a mixture of a binomial and a beta-binomial, a mixture of three binomials and a mixture of two beta-binomials. Shown below are his posterior model probabilities. The coding used for the data sets is the same as that used in Brooks et al. (1997), and data set E1 is the data presented in Exercise 3.3. BBin denotes beta-binomial, etc.

Data	BBin	2-Bin mixture	Bin/BBin mixture	3-Bin mixture	2-BBin mixture
E1	0.028	**0.570**	0.188	0.204	0.004
E2	0.066	**0.424**	0.237	0.258	0.008
HS1	0.014	0.000	0.201	0.047	**0.725**
HS2	0.000	0.000	**0.936**	0.009	0.054
HS3	0.000	0.054	**0.802**	0.124	0.017
AVSS	**0.252**	**0.362**	0.151	0.206	0.019

Discuss these results. In any mixture, the *main* component is the one with the largest associated probability in the mixture. Shown below are the mean probabilities of response for the main components of each of the six data sets, both from the model with the highest posterior model probability, shown in bold, and averaged over all models (Brooks, 2001).

Data	Highest post. prob.		Model averaged	
	Mean	SD	Mean	SD
E1	0.066	0.007	0.067	0.031
E2	0.067	0.007	0.073	0.034
HS1	0.076	0.019	0.079	0.022
HS2	0.072	0.036	0.074	0.035
HS3	0.060	0.004	0.056	0.006
AVSS	0.033	0.012	0.055	0.020

Discuss these results.

7.21   Consider how you would provide a Bayesian formulation and a Bayesian analysis of the fecundability and polyspermy models of Chapter 2.

7.22* (Brooks et al., 2000b.) In Examples 4.6 and 5.5 we have discussed a model for bird survival in which the likelihood surface has a completely flat ridge. Consider whether a Bayesian analysis with independent $U(0, 1)$ priors for all of the model probabilities is likely to produce useful results.

7.23   Data, $\mathbf{x} = (x_1, x_2, x_3)$, arise with the multinomial probability,

$$f(\mathbf{x}|\theta, \eta) \propto \theta^{x_1} \eta^{x_2} (1 - \theta - \eta)^{x_3},$$

for $0 < \theta$, $\eta < 1$ and $\theta + \eta < 1$.

The joint prior distribution for the pair of parameters (θ, η) is said to be *Dirichlet*, with probability density function

$$\pi(\theta, \eta) \propto \theta^{\alpha_1} \eta^{\alpha_2} (1 - \theta - \eta)^{\alpha_3},$$

for given values of α_1, α_2 and α_3.

Show that the posterior distribution $\pi(\theta, \eta | \mathbf{x})$ is also Dirichlet. Find the conditional posterior distributions for θ and η, each given the other, and outline how these might be used in Gibbs sampling.

7.24* (Examination question, University of Nottingham.)

Consider a long document that consists of $n(> 2)$ pages. It is known that one secretary started typing the document but it was completed by another. However, exactly where in the document the second secretary took over from the first is unknown. It is proposed to model this by supposing that the first r pages were typed by the first secretary and the final $n - r$ pages by the second where r is an unknown integer between 1 and $n - 1$ inclusive.

The two secretaries make typing errors at different rates θ_1 and θ_2 per page. Let X represent the number of errors on the ith page and assume that X has a Poisson distribution, $i = 1, 2, \dots, n$.

Then for $1 \le i \le r$

$$pr(X_i = x_i) = \frac{\theta_1^{x_i} e^{-\theta_1}}{x_i!} \quad x_i = 0, 1, 2, \dots$$

and for $r + 1 \le i \le n$

$$pr(X_i = x_i) = \frac{\theta_2^{x_i} e^{-\theta_2}}{x_i!} \quad x_i = 0, 1, 2, \dots$$

It is proposed to use a Bayesian approach to make inferences about r. Prior information about θ_i is modelled by a gamma, $\Gamma(\alpha_i, \beta_i)$, distribution. Little is known about when the secretaries changed, so that the prior for r is taken to be uniform over the integers $1, 2, \dots, n - 1$.

(i) Show that the joint posterior distribution of θ_1, θ_2 and r is proportional to

$$\theta_1^{S_r + \alpha_1 - 1} \theta_2^{T_r + \alpha_2 - 1} e^{-\theta_1(\beta_1 + r)} e^{-\theta_2(\beta_2 + n - r)}, \quad \text{for } 1 \le r \le n - 1,$$

where

$$S_r = \sum_{i=1}^{r} x_i \quad \text{and} \quad T_r = \sum_{i=r+1}^{n} x_i.$$

(ii) Hence identify the conditional posterior distributions of each of the three parameters, conditional on the remaining two.

(iii) Outline how you would sample from the full conditional posterior probability function of r.

(iv) Explain how you would use Gibbs sampling in this situation.

(v) Describe how you would estimate the posterior marginal probability function of r, its mean and variance.

(vi) How would you adapt your algorithm if it were known that the second secretary was more error prone than the first?

7.25 (Examination question, University of Nottingham.)

A model for the spread of a measles epidemic initiated by one infective in a household of size three predicts that the total number of people infected in the household has the following distribution

Total number infected	1	2	3
Probability	$(1-\theta)^2$	$2\theta(1-\theta)^2$	$\theta^2(3-2\theta)$

where θ, the probability of adequate contact, is an unknown parameter. The following data give the frequency distribution of 334 such measles epidemics,

Total number infected	1	2	3
Frequency	34	25	275

Obtain a quadratic equation that is satisfied by the maximum-likelihood estimate $\hat{\theta}$, and deduce that $\hat{\theta} = 0.728$.

Carry out a Pearson goodness-of-fit test to examine if the model provides a good fit to these data.

A more refined model splits those epidemics in which three people are infected into two types: 3(a), in which the initial infective infected both the other individuals in the household and 3(b), in which the initial infective only infects one of the other individuals in the household, who then infects the third individual. The corresponding probability distribution is

Total number infected	1	2	3(a)	3(b)
Probability	$(1-\theta)^2$	$2\theta(1-\theta)^2$	$2\theta^2(1-\theta)$	θ^2

Suppose that n independent epidemics are observed and their final outcomes, $y_1, y_2, \ldots, y_n$ say, recorded, yielding the following frequency distribution.

Final outcome	1	2	3(a)	3(b)
Frequency	a	b	c	d

Derive the posterior distribution $f(\theta|y)$ when the prior distribution of θ is $Be(\alpha, \beta)$. Is the beta distribution a conjugate prior for the case when the final outcomes 3(a) and 3(b) are not separated, as in part (a)?

If the values of c and d are not known, show how the breakdown into types

3(a) and 3(b) in part (b) may be used to provide an EM algorithm for a classical maximisation of the likelihood. You can use the example of (a) as an illustration, but you are not expected to carry out extensive iterations.

7.26   For the model of Exercise 7.25, suppose that the total number infected, i, occurs with frequency X_i, $i = 1, 2, 3$. Obtain the expected Fisher information. Write down the asymptotic distribution of the maximum-likelihood estimator $\hat{\theta}$, and use it to obtain an approximate 99.5% confidence interval for θ when $X_1 = 6, X_2 = 11, X_3 = 13$ and $\hat{\theta} = 0.4765$. Compare this interval with that resulting from taking an appropriate section of the log-likelihood.

In standard notation, an iterative procedure for evaluating $\hat{\theta}$ proceeds according to the iteration:

$$\theta^{(r+1)} = \theta^{(r)} + d\ell/d\theta \frac{(\theta^{(r)^2} - 4\theta^{(r)} - 3)}{\{\theta^{(r)}(2\theta^{(r)} - 3)\}}.$$

Explain whether this is the Newton-Raphson method, or the method of scoring. For the data of (a), one of these methods produces the sequence:

A: $0.01 \rightarrow 0.608 \rightarrow 0.464 \rightarrow 0.477 \rightarrow 0.4765$.

The other produces the sequence:

B: $0.01 \rightarrow 0.02 \rightarrow 0.04 \rightarrow 0.08 \rightarrow 0.146 \rightarrow 0.259 \rightarrow 0.398 \rightarrow 0.471 \rightarrow 0.4765$.

Explain which sequence you think comes from which method.

7.27   Devise a Metropolis method to simulate random variables from the Cauchy pdf,

$$f(x) = \frac{1}{\pi(1 + x^2)}, \quad -\infty < x < \infty.$$

Provide a MATLAB program for your procedure.

7.28   A simple probability model for the data of Table 7.1 has two probabilities as parameters, p, the probability of capture in any year and ϕ, the annual dipper survival probability. For the first five rows of the table the data are multinomial, with first two multinomial probabilities, respectively, ϕp and $\phi^2 p(1 - p)$.

Describe fully the probability distribution for the 80 dippers released in 1984, noting that the study stopped in 1987. Figure 7.12 contains several panels. In one is shown the likelihood for this application. In two are shown the marginal posterior distributions for each of the parameters, corresponding to independent $U(0, 1)$ prior distributions for the two parameters.

Explain in detail how the marginal distributions are obtained. In the last two figure panels are shown histograms resulting from simulations from the two marginal posterior distributions. The simulations have been obtained by means of an MCMC procedure, the MATLAB code for which is shown in the program below.

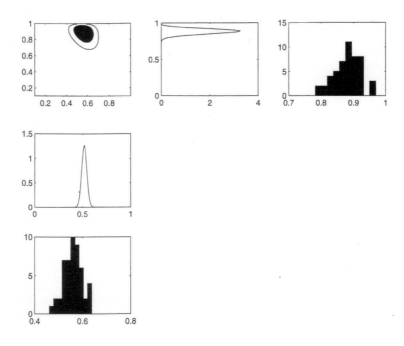

Figure 7.12 *Dipper likelihood, marginals and simulations.*

```
m=500; k=100;
for j=1:k
    b=[.9,.4];
    for i=2:m
        a=[rand, rand];
        alpha=min(1,dipper(a)/dipper(b));
        if alpha > rand;
            b=a;
        else
        end
    end
    x(j)=b(1);y(j)=b(2);
end
subplot(3,3,3), hist(y); subplot(3,3,7), hist(x);
```

Provide a full description of how the method of the MATLAB program operates. Include in your answer an evaluation of whether you think this MCMC procedure might be made more efficient, and of the method used to obtain the samples described by the histograms.

7.29 The Hardy-Weinberg law predicts that the proportions of genotypes AA, Aa and aa in a population are θ^2, $2\theta(1-\theta)$ and $(1-\theta)^2$, respectively, where $\theta \in (0,1)$ is an unknown parameter. Suppose that the genotypes of n independent individuals are observed, yielding the following frequency distribution:

Genotype	AA	Aa	aa
Frequency	n_1	n_2	n_3

1. Derive the posterior distribution $f(\theta \mid \mathbf{y})$, where $\mathbf{y} = (n_1, n_2, n_3)$, when the prior distribution of θ is $\mathrm{Be}(\alpha, \beta)$.

2. Suppose now that the allele a is dominant, so the genotypes Aa and aa cannot be distinguished and rather than observing n_2 and n_3, only the total number of individuals having genotypes Aa or aa, $(n_2 + n_3)$, is observed. Explain whether the beta distribution is a conjugate prior for θ.

7.30 In the genetic linkage study of Example 5.6, 197 animals are divided into four categories with frequencies of $125, 18, 20, 34$, with corresponding probabilities given below

$$\frac{1}{2} + \frac{\pi}{4}, \frac{1}{4}(1 - \pi), \frac{1}{4}(1 - \pi), \frac{\pi}{4}.$$

With a uniform prior density for π, write down the posterior density for π. Explain how you would use an MCMC method for sampling from this posterior distribution.

7.31 (Millar, 2004.) Consider the case of a random variable X which has a $N(\mu, \sigma^2)$ distribution, where σ^2 is known, and the parameter μ has a $N(\nu, \tau^2)$ distribution.

An important issue in Bayesian analysis is the sensitivity of posterior distributions to assumptions made regarding prior distributions. Let the posterior distribution for μ have mean ν^*. One way to evaluate the sensitivity of the posterior distribution to the prior is to form the derivative, $\frac{\partial \nu^*}{\partial \nu}$.

Show for the example of this question that the above derivative is the ratio of posterior variance to prior variance, and discuss whether you think this is a sensible result.

7.32 Write a MATLAB program in order to derive model probabilities for the geometric and beta-geometric models of Example 7.4, applied to the smokers data of Table 1.1. Interpret the results in the light of a likelihood-ratio test comparison of the two models.

7.33* Devise a RJMCMC procedure for estimating model probabilities for Example 7.3 on the survival of dippers. Compare the results with those of Table 7.2.

7.34 Modify and run the program of Figure 7.2, so that it corresponds to the trivariate distribution of Example 7.1.

7.35 Experiment with the program of Figure 7.10.

CHAPTER 8

General Families of Models

8.1 Common structure

Many of the models that we have considered in this book have been tailored to a particular situation and fitted using general methods of function optimisation. It is important, however, to realise that examples which appear initially to be quite different often share a common structure, which can be exploited to remove much of the labour of non-linear model-fitting. Two apparently different models considered so far are logit models for quantal response data, and multinomial models for *contingency* tables, which are rectangular tables of counts. These can both be fitted to data using a computerised procedure for fitting *generalised linear models* (GLMs). Procedures for fitting GLMs may also be used more widely — for example, to fit models to data on survival times.

An advantage of being able to use a general procedure to fit a range of different models is that for each model the same basic tools are readily available, for instance for calculating and plotting different types of *residual* which can be used for assessing goodness-of-fit, and for judging whether certain features of the data are exerting a particular influence on the conclusions of the analysis.

Some of the models we have considered so far in this book have involved ideas of *regression* and of *covariates*. This is the topic of Section 5.5, and further examples are provided in Exercise 5.10, in which proportions of dead aphids are related to dose of nicotine, in Exercise 5.11, in which failure times of motorettes are related to temperature, and elsewhere. Most of the examples of this chapter will involve regressions of various kinds. In describing a range of modern useful statistical procedures, the work of this chapter makes use of all the work that has gone before and helps to unify all of the material of the book.

8.2 Generalised linear models (GLMs)

8.2.1 The link function

The simplest linear model that we shall consider is the regression model in which

$$Y = \alpha + \beta x + \epsilon, \tag{8.1}$$

where x is a dependent variable, and Y is a response variable, α, β are parameters to be estimated and ϵ is a random error term with a $N(0, \sigma^2)$ distribution.

Consider now the logit model for quantal response data. In this case, the random variation is described by a binomial distribution, and not a normal distribution. Furthermore, the binomial probability of response to dose x is

$$P(x) = \frac{1}{1 + e^{-(\alpha + \beta x)}}. \tag{8.2}$$

What do expressions (8.1) and (8.2) have in common? We can see from Equation (8.2) that

$$\text{logit}\,(P(x)) = \log\left(\frac{P(x)}{1 - P(x)}\right) = \alpha + \beta x.$$

Suppose that in the logit model, n individuals are given dose x, and R of these individuals respond, each independently with probability $P(x)$. Then the random variable R has the binomial, $\text{Bin}(n, P(x))$ distribution, and $\mathbb{E}[R] = nP(x)$. Hence,

$$\left.\begin{aligned} \log\left(\frac{\mathbb{E}\,[R]}{n - \mathbb{E}\,[R]}\right) &= \alpha + \beta x, \\ \text{whereas from Equation (8.1), for comparison} \qquad\qquad & \\ \mathbb{E}\,[Y] &= \alpha + \beta x. \end{aligned}\right\} \tag{8.3}$$

The expressions of Equation (8.3) demonstrate what it is that these two models have in common: in each case a function of the expectation of the response variable is a linear function of the dependent variable x. The function used to link the mean response to the linear function of the dependent variable (or variables, in the case of many of these) is simply called the *link* function. A formal definition is given later. The two important components of GLMs are the link function and a general family of distributions which includes the normal distribution and the binomial distribution as special cases.

The year 1972 was an important one for the development of useful statistical methodology. It saw the publication of Cox (1972) in which an elegant way was proposed for analysing survival data with covariates, providing a semiparametric and robust formulation of a proportional hazards model. We shall discuss proportional hazards models later. It also saw the publication of the paper by Nelder and Wedderburn (1972), in which the family of GLM distributions considered was the exponential family. Since then, there have been further extensions of the theory (see, e.g., Green, 1984) but here we shall only consider the standard cases.

8.2.2 The exponential family

This general family of distributions includes the exponential and normal pdfs and also covers important discrete distributions. It has already been encountered in Section 5.3.2. Thus if here we write $f(w; \theta)$ as a pdf or a probability function with a single parameter θ, then for the exponential family we can write

$$f(w; \theta) = \exp \{a(w) \, b(\theta) + c(\theta) + d(w)\}, \tag{8.4}$$

for quite general functions, a, b, c, and d. If $a(w) = w$, then we are said to have the *canonical*, or *standard*, form of the exponential family, and $b(\theta)$ is then called the *natural parameter*. It is the canonical form of the exponential family that is used in GLMs.

Note that

$$\ell = \log f(w; \theta) = a(w)b(\theta) + c(\theta) + d(w)$$

is the log-likelihood corresponding to a single value w from the exponential family. We know, from Section 4.1.1, that $\mathbb{E}[\frac{d\ell}{d\theta}] = 0$. Hence in this instance,

$$\frac{d\ell}{d\theta} = a(w)b'(\theta) + c'(\theta),$$

$$\text{where} \quad b'(\theta) = db/d\theta, \text{ etc.,}$$

and then taking expectations and setting $\mathbb{E}\left[\dfrac{d\ell}{d\theta}\right] = 0$ provides us with the result that

$$\mathbb{E}[a(W)] = -\frac{c'(\theta)}{b'(\theta)}. \tag{8.5}$$

An alternative expression for the form of the exponential family is given in Exercise 8.15.

Example 8.1: The Poisson distribution

The Poisson probability function, $P_o(\theta)$, is given by

$$f(w; \theta) = \frac{\theta^w e^{-\theta}}{w!} \quad \text{for } w = 0, 1, 2 \dots.$$

Here, $f(w; \theta) = \exp(w \log \theta - \theta - \log w!)$, which is one example of the canonical form of the exponential family, with $\log \theta$ as the natural parameter. □

Example 8.2: The binomial distribution

The binomial Bin (n, θ) probability function is given by

$$f(w; \theta) = \binom{n}{w} \theta^w (1 - \theta)^{n-w}, \quad 0 \le w \le n,$$

in which, as usual, n is supposed to be known. We see that we can write

$$f(w; \theta) = \exp\left\{ w \log \theta - w \log(1 - \theta) + n \log(1 - \theta) + \log \binom{n}{w} \right\}.$$

This is therefore another example of the canonical form of the exponential family, in this case with natural parameter, $\log\{\theta/(1 - \theta)\}$. □

Example 8.3: The normal distribution, $N(\theta, \sigma^2)$

Here the pdf is

$$f(w; \theta, \sigma^2) = \frac{1}{\sigma\sqrt{2\pi}} \exp\left\{-\frac{1}{2\sigma^2}(w - \theta)^2\right\}$$

$$= \exp\left\{-\frac{w^2}{2\sigma^2} + \frac{w\theta}{\sigma^2} - \frac{\theta^2}{2\sigma^2} - \frac{1}{2}\log\left(2\pi\sigma^2\right)\right\}.$$

Unlike the last two examples, the normal pdf does not fit directly into the form of Equation (8.4) because of the presence of the additional parameter σ^2. However, we can see that if we regard σ^2 as *known*, then once again we get the canonical form of the exponential family, with natural parameter (θ/σ^2).

□

If $W_1, \ldots, W_n$ are independent identically distributed random variables, all with the exponential family distribution of Equation (8.4), then the joint pdf is given by:

$$f(w_1, \ldots, w_n) = \exp\left\{b(\theta)\sum_{i=1}^{n} a(w_i) + nc(\theta) + \sum_{i=1}^{n} d(w_i)\right\}.$$

We can see that the log-likelihood $l(\theta; \mathbf{w})$ is then

$$\ell(\theta; \mathbf{w}) = b(\theta)\sum_{i=1}^{n} a(w_i) + nc(\theta) + \sum_{i=1}^{n} d(w_i).$$

The only way in which the data, $\mathbf{w}$, influence the maximum-likelihood estimate $\hat{\theta}$ is through the sum, $\sum_{i=1}^{n} a(w_i)$, as the term $\sum_{i=1}^{n} d(w_i)$ vanishes when we differentiate $\ell(\theta; \mathbf{w})$ with respect to the components of θ. Thus $\sum_{i=1}^{n} a(w_i)$ is said to be *sufficient* for the maximum-likelihood estimation of θ.

Sufficiency is clearly an important consideration in dealing with likelihoods in general.

8.2.3 Maximum-likelihood estimation

In GLMs we use the canonical form of the exponential family. We observe n random variables, $Y_1, \ldots, Y_n$, all from the same member of the exponential family, but with the ith random variable, Y_i, depending on parameter θ_i; thus we allow for the possibility that the $\{\theta_i\}$ may be different. Then, from Equation (8.4), the log-likelihood is given by:

$$\ell(\boldsymbol{\theta}; \mathbf{y}) = \sum_{i=1}^{n} y_i b(\theta_i) + \sum_{i=1}^{n} c(\theta_i) + \sum_{i=1}^{n} d(y_i),$$

where $\mathbf{y}$ is a realisation of the random variable $\mathbf{Y}$.

We let $\mu_i = \mathbb{E}[Y_i]$, $1 \leq i \leq n$. The link function is then defined by:

$$g(\mu_i) = \mathbf{x}_i'\boldsymbol{\beta} = \xi_i,$$

say, where the column vector $\mathbf{x}_i$ contains the set of p independent variables for observation y_i, and $g(\cdot)$ is a monotonic differentiable function.

In order to obtain the maximum-likelihood estimate, $\hat{\boldsymbol{\beta}}$, numerical optimisation is usually necessary. An attractive feature of the exponential family is that there is a unique maximum-likelihood estimate (see Cox and Hinkley, 1974, p.286). If we obtain $\hat{\boldsymbol{\beta}}$ by using the method of scoring then a surprise is in store, as the iterative procedure can be expressed in terms of weighted least squares, as we shall now see.

We let $\mathbf{X}$ denote the $n \times p$ matrix with ith row $\mathbf{x}_i'$, $1 \leq i \leq n$. We let $\mathbf{W}$ denote the $n \times n$ diagonal matrix, with zeroes everywhere except on the main diagonal, and with $w_{ii} = \dfrac{1}{\text{var}(Y_i)} \left(\dfrac{\partial \mu_i}{\partial \xi_i} \right)^2$.

Then the rth iterative estimate of $\boldsymbol{\beta}$ using the method of scoring is given as:

$$\hat{\boldsymbol{\beta}}^{(r)} = (\mathbf{X}'\mathbf{W}\mathbf{X})^{-1}\mathbf{X}'\mathbf{W}\mathbf{z},$$

for suitable $\mathbf{z}$, specified below (Exercise 8.1).

We know from Equation (5.22) that $\hat{\boldsymbol{\beta}}^{(r)}$ is therefore the weighted least squares estimate of $\boldsymbol{\beta}$ when the random variable $\mathbf{z}$ has expectation, $\mathbf{X}\boldsymbol{\beta}$, and $\mathbf{W}^{-1}$ is the variance-covariance matrix of $\mathbf{z}$. In our case, we have

$$z_i = \sum_{k=1}^{p} x_{ik}\hat{\beta}_k^{(r-1)} + (y_i - \mu_i)\frac{\partial \xi_i}{\partial \mu_i},$$

and μ_i, $\dfrac{\partial \xi_i}{\partial \mu_i}$ and $\mathbf{W}$ are evaluated at $\hat{\boldsymbol{\beta}}^{(r-1)}$. Hence we obtain the maximum-likelihood estimate $\hat{\boldsymbol{\beta}}$, by the method-of-scoring, by iterating the solution to a generalised weighted least squares estimation procedure. It is therefore possible to fit GLMs using a computerised procedure for generalised weighted least squares — see, e.g., Morgan (1992, p.52) and Exercise 8.2. The iteration is frequently termed IRLS — iteratively reweighted least squares.

We know from Chapter 4 that we may choose between alternative models for a particular data set using likelihood-ratio tests or score tests. We know also that score tests only require the simpler of the two models being compared to be fitted. Pregibon (1982) shows that the algebra of score tests may be avoided in GLMs, by taking *one step* of the iterations for fitting the more complicated of the two models.

We have already defined the *deviance* in Section 4.7. For a GLM with parameter vector $\boldsymbol{\beta}$ the deviance is defined as

$$D = 2\{\ell(\hat{\boldsymbol{\beta}}; \mathbf{y}) - \ell_{sat}\},$$

where ℓ_{sat} is the value of the log-likelihood when we set $\mu_i = y_i$ for all i. In the case of ℓ_{sat}, the model is *saturated* with parameters, as there is one for each observation. The log-likelihood ℓ_{sat} does not involve any parameters. For different nested models, with different dimensions of the parameter vector $\boldsymbol{\beta}$, deviance differences correspond to likelihood-ratio test statistics.

Examples 8.4: Common Birds Census

The Common Birds Census (CBC) is a scheme for monitoring the changes in abundance of birds. The CBC is carried out by volunteers, and co-ordinated by the British Trust for Ornithology. For each of a range of bird species, CBC data result in counts, y_{it}, of the number of individuals of that species observed at site i in year t. As we are dealing with counts, we may use a Poisson distribution to describe y_{it}, with mean μ_{it}. We would expect μ_{it} to vary with respect to site and time, and a possible model therefore is to use the canonical link function, and set

$$\log(\mu_{it}) = \alpha_i + \beta_t,$$

where α_i is the site effect for site i, and β_t is the year effect for year t.

This GLM is one of a range of models that may be fitted to ecological data using the TRIM software package (Pannekoek and van Strien, 1996). We continue the discussion of this example in Section 8.4.2 and in Exercise 8.6.   □

Example 8.5*: Proportional hazards

Models for survival data have been encountered in Section 5.6.4. A non-negative function $h(t)$ such that for any time-interval $(t, t+\delta t)$ the probability of a death of an individual can be written as $h(t)\delta t + o(\delta t)$ is called the *hazard* function. The Weibull model is more flexible than the exponential model considered earlier, as it has a hazard function which may increase or decrease with time, in addition to possibly being constant, which corresponds to the exponential model. Frequently survival data not only provide times to response, but in addition each individual involved has a vector $\mathbf{x}$ of covariates, and there is interest in modelling the hazard function in terms of $\mathbf{x}$. One example is provided by Prentice (1973) in which a single covariate x describes tumour type for individuals in a lung cancer trial — see Exercise 8.10. A convenient way to model the hazard for an individual with covariate vector $\mathbf{x}$ is as

$$h(t|\mathbf{x}) = \lambda(t)e^{\boldsymbol{\beta}'\mathbf{x}},$$

for some positive function of time $\lambda(t)$. The use of the exponential term ensures that the hazard is positive, as it must be.

This is called a *proportional hazards* model, since at time t, two individuals with covariates $\mathbf{x}_1$ and $\mathbf{x}_2$ will have the hazard ratio:

$$\frac{h(t|\mathbf{x}_1)}{h(t|\mathbf{x}_2)} = e^{\boldsymbol{\beta}'(\mathbf{x}_1 - \mathbf{x}_2)}$$

which does not involve t — whatever this hazard ratio is, it is a constant, determined by the covariate vectors, and does not vary over time. The hazards are said to be proportional. Two points are worth noting here, viz., that for some data this could be an inappropriate model (see Exercise 8.10), and the way in which $\mathbf{x}$ enters $h(t|\mathbf{x})$ has been chosen for convenience, and many other alternative expressions could have also been used.

$$\text{Let} \qquad \Lambda(t) = \int_0^t \lambda(\tau)d\tau,$$

and suppose that the time to response has pdf $f(t)$ and cdf, $F(t) = 1 - S(t)$. We call $S(t)$ the *survivor function*, already familiar from Equation (5.17). From its definition, we see that we can write

$$h(t) = \frac{f(t)}{S(t)},$$

and as $f(t) = \dfrac{d}{dt}F(t)$, and $F(t) = 1 - S(t)$,

$$h(t) = -\frac{dS(t)}{dt}/S(t).$$

When $h(t|\mathbf{x}) = \lambda(t)e^{\boldsymbol{\beta}'\mathbf{x}}$, then by integrating this equation, we get

$$\Lambda(t)e^{\boldsymbol{\beta}'\mathbf{x}} = -\log S(t),$$

so that $S(t) = \exp\{-\Lambda(t)e^{\boldsymbol{\beta}'\mathbf{x}}\}$, and then by differentiating, as $f(t) = -\dfrac{dS(t)}{dt}$, we obtain the expression for $f(t)$:

$$f(t) = \lambda(t)e^{\boldsymbol{\beta}'\mathbf{x}}\exp\{-e^{\boldsymbol{\beta}'\mathbf{x}}\Lambda(t)\}.$$

Presented with a set of n survival times, some of which may be censored, and for each of which there is a corresponding covariate vector, we are interested in estimating $\boldsymbol{\beta}$, and possibly also $\lambda(t)$. Although we can see that the model contains the linear function of covariates, $\boldsymbol{\beta}'\mathbf{x}$, the tie-in with GLMs is not yet clear. In order to provide it, we introduce indicator variables which denote right-censored values if they occur in the data:

$$\delta_i = 1 \text{ if the } i\text{th individual is not censored,}$$
$$\delta_i = 0 \text{ if the } i\text{th individual is censored.}$$

We can therefore write the log-likelihood of the data as:

$$\ell = \sum_{i=1}^n \{\delta_i \log f(t_i) + (1 - \delta_i)\log S(t_i)\},$$

i.e.,

$$\ell = \sum_{i=1}^n \{-\delta_i\Lambda(t_i)e^{\boldsymbol{\beta}'\mathbf{x}_i} + \delta_i\boldsymbol{\beta}'\mathbf{x}_i + \delta_i \log \lambda(t_i) - (1 - \delta_i)\Lambda(t_i)e^{\boldsymbol{\beta}'\mathbf{x}_i}\},$$

which simplifies to give

$$\ell = \sum_{i=1}^n \{\delta_i\boldsymbol{\beta}'\mathbf{x}_i + \delta_i \log \lambda(t_i) - \Lambda(t_i)e^{\boldsymbol{\beta}'\mathbf{x}_i}\}.$$

If we now set $\mu_i = \Lambda(t_i)e^{\boldsymbol{\beta}'\mathbf{x}_i}$, then the above expression can be written more simply as

$$\ell = \sum_{i=1}^{n} \{\delta_i \log \mu_i - \mu_i\} + C, \qquad (8.6)$$

in which C does not involve $\boldsymbol{\beta}$.

We recognise the log-likelihood of Equation (8.6) as that which would arise if the $\{\delta_i\}$ were Poisson random variables with mean μ_i — see for example Exercise 8.4. Additionally, $\log \mu_i$ involves the linear function $\boldsymbol{\beta}'\mathbf{x}$ of the covariates. Of course, the $\{\delta_i\}$ are simply indicator variables, but by treating them as Poisson we are able to fit this model using computer software designed to fit GLMs. We need to consider further what assumptions we may make concerning $\lambda(t)$, and the discussion of this example is continued in Exercises 8.8 and 8.11.   □

8.2.4 Extensions

We know from the expression of (8.4) that if $a(w) = w$ then we have the canonical form of the exponential family, and that then $b(\theta)$ is the natural parameter. We also know from Equation (8.5) that in this case, $\mu = \mathbb{E}[W] = -\dfrac{c'(\theta)}{b'(\theta)}$. The general definition of the link function tells us that $g(\mu) = \mathbf{x}'\boldsymbol{\beta}$. If the link function is chosen so that $g(\mu) = b(\theta)$, then it is said to be the *canonical* link function. For example, when W has a binomial distribution, the canonical link function is logistic, while when W has a Poisson distribution, the canonical link function is log — see Exercise 8.13. When the link function used is the canonical link function, then the Newton-Raphson method and the method of scoring for fitting the model to data by maximum likelihood are equivalent — see Exercise 8.16. Quite complex link functions may be adopted, as we can see from the next example.

Example 8.6: An extended link function

We have seen that the logit model for quantal response data gives as the expression for the probability of response to dose d,

$$P(d) = \frac{1}{1 + e^{-(\alpha+\beta d)}}.$$

This is an example of a GLM, with a binomial random variable and a logistic link function. This model was extended by Aranda-Ordaz (1983), to give:

$$\begin{aligned} P(d) &= 1 - (1 + \lambda e^{\alpha+\beta d})^{-\frac{1}{\lambda}}, \quad \text{for } \lambda e^{\alpha+\beta d} > -1, \\ &= 0 \qquad\qquad\qquad\qquad \text{otherwise.} \end{aligned}$$

The additional parameter λ allows greater flexibility for fitting the model to data. It is readily verified that the logit model arises when $\lambda = 1$. For a fixed value of λ this model corresponds to a GLM. It may therefore be fitted to data by means of a computer program for fitting GLMs, and nesting the optimisation with respect to λ. In this case, we maximise with respect to the pair of parameters, λ and β, at each of a number of values, λ_i, say, selected for λ. This allows us to construct a profile log-likelihood with respect to λ, and hence to derive the maximum-likelihood estimates of all three parameters, α, β and λ. (See Section 4.3.) For further discussion of this model, see Exercise 8.14 and Morgan (1992, Chapter 4). □

We have emphasised in Section 5.5.2 the importance of checking whether the conclusions from model fitting may be influenced by certain aspects of the data. For linear regression models the procedures that may be used are described by Cook and Weisberg (1982). Pregibon (1981) has elegantly demonstrated how one-step approximations to iterative Newton-Raphson for obtaining maximum-likelihood estimates allow the linear model ideas to be applied to general logistic regression, which extends the logit model of Example 8.6 to the case of several explanatory covariates.

8.3 Generalised linear mixed models (GLMMs)

8.3.1 Overdispersion

We saw in Chapter 2 that simple models may be inadequate for describing the variation in a data set. Any overdispersion relative to a simple model may be accommodated by making the model more complex, as in moving from a geometric distribution to a beta-geometric distribution, or from a Poisson distribution to a negative-binomial distribution, for example.

Overdispersion may be the result of the data being grouped in some way which is not accounted for by the model. For instance, surveys may be administered by a number of different interviewers; perinatal deaths of animals may be distributed over a number of different litters; there may be differing environmental conditions which affect the survival of wild animals differentially from year to year. For illustration we shall consider the case of animal litters.

If we consider modelling a litter effect, within any litter of size n, the number of deaths, X, may be regarded as binomial, $\text{Bin}(n, p)$, say, and with $\text{Var}(X|p) = np(1-p)$. If p is then allowed to vary between litters according to the beta, $\text{Be}(\alpha, \beta)$, distribution, then X will have a beta-binomial distribution. The probability function of the beta-binomial distribution has been given in Exercise 3.3. Unconditionally, $\text{Var}(X) = n\phi\mu(1-\mu)$, where $\mu = \alpha/(\alpha+\beta)$ and $\phi > 1$. The beta-binomial model induces a correlation between the responses of animals in the same litter (Altham, 1978). It thereby appropriately models the feature resulting in overdispersion. For further discussion, see Morgan (1992, Chapter 6), and Brooks et al. (1997). The method of *quasi-likelihood* may be used to analyse overdispersed data without making distributional assumptions, but only using the specification of the mean and the variance. This is rather like using a method of least squares in that in both cases a model is not fully specified. Model fitting using quasi-likelihood may be conveniently carried out by suitably adapting the IRLS procedure used to fit GLMs (Morgan, 1992, p.257). An alternative approach to dealing with overdispersion, which may be suitable for some applications, is to include additive random effects, which extend GLMs to generalised linear mixed models, denoted by GLMMs, to which we shall now turn. Williams (1988) provided a comparison between the beta-binomial model and a particular GLMM.

8.3.2 Model specification

In a linear model, correlation between individual responses, as between littermates in the illustration discussed above, is straightforwardly produced by the addition of suitable random effects. For example, we may have two individual responses given by:

$$y_1 = \boldsymbol{\beta}' \mathbf{x}_1 + u + \epsilon_1$$
$$y_2 = \boldsymbol{\beta}' \mathbf{x}_2 + u + \epsilon_2. \tag{8.7}$$

Here the individuals have respective covariate vectors $\mathbf{x}_1$ and $\mathbf{x}_2$, and ϵ_1 and ϵ_2 are appropriate independent random error terms, of variance σ_0^2. The vector of coefficients, $\boldsymbol{\beta}$, corresponds to the *fixed* effects, while the common independent random variable u is the *random* effect, with mean zero and variance σ_1^2. Typically u would be assumed to have a normal distribution. The correlation between y_1 and y_2 is then readily shown to be $\rho = \sigma_1^2/(\sigma_0^2 + \sigma_1^2)$. For some data it is useful to have several independent random effects. In such a case the variances of the random effects are called the *variance components*. The formulation of Equations (8.7) is called a linear mixed model (LMM) and it generalises simply to produce a GLMM. In a GLM, the linear predictor term, ξ, is written as $\xi = \boldsymbol{\beta}' \mathbf{x}$. In a GLMM, we write

$$\xi = \boldsymbol{\beta}' \mathbf{x} + \boldsymbol{\gamma}' \mathbf{u}. \tag{8.8}$$

There is now a vector of random effects, $\mathbf{u}$, the contribution of which is determined by the parameter vector $\boldsymbol{\gamma}$. For instance, if we have two correlated Bernoulli random variables, Y_1 and Y_2, with respective means, $\mathbb{E}[Y_1] = \mu_1, \mathbb{E}[Y_2] = \mu_2$, and respective covariate vectors, $\mathbf{x}_1$ and $\mathbf{x}_2$, then we obtain the following GLMM from using a logit link:

$$\text{logit} (\mu_1) = \boldsymbol{\beta}' \mathbf{x}_1 + u,$$

$$\text{logit} (\mu_2) = \boldsymbol{\beta}' \mathbf{x}_2 + u,$$

where once again u is the random effect, with mean zero and variance σ_1^2.

8.3.3 Maximum likelihood estimation

Conditional upon a set of random effects in the vector $\mathbf{u}$, we may write the log-likelihood as $\ell(\boldsymbol{\beta}, \boldsymbol{\sigma}; \mathbf{y} | \mathbf{u})$, where $\boldsymbol{\sigma}$ is a vector containing the variance components. If the random variable $\mathbf{u}$ has the multivariate pdf $f(\mathbf{u})$, which is usually assumed to be multivariate normal, then in order to obtain maximum-likelihood estimates for $\boldsymbol{\beta}$ and $\boldsymbol{\sigma}$, it is first necessary to form the unconditional log-likelihood,

$$\ell(\boldsymbol{\beta}, \boldsymbol{\sigma}; \mathbf{y}) = \int \ldots \int \ell(\boldsymbol{\beta}, \boldsymbol{\sigma}; \mathbf{y}|\mathbf{u}) f(\mathbf{u}) \mathrm{d}\mathbf{u}. \tag{8.9}$$

In some cases it is possible to evaluate this integral explicitly – for example when the log-likelihood results from a multivariate normal distribution and $f(\mathbf{u})$ is also multivariate normal — but this is rare (see also van Duijn, 1993). Jansen (1993) has shown how numerical analysis may be used effectively for approximating the integrals in Equation (8.9). However for complex problems the numerical analysis approach is not a practical proposition. The approach of approximate Laplacian integration has been used by Breslow and Clayton (1993), Wolfinger (1993), and Wolfinger and O'Connell (1993).

For LMMs, parameters are estimated by a combination of maximum likelihood and *residual maximum likelihood* (REML). Due to Patterson and Thompson (1971), REML is a procedure which may reduce the bias in the estimation of the variance components. Schall (1991) and Engel and Keen (1994) produced an integration of REML and IRLS, to give IRREML, as a means of fitting GLMMs. As pointed out by Engel (1997, p.7), IRREML shares the robustness property of quasi-likelihood, in that it does not fully specify distributions.

In the paper by Anderson and Aitkin (1985), for a survey on consumer purchasing, there were 1265 respondents spread over 32 different areas. In all there were 64 interviewers, modelled as a random effect. In Engel and Buist (1996), cattle carcasses from 23 slaughterhouses were classified into five main groups by 56 different individuals; analyses in this case involved a number of different random effects.

8.3.4 The Bayesian approach

We know from the work of the last chapter that integrals that are similar to that of Equation (8.9) may be readily evaluated using MCMC methodology. The Bayesian approach to GLMMs is to give the vector of random effects, $\mathbf{u}$, in Equation (8.8) a prior distribution. In Zeger and Karim (1991) a multivariate normal distribution is adopted with mean $\mathbf{0}$ and dispersion matrix $\boldsymbol{\Sigma}$. The full parameter set for the model is then $(\boldsymbol{b}, \boldsymbol{\Sigma})$, which is given a prior distribution. Zeger and Karim (1991) assume that $\boldsymbol{b}$ and $\boldsymbol{\Sigma}$ are independent in the prior distribution. They show that the model is an hierarchical Bayes model, and that the necessary conditional distributions are readily simulated, allowing Gibbs sampling to take place. Further discussion of the Bayesian approach to GLMMs is provided by McGilchrist (1994).

8.4 Generalised additive models (GAMs)

In some cases GLMs are insufficiently flexible to describe the variation in the data. The generalised additive model (GAM) provides a flexible and useful extension of GLMs. Before discussing GAMs, we provide two examples of 'smoothers' at work, as these form the components of GAMs.

8.4.1 Smoothers

Example 8.7: University performance

Plotted in Figure 8.1 are scores in a first-year algebra examination taken in the Australian Defence Force Academy in Canberra and a measure of performance at the end of secondary school (NTER). The individuals are categorised as those from the state of Victoria, and those from the state of New South Wales. Instead of using parametric models to demonstrate relationships between the variables, Anderson and Catchpole (1998) used a particular nonparametric smoother, *loess* (see Cleveland et al., 1992, for a description of this technique). □

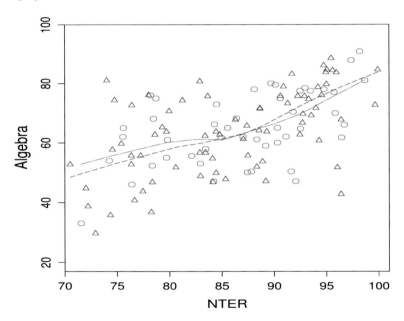

Figure 8.1 *A plot of first-year algebra examination marks vs a measure of secondary school performance (NTER) for students at the Australian Defence Force Academy.* Δ *denotes students from New South Wales (NSW), O denotes students from Victoria. Also shown are loess smoothers (dashed line) NSW; (solid line) Victoria, from (Anderson and Catchpole, 1998). Reprinted with permission from the* Australian & New Zealand Journal of Statistics.

Example 8.8: Simulated data

The data of Figure 8.2 are the result of adding random disturbances to points on a parabola. Shown also are the results of fitting a smoothing spline

with two alternative forms of roughness. The programs to produce the data
and the smoothing spline are to be found later in Figure 8.4.   □

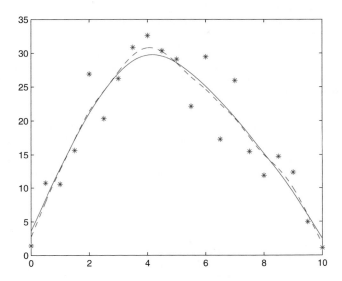

Figure 8.2 *An example of a cubic smoothing spline fitted to simulated data, with two
alternative values for the roughness parameter.*

Smoothers are essentially non-parametric ways of doing regression, which
are useful when the variation in the data is too complex to be described
satisfactorily by means of a parametric model. We have already encountered
a simple example of a smoother in the moving average of Figure 6.7. Kernel
methods, described in Appendix C, may be used to produce smoothers — see
Exercise 8.17. The detail for cubic spline smoothers will be given in Section
8.4.3. There are many further alternatives which may be used. The differences
in practice are unlikely to be substantial, and what is used in any application
will probably be due to a combination of convenience and personal preference.

In GLMs, we have seen that the mean, μ, of a response variate Y is expressed
through the link function as a linear function of covariates:

$$g(\mu) = \boldsymbol{\beta}'\mathbf{x} = \sum_{i=0}^{p} \beta_i x_i.$$

In GAMs, we have instead

$$g(\mu) = \alpha + \sum_{i=1}^{d} f_i(x_i),$$

in which the functions $f_i(\cdot)$ are suitably fitted smoothers. The simplest case is when $d = 1$ and we return now to Example 8.4 to illustrate this case.

Example 8.4 continued

The model used in Example 8.4 for the mean bird count of a particular species, μ_{it}, at site i in year t, was of the form:

$$\log(\mu_{it}) = \alpha_i + \beta_t.$$

An alternative description by means of a GAM was investigated by Fewster et al. (2000). They set:

$$\log(\mu_{it}) = \alpha_i + f(t) \tag{8.10}$$

in which $f(t)$ is a smoothing spline. An illustration of the result is shown in Figure 8.3 for CBC data for the corn bunting, *Emberiza calandra*. The GAM is evidently useful in portraying the general changes over time.

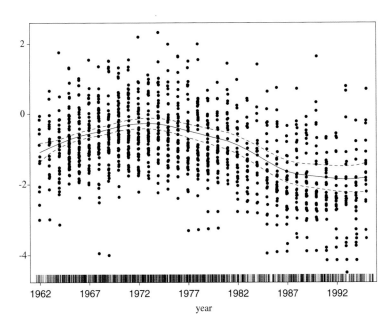

Figure 8.3 *Plot by (Fewster et al., 2000) of the year effect function $f(t)$, from Equation (8.10) from a GAM fitted to CBC data for the corn bunting (solid curve). Also shown are twice standard error bands (dashed curve), as well as partial deviance residuals (●) and a rug plot. See Chambers and Hastie (1993) for explanation of these last two topics, which are not discussed here. The rug plot indicates the density of the observations. The amount of smoothing adopted is indicated in the source paper. Figure reprinted with permission from* Ecology.

□

8.4.2 Backfitting

Additive models are a special case of GAMs, with the response variate Y given directly as:

$$Y = \alpha + \sum_{j=1}^{d} f_j(x_j) + \epsilon,$$

where ϵ denotes random error, and the $f_j(\cdot)$ are estimated by appropriately chosen smoothers. Additive models may be fitted by means of an iterative, *backfitting* algorithm, which proceeds as follows: initial values are taken for α and the $\{f_j\}$; for instance, one might set as the initial value for α, $\alpha^{(0)} = \bar{y}$ and the $f_j^{(0)}$ could result from a multiple linear regression of Y on $\mathbf{x}$. Then $f_1^{(0)}$ is updated by smoothing the partial residuals resulting from $Y - \alpha^{(0)} - \sum_{j=2}^{d} f_j^{(0)}(x_j)$, to give $f^{(1)}(x_1)$; $f_2^{(0)}$ is then updated by smoothing the partial residuals resulting from $Y - \alpha - f_1^{(1)}(x_1) \sum_{j=3}^{d} f_j^{(0)}(x_j)$, etc, and the process continues until convergence is obtained.

In some cases convergence is certain, as discussed by Hastie and Tibshirani (1990, p.91). GAMs are fitted using a combination of an adaptation of the method of scoring and a modified backfitting algorithm. Hastie and Tibshirani (1990, p.141) provide the details corresponding to exponential family models.

We now explain how one smoother, based on cubic splines, works.

*8.4.3 *Smoothing splines*

The lucid account of smoothing splines provided by Green and Silverman (1994) provides the basis for this section. As in Figure 8.2, we have a response variable Y measured at each of the points, a, $\{t_i\}$, b, with $a < t_1 < t_2 < \ldots < t_n < b$. The points $\{t_i\}$ are known as the *knots*. A cubic spline is a function which is a cubic polynomial on each of the intervals, $[a, t_1], (t_1, t_2], \ldots, (t_n, b]$ and at the joins the function and its first two derivatives are continuous. It may possibly be noticed that the first and last segments of the cubic splines in Figure 8.2 are exactly straight lines. This is because these splines are *natural cubic splines* which we denote by NCS, which have second and third derivatives zero at a and b.

We choose the smoother $f(t)$ to minimise the function,

$$S(f) = \sum_{i=1}^{n} (y_i - f(t_i))^2 + \nu \int_a^b (f''(x))^2 dx, \tag{8.11}$$

where ν is a smoothing parameter which needs to be chosen. The second term in Equation (8.11) measures the variation in $S(f)$. Thus that term places a brake on making $f(t)$ too variable when minimising $S(f)$, and the effect of this depends on the size of ν. The effect of changing ν has already been seen in the particular illustration of Figure 8.2.

It is a beautiful result that amongst the set of functions that are differentiable on the interval $[a, b]$ and which have absolutely continuous first derivative, a particular NCS minimises $S(f)$ — see Green and Silverman (1994, p.19) and Exercise 8.18. We now show how we can compute this NCS.

$$\begin{aligned}
\text{Let} \quad & f(t) = d_i(t - t_i)^3 + c_i(t - t_i)^2 + b_i(t - t_i) + a_i, \\
\text{for} \quad & t_i \leq t \leq t_{i+1}. \\
\text{Let} \quad & h_j = t_{j+1} - t_j \quad \text{for} \quad 1 \leq j \leq (n-1) \\
\text{and} \quad & f_j = f(t_j); \ \gamma_j = f''(t_j) \quad \text{for} \quad 1 \leq j \leq n, \\
& \mathbf{f} = [f_1 \ldots f_n]'; \ \boldsymbol{\gamma} = [\gamma_2 \ldots \gamma_{n-1}]'.
\end{aligned}$$

The $\{h_j\}$ are used to build two tridiagonal matrices, $\mathbf{V}$, of dimensions $n \times (n-2)$ and $\mathbf{R}$ of dimensions $((n-2) \times (n-2))$, which are zero except for the main diagonal and the diagonals above and below the main diagonal. In detail, we have

$$\left. \begin{aligned}
v_{j-1,j} &= h_{j-1}^{-1} \\
v_{j,j} &= -h_{j-1}^{-1} - h_j^{-1} \\
v_{j+1,j} &= h_j^{-1}
\end{aligned} \right\} \quad \text{for} \quad j = 2, \ldots, n-1,$$

$$v_{ij} = 0 \qquad \text{for} \quad |i - j| \geq 2;$$

$$\begin{aligned}
r_{ii} &= \tfrac{1}{3}(h_{i-1} + h_i), & \text{for} \quad i = 2, \ldots, n-1, \\
r_{i,i+1} &= r_{i+1,i} = \tfrac{1}{6}h_i, & \text{for} \quad i = 2, \ldots, n-2, \\
r_{ij} &= 0 & \text{for} \quad |i - j| \geq 2.
\end{aligned}$$

There are obvious constraints on the elements of $\mathbf{f}$ and $\boldsymbol{\gamma}$ (Exercise 8.19), and these are expressed through as

$$\mathbf{V}' \, \mathbf{f} = \mathbf{R} \, \boldsymbol{\gamma}. \qquad (8.12)$$

As $\mathbf{R}$ is positive definite, we can define the matrix $\mathbf{K}$ as:

$$\mathbf{K} = \mathbf{V} \, \mathbf{R}^{-1} \mathbf{V}'. \qquad (8.13)$$

The NCS which minimises $S(f)$ in Equation (8.11) satisfies

$$\mathbf{f} = (\mathbf{I} + \nu \, \mathbf{K})^{-1} \, \mathbf{y} \qquad (8.14)$$

(see Exercise 8.18). We could now obtain $\mathbf{f}$ from Equation (8.14) and even, crudely, join up the components of $\mathbf{f}$ with straight-line segments. However Reinsch (1967) improved on the efficiency of Equation (8.14), and his approach is as follows: From Equation (8.14),

$$\mathbf{f} + \nu \, \mathbf{K} \, \mathbf{f} = \mathbf{y}, \quad \text{and by Equation (8.13)},$$

$$\begin{aligned}
\mathbf{f} &= \mathbf{y} - \nu \, \mathbf{V} \, \mathbf{R}^{-1} \, \mathbf{V}' \, \mathbf{f} \\
&= \mathbf{y} - \nu \, \mathbf{V} \, \boldsymbol{\gamma}, \quad \text{by Equation (8.12)}.
\end{aligned}$$

Hence,

$$\mathbf{V}' \, \mathbf{f} = \mathbf{V}' \, \mathbf{y} - \nu \, \mathbf{V}' \, \mathbf{V} \, \boldsymbol{\gamma},$$

and by Equation (8.12),

$$\mathbf{R}\,\boldsymbol{\gamma} = \mathbf{V}'\,\mathbf{f} = \mathbf{V}'\mathbf{y} - \nu\,\mathbf{V}'\,\mathbf{V}\,\boldsymbol{\gamma},$$

so that

$$(\mathbf{R} + \nu\,\mathbf{V}'\,\mathbf{V})\,\boldsymbol{\gamma} = \mathbf{V}'\,\mathbf{y}. \qquad (8.15)$$

This equation may be solved efficiently to give $\boldsymbol{\gamma}$ and then $\mathbf{f}$ is obtained, from above, as $\mathbf{f} = \mathbf{y} - \nu\,\mathbf{V}\,\boldsymbol{\gamma}$.

As $f(t)$ is a cubic, we can use $\mathbf{f}$ and $\boldsymbol{\gamma}$ to give the functions $f(t)$ for plotting. For $t_i \le t \le t_{i+1}$, and $1 \le i \le (n-1)$,

$$f(t) = \frac{(t - t_i)f_{i+1} + (t_{i+1} - t)f_i}{h_i} - \frac{1}{6}(t - t_i)(t_{i+1} - t)\times$$

$$\left\{\left(1 + \frac{(t - t_i)}{h_i}\right)\gamma_{i+1} + \left(1 + \frac{(t_{i+1} - t)}{h_i}\right)\gamma_i\right\}.$$

The MATLAB program of Figure 8.4 plots a cubic smoothing spline, without solving efficiently for $\boldsymbol{\gamma}$ (though this is easily done via a Cholesky decomposition of $\mathbf{R} + \nu\,\mathbf{V}'\mathbf{V}$).The program selects ν by means of a *cross-validation* procedure, which operates as follows. Suppose we write $\hat{f}^{(-i)}(t;\nu)$ for the estimate of the smoothing spline when the value y_i at t_i is omitted. For any ν, we can consider how well $\hat{f}^{(-i)}(t_i;\nu)$ predicts y_i. This then results in the cross-validation criterion,

$$cv(\nu) = \frac{1}{n}\sum_{i=1}^{n}\left\{y_i - \hat{f}^{(-i)}(t_i;\nu)\right\}^2,$$

and we choose ν to minimise $cv(\nu)$.

We recall from Equations (8.13) and (8.14) that

$$\mathbf{f} = \mathbf{A}(\nu)\,\mathbf{y}, \qquad (8.16)$$

where $\mathbf{A}(\nu)$ is what we call the *hat* matrix,

$$\mathbf{A}(\nu) = (\mathbf{I} + \nu\,\mathbf{V}\,\mathbf{R}^{-1}\,\mathbf{V}')^{-1},$$

so called because we may regard the operation of Equation (8.16) as producing an estimated value of $\mathbf{y}$, or $\hat{\mathbf{y}}$.

Conveniently, as shown in Green and Silverman (1994, p.31), we can write

$$cv(\nu) = \frac{1}{n}\sum_{i=1}^{n}\left\{\frac{y_i - \hat{f}(t_i;\nu)}{1 - a_{ii}(\nu)}\right\}^2,$$

when the terms $\{a_{ii}(\nu)\}$ are the diagonal entries of the hat matrix. This means that it is not necessary to fit $(n-1)$ separate smoothed splines corresponding to leaving out each of the data values in turn. Although we have used *fminbnd* in Figure 8.4(a) to minimise $cv(\nu)$, we also plot a graph of $cv(\nu)$, and we show in Figure 8.5(b) the result of one such plot.

(a)

```
% Program to fit a smoothed spline to a set of data, including
% the cross-validation method for choosing the smoothing parameter nu.
% Global variables are : Q, R and Y.
% Variables Y and t are respectively the 'y-' and 'x-' values of the data.
% In this case they are set by a separate program.
% Calls CVSEARCH and FMINBND.
% Code due to Paul Terrill.
%_____
global Q R Y                     % we start by forming the Q and
h=t(:,2:m)-t(:,1:m-1);           % R matrices
ih=1./h;
Q=zeros(m-2,m-2)+diag(ih(:,2:m-2),1)+diag(ih(:,2:m-2),-1)+...
   diag(-ih(:,1:m-2)-ih(:,2:m-1),0);
Q=[ih(1) zeros(1,m-3);Q;zeros(1,m-3) ih(m-1)];
R=zeros(m-2,m-2)+diag(1/6*h(:,2:m-2),1)+diag(1/6*h(:,2:m-2),-1)+...
   diag(1/3*(h(:,1:m-2)+h(:,2:m-1)),0);

                                 % We now need to choose an alpha that
                                 % minimises the cross-validation
                                 % criterion, CV(nu)
nu=0.01:0.005:5;
for i=1:length(nu)
  cv(i)=cvsearch(nu(i));
end
figure(2)
plot(nu,cv);                     % this plots the graph of CV(nu)
xlabel('\nu')
lv=input('Enter lower value for fminbnd and press return: ');
uv=input('Enter upper value for fminbnd and press return: ');
alpha=fminbnd('cvsearch',lv,uv)
                                 % we use 'fminbnd' to select nu
gamma=(R+nu*Q'*Q)\(Q'*Y);
g=Y-nu*Q*gamma;                  % next we plot the splines
figure(1)
hold on
y1=1+9*t'-t'.*t'+6;
plot(t',y1,'.')
gamma=[0;gamma;0];
for i=1:m-1
  x=[t(i):0.01:t(i+1)]';
  q1=ones(size(x));
  t1=x-t(i)*q1;
  t2=(t(i+1)*q1-x);
  w1=(t1*g(i+1)+t2*g(i))/h(i);

                                 % now we have to account for the unusual
                                 % indexing used in gamma, and the fact
                                 % that the second-order derivatives are
                                 % zero at each end.

  w2=t1.*t2.*(gamma(i+1)*(q1+t1/h(i))+gamma(i)*(q1+t2/h(i)));
  w=w1-w2/6;
  plot(x,w);
end
hold off
```

Figure 8.4 MATLAB *programs (a) to produce a cubic smoothing spline; (b) to estimate the roughness parameter ν by means of cross-validation; and (c) to simulate data to illustrate the performance of a cubic smoothing spline.*

```
(b) function x=cvsearch(nu)
%
%CVSEARCH calculates the cross-validation criterion, CV(nu)
%    used in fitting a cubic spline, and which is then minimised.
%    global matrices are, Q, R, specified by the cubic-spline program.
%    The 'y-values', Y, produced by the data generation program are global.
%    Code due to Paul Terrill.
%_____
global Q R Y
A=inv(eye(length(Y))+nu*Q*inv(R)*Q');    % eye : identity matrix
gamma=(R+nu*Q'*Q)\(Q'*Y);
g=Y-nu*Q*gamma;
x=((Y-g)./(1-diag(A))).^2;
x=sum(x)/length(Y);


(c)
% Program to generate and plot a set of data for use with the cubic spline
% program; the data are a simple random disturbance added to a parabola
% the y- coordinates of the points are a global variable, needed by
% the function, 'cvsearch'. This program is
% designed to be followed directly by the 'cubicspline' program.
%_____
global Y t=0:0.2:10; m=length(t);
rand('state',8);                          % sets the seed for the
Y=1+9*t'-t'.*t'+12*rand(m,1);             % random number generator
figure(1)
plot(t',Y,'*');
```

Figure 8.4 *Continued.*

8.5 Discussion

The work of this chapter builds on the material in the earlier chapters of the book. In this chapter we have focussed on three general families of non-linear models. Simpler linear models result as special cases, as we have seen, for example, in Section 8.3.2. The models of this chapter are clearly computationally intensive and need computerised algorithms in order to be fitted to data. What has been presented here is only introductory, and the richness of the subjects can be appreciated from books such as Green and Silverman (1994), Hastie and Tibshirani (1990), and McCullagh and Nelder (1989). Recent developments for GAMs are described in Wood (2006), and for GLMMs by Molenberghs and Verbecke (2005). Splines are used in models for including covariates in models for capture-recapture by Gimenez et al. (2006).

Liang and Zeger (1986) derive generalised estimating equations (GEEs) for when there is time-dependence among repeated measures on individuals, while Zeger et al. (1988) present a set of GEEs for GLMM models for longitudinal data.

In this book, we have used MATLAB computer programs to illustrate the

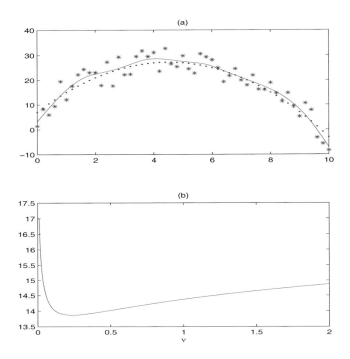

Figure 8.5 *(a) The cubic smoothing spline obtained as a result of cross-validatory choice of ν. Also shown is the parabola (dotted curve) used to generate the data. (b) Graph of $cv(\nu)$ against ν, showing the minimising value of ν.*

theory. For many applications MATLAB programs will provide a convenient answer, especially if the problem at hand is atypical. However, in practice, for much applied statistical work, it is efficient to make use of computer packages if they have routines which perform the desired analyses. For example, GEN-STAT is useful for REML and IRREML; the book by Lloyd (1999) makes use of GLIM, MINITAB, S-PLUS, LogXact and StatXact; SAS is used extensively, especially in the pharmaceutical industry.

8.6 Exercises

8.1 Verify that the rth iterative estimate of b in GLMs, using the method of scoring, is given as

$$\hat{\beta}^{(r)} = (\mathbf{X}' \, \mathbf{W} \, \mathbf{X})^{-1} \, \mathbf{X}' \, \mathbf{W} \, \mathbf{z},$$

in the notation of Section 8.2.3.

8.2 Write a MATLAB program to fit a logit model to the age-of-menarche data of Exercise 4.14, by iterating generalised least squares.

8.3 The contingency table below presents data from Roberts et al. (1981). Individuals with malignant melanoma, which is a skin cancer, are classified according to tumour type and site.

	Site			
	Head and neck	Trunk	Extremities	Total
Tumour type				
Hutchinson's melanotic freckle	22	2	10	32
Superficial spreading melanoma	16	54	115	185
Nodular	19	33	73	125
Indeterminate	11	17	28	56
Total	68	106	226	400

Indicate how you would analyse these data within the general framework of GLMs. (Dobson, 1990, p.123).

8.4 Assume that random variables Y_i are independent, and have the distribution, $Y_i \sim P_o(\alpha + \beta x_i)$, where the $\{x_i\}$ are known covariates and α, β are parameters to be estimated, $1 \le i \le n$. Write down the likelihood, and show that the maximum-likelihood estimates of α and β satisfy the equations:

$$\sum_{i=1}^{n} \frac{y_i}{(\hat{\alpha} + \hat{\beta} x_i)} = n; \quad \sum_{i=1}^{n} \frac{x_i y_i}{(\hat{\alpha} + \hat{\beta} x_i)} = n\bar{x},$$

where $\{y_i\}$ are the observed values of $\{Y_i\}$, and $\bar{x} = \dfrac{\sum_{i=1}^{n} x_i}{n}$.

For the following data,

i	1	2	3	4	5	6
x_i	0	0	1	1	2	2
y_i	1	0	5	7	20	26

show that $\hat{\beta} = 22.5\hat{\alpha}$, and calculate $\hat{\alpha}$ and $\hat{\beta}$.

8.5 (Dobson, 1990, p.33.) The data below are the numbers of deaths from AIDS in Australia per quarter for the period, 1983–1986. Plot the data and describe the salient features. Explain how you would analyse these data using a GLM.

Quarter	1	2	3	4	5	6	7	8	9	10	11	12	13	14
No. dead	0	1	2	3	1	4	9	18	23	31	20	25	37	45

8.6 The Common Birds Census data are described in Example 8.4, where a
GLM is proposed for the analysis. *The Independent* newspaper published
in Britain on 12 August 1999 comments on the dramatic decline in many
farmland bird species over the past 30 years. Discuss how you would use
the results of the GLM analysis to represent these drops in abundance.

8.7 The zero-truncated Poisson distribution has probabilities:

$$pr(Y = j) = \frac{e^{-\lambda}\lambda^j}{j!(1 - e^{-\lambda})} = \frac{\lambda^j}{j!(e^{\lambda} - 1)}, \quad \text{for } j = 1, 2. \ldots.$$

Show that Y is a member of the exponential family distribution with canonical parameter $\log(\lambda)$. Discuss how you would fit this model to data when
there is a set, $(x_1, x_2, \ldots, x_n)$, of covariates, corresponding to n observations from the zero-truncated Poisson. A botanical application is provided
by Ridout and Demétrio (1992).

8.8 In the Weibull model for survival data, the time to response, T, has pdf

$$f(t) = \alpha\lambda t^{\alpha-1} e^{-\lambda t^{\alpha}} \quad \text{for } t > 0.$$

Derive the forms for the hazard function and the survivor function. Investigate how changing the shape parameter α affects the shape of the hazard
function. Suppose all individuals in a study share the same shape parame-
ter α, but differ in a set of covariates, $\mathbf{x}$. For the ith individual we assume
that the λ-parameter, given by λ_i, is related to the covariates $\mathbf{x}_i$ by means
of the equation,

$$\log \lambda_i = \boldsymbol{\beta}'\mathbf{x}_i.$$

Show that the model is a proportional hazards model. Show how it may be
fitted to the data as a GLM. Compare and contrast this model with that
of Example 8.5.

8.9* If T has the Weibull distribution of Exercise 8.8, then $Y = \log T$ has the
extreme value distribution. Derive the form of the pdf of Y. Consider how
the extreme value distribution may be fitted to survival data as a GLM,
when the covariates enter the model in the same way as in Exercise 8.8.

8.10 A group of 97 patients with inoperable lung cancer were given chemother-
apy, resulting in observed survival times ranging from 1 to 587 days. On
each patient the covariates recorded were: tumour type (adeno, small, large
or squamous) and a measure of fitness. This is a subset of a data set presented by Prentice (1973). Discuss whether you think that a proportional
hazards model would be appropriate for the data. Bennett (1983) presented
an alternative model which we shall now describe. The survival time T has
a log-logistic distribution, with cdf,

$$F(t) = \frac{1}{1 + t^{-\phi}e^{-\theta}}, \quad t \geq 0, \quad \phi > 0.$$

Derive the hazard function, and consider how it varies as a function of time.
Compare and contrast this behaviour with that of the hazard functions of

Exercises 8.8 and 8.11. It is assumed that ϕ is a constant, but that for an individual with covariates $\mathbf{x}$, we can write $\theta = \boldsymbol{\beta}'\mathbf{x}$. By considering the ratio, $F(t)/(1 - F(t))$, show that this is a *proportional odds* model. Write down the likelihood function, and show how the model may be fitted as a GLM. Discuss the maximum likelihood estimates shown below, together with estimates of standard error, and their implications for survival.

$\hat{\phi}$		1.841	(0.163)
$\hat{\beta}_0$	(constant)	−5.293	(0.860)
$\hat{\beta}_1$	(fitness)	−0.055	(0.010)
$\hat{\beta}_2$	(adeno vs large)	1.295	(0.552)
$\hat{\beta}_2$	(small vs large)	1.372	(0.521)
$\hat{\beta}_2$	(squamous vs large)	−0.138	(0.576)

8.11* In the Cox proportional hazards model (Cox, 1972), the multiplicative function $\lambda(t)$ in the hazard:

$$h(t|\mathbf{x}) = \lambda(t)e^{\boldsymbol{\beta}'\mathbf{x}}$$

is not specified. Consider how you might fit such a model to data. Might a GLM approach be feasible? See Aitkin et al. (1989, Section 6.15) for detailed discussion.

8.12 Discuss whether the model of Exercise 5.10 is a GLM.

8.13 In GLMs, verify that for a binomial distribution the canonical link function is logistic, while for a Poisson distribution the canonical link function is log.

8.14 Discuss the power logit model for quantal response data, in which the probability of response to dose d is given by:

$$P(d) = (1 + e^{-\beta(d-\theta)})^{-m},$$

where β, θ and m are parameters.

8.15 As in Section 5.3.2, the exponential family is often presented as:

$$f(w; \theta) = \exp\{(w\theta - b(\theta))/a(\phi) + c(w, \phi)\},$$

for specific functions, $a(\ )$, $b(\ )$ and $c(\ )$, and known parameter ϕ. Verify that this is identical to the canonical form of the expression of (8.4).

8.16* Show that when a canonical link function is used in a GLM, then the method of scoring is equivalent to the Newton-Raphson method for obtaining maximum-likelihood estimates.

8.17 A *kernel smoother* uses a kernel function to weight local information in a scatter plot. Suppose we have a set of data, $(x_i, y_i, 1 \le i \le n)$. For any value x, the smoothed y-value is given by:

$$s(x) = \frac{\sum_{i=1}^{n} K\left(\frac{x - x_i}{\lambda}\right) y_i}{\sum_{i=1}^{n} K\left(\frac{x - x_i}{\lambda}\right)}$$

where $K(\cdot)$ is a suitable kernel function and λ is the kernel window width. Experiment with kernel smoothing. For example, generate random data using the simulation program of Figure 8.4(c) and then, for a chosen kernel, obtain the smoothed function, $s(x)$, for a number of different window widths. For extensive discussion of kernel regression, see Chu and Marron (1991).

8.18* Show, using integration by parts, that the penalty term of Equation (8.11) can be written as:

$$\int_a^b (f''(x))^2 dx = \mathbf{f}'\mathbf{K}\mathbf{f}.$$

Show further that the function $S(f)$ of Equation (8.11) can be written as:

$$S(f) = (\mathbf{y} - \mathbf{f})'(\mathbf{y} - \mathbf{f}) + \nu\mathbf{f}'\mathbf{K}\mathbf{f},$$

and deduce that the minimising value of $\mathbf{F}$ is

$$\mathbf{F} = (\mathbf{I} + \nu\mathbf{K})^{-1}\mathbf{y}.$$

8.19* Explain why the constraints of Equation (8.12) hold.

8.20* Consider how you would use numerical integration in MATLAB in order to evaluate the unconditional GLMM log-likelihood of Equation (8.9).

8.21* In Example 7.2, the annual reporting probabilities of dead birds are assumed to vary from year-to-year. The model adopted there has a separate reporting probability, λ_i, for the ith year of the study. An alternative approach to describing variation in the reporting probability is to treat it as a random effect, and give it a suitable distribution. Investigate how you would write down the likelihood for a model in which the reporting probability λ is treated as a random effect over time. Discuss whether random effects might be used elsewhere in models for the annual survival of wild animals.

8.22 Discuss the results of Figure 8.6, in which the same GAM has been fitted to the data from each state, but the state responses differ by a constant term. Compare the results with those of Figure 8.1. Suggest alternative breakdowns of the data which may be important.

8.23 Use the MATLAB program of Figure 8.4(a) to smooth the approximate likelihoods from Monte Carlo inference, and displayed in Figure 6.7.

8.24 Write down the link function for the model of Example 8.6, for fixed λ. If

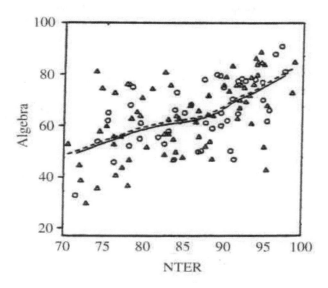

Figure 8.6 *The results of a GAM fitted to the same data as in Figure 8.1. From Anderson and Catchpole (1998). Reprinted with permission from the* Australian & New Zealand Journal of Statistics.

a computer package for fitting GLMs is used for the Aranda-Ordaz model, by nesting the optimisation with respect to λ, the estimated standard errors of $\hat{\alpha}$ and $\hat{\beta}$ will be incorrect, as they do not take account of the estimation of λ. Consider how you would derive the correct standard errors.

Modify the logit function of Exercise 3.10, to allow the Aranda-Ordaz model to be fitted. Investigate whether the Aranda-Ordaz model provides a better fit to the data of Exercise 4.14, which describe the incidence of menarche in groups of Polish girls, using a likelihood-ratio test. Compare your conclusions with those resulting from the score test of Exercise 4.8.

Index of Data Sets

Index of MATLAB Programs

Appendix A: Probability and Statistics Reference

A.1: Distributions

For convenience of reference, we list here the forms of the pdf's and probability functions used in the text.

Multinomial:

$$pr(x_1, x_2, \ldots, x_k) = \frac{(\Sigma x_i)!}{x_1! x_2! \ldots x_k!} \prod_{i=1}^{k} p_i^{x_i}, \quad \sum_{i=1}^{k} p_i = 1.$$

Binomial:

$$pr(X = k) = \binom{n}{k} p^k (1-p)^{n-k}, \quad k = 0, 1, \ldots, n.$$

Geometric:

$$pr(X = k) = (1-p)^{k-1} p, \quad k = 1, 2, \ldots.$$

Poisson:

$$pr(X = k) = \frac{e^{-\lambda} \lambda^k}{k!}, \quad p = 0, 1, 2, \ldots.$$

Zero-inflated Poisson:

$$pr(X = k) = \begin{cases} \frac{(1-w)e^{-\lambda}\lambda^k}{k!}, & k = 1, 2, \ldots, \\ w + (1-w)e^{-\lambda}, & k = 0. \end{cases}$$

Zero-truncated Poisson:

$$pr(X = k) = \frac{\lambda^k}{k!(e^\lambda - 1)}, \quad k = 1, 2, \ldots.$$

Negative binomial:

$$pr(X = n + k) = \binom{n+k-1}{k} p^k (1-p)^n, \quad k = 0, 1, 2, \ldots.$$

Beta-geometric:

$$pr(X = k) = \frac{\mu \prod_{i=1}^{k-1}\{1 - \mu + (i-1)\theta\}}{\prod_{i=1}^{k}\{1 + (i-1)\theta\}}, \quad k = 1, 2, \ldots.$$

Beta-binomial:

$$pr(X = k) = \frac{\binom{n}{k} \prod_{r=0}^{k-1}(\mu + r\theta) \prod_{r=0}^{n-k-1}(1 - \mu + r\theta)}{\prod_{r=0}^{n-1}(1 + r\theta)}, \quad k = 0, 1, 2, \ldots, n.$$

Uniform:

$$f(x) = \frac{1}{(b-a)}, \quad a < x < b,$$
$$= \quad 0 \quad \text{otherwise.}$$

Exponential:

$$f(x) = \lambda e^{-\lambda x}, \quad x \ge 0.$$

Gamma:

$$f(x) = \frac{e^{-\lambda x}\lambda^n x^{n-1}}{\Gamma(n)}, \quad x \ge 0.$$

Chi-square with ν degrees of freedom:

$$f(x) = \frac{e^{-x/2}x^{\nu/2-1}}{\Gamma(\nu/2)2^{\nu/2}}, \quad x \ge 0.$$

t_ν:

$$f(x) = \frac{\Gamma\left(\frac{\nu}{2}+\frac{1}{2}\right)}{\sqrt{\nu\pi}\;\Gamma\left(\frac{\nu}{2}\right)}\left(1+\frac{x^2}{\nu}\right)^{-\frac{1}{2}(\nu+1)}.$$

Cauchy:

$$f(x) = \frac{\beta}{\pi\{\beta^2+(x-\alpha)^2\}}, \quad -\infty < x < \infty.$$

Weibull:

$$f(x) = \kappa\rho(\rho x)^{\kappa-1}\exp\{-(\rho x)^\kappa\}, \quad x \ge 0.$$

Normal:

$$f(x) = \frac{e^{-(x-\mu)^2/(2\sigma^2)}}{\sigma\sqrt{2\pi}}, \quad -\infty < x < \infty.$$

Half-normal:

$$f(x) = \frac{\sqrt{2}e^{-(x-\mu)^2/(2\sigma^2)}}{\sigma\sqrt{\pi}}, \quad x \ge 0.$$

Multivariate normal:

$$\phi(\mathbf{x}) = (2\pi)^{-p/2}|\mathbf{\Sigma}|^{-1/2}\exp\left\{-\frac{1}{2}(\mathbf{x}-\boldsymbol{\mu})'\mathbf{\Sigma}^{-1}(\mathbf{x}-\boldsymbol{\mu})\right\},$$
$$-\infty < x_i < \infty, \quad i = 1, 2, \ldots, p.$$

Logistic:

$$f(x) = \frac{\beta e^{-(\alpha+\beta x)}}{(1+e^{-(\alpha+\beta x)})^2}, \quad -\infty < x < \infty.$$

Log-logistic:

$$f(x) = \frac{\beta e^{-\alpha}x^{-(\beta+1)}}{(1+e^{-\alpha}x^{-\beta})^2}, \quad 0 \le x < \infty.$$

Inverse Gaussian:

$$f(x) = \left(\frac{\lambda}{2\pi x^3}\right)^{\frac{1}{2}}\exp\left\{-\frac{\lambda}{2}\left(\frac{x}{\mu^2}-\frac{2}{\mu}+\frac{1}{x}\right)\right\} \quad 0 \le x < \infty, \quad \lambda > 0, \quad \mu > 0.$$

Extreme-value:

$$f(x) = \kappa \rho^{\kappa} \exp(\kappa x - \rho^{\kappa} e^{\kappa x}), \quad -\infty < x < \infty.$$

A.2: The Poisson process

We shall just consider the case of one-dimension, which for simplicity we shall refer to as *time*.

The Poisson process is a model for the random occurrence of events in time. If the rate of the Poisson process is a positive constant λ, then in a small interval of time $(t, t + \delta t]$, the rules of the process are:

$$Pr(\text{ one event in } (t, t + \delta t]) = \lambda \delta t + o(\delta t),$$
$$Pr(\text{ no event in } (t, t + \delta t]) = 1 - \lambda \delta t + o(\delta t),$$
$$Pr(> 1 \text{ event in } (t, t + \delta t]) = o(\delta t).$$

We assume that the number of events occurring in $(t, t + \delta t]$ is independent of occurrences in $(0, t]$. The consequences of these rules are that:

(i) $Pr(k \text{ events in time interval } (0, t]) =$

$$\frac{e^{-\lambda t}(\lambda t)^k}{k!}, \quad k = 0, 1, 2, \ldots.$$

(ii) At any time τ, the time to the next event, T, is an exponential random variable, with pdf:

$$f(t) = \lambda e^{-\lambda t}, \quad \text{for } t \geq 0.$$

We can see that $f(t)$ does not involve τ, so that the Poisson process lacks memory. This is a property shared with Markov chains, the subject of A.4. See for instance, Cox and Miller (1965, p.146). There are many generalisations of the Poisson process — see for example Cox and Miller (1965, p.153).

A.3: Normal quadratic forms

Suppose $\mathbf{X} \sim N_d(\boldsymbol{\mu}, \boldsymbol{\Sigma})$, and that $\boldsymbol{\Sigma}$ is non-singular.

Let us define the scalar random-variable Z by:

$$Z = (\mathbf{X} - \boldsymbol{\mu})' \boldsymbol{\Sigma}^{-1} (\mathbf{X} - \boldsymbol{\mu}).$$

The matrix $\boldsymbol{\Sigma}$ is positive-semi-definite, so we can find a non-singular matrix $\mathbf{B}$ such that

$$\mathbf{B} \boldsymbol{\Sigma} \mathbf{B}^{-1} = \mathbf{I}. \tag{A.1}$$

See, for example, Carroll et al. (1997 p.229).

Let $\mathbf{Y} = \mathbf{B}(\mathbf{X} - \boldsymbol{\mu})$, so that

$$\mathbf{Y} \sim N_d(\mathbf{0}, \mathbf{B}\boldsymbol{\Sigma}\mathbf{B}') = N_d(\mathbf{0}, \mathbf{I}), \quad \text{by Equation (A.1).} \tag{A.2}$$

Now we make use of the fact that for appropriate matrices $\mathbf{A}$ and $\mathbf{B}$, $(\mathbf{A}^{-1})' = (\mathbf{A}')^{-1}$ (Carroll et al., 1997, p.165) and $(\mathbf{AB})' = \mathbf{B}'\mathbf{A}'$ (Carroll et al., 1997, p.50).

$$
\begin{aligned}
Z &= (\mathbf{X} - \boldsymbol{\mu})'\boldsymbol{\Sigma}^{-1}(\mathbf{X} - \boldsymbol{\mu}) \\
&= (\mathbf{B}^{-1}\mathbf{Y})'\boldsymbol{\Sigma}^{-1}(\mathbf{B}^{-1}\mathbf{Y}) \\
&= \mathbf{Y}'(\mathbf{B}^{-1})'\boldsymbol{\Sigma}^{-1}\mathbf{B}^{-1}\mathbf{Y} \\
&= \mathbf{Y}'(\mathbf{B}')^{-1}\boldsymbol{\Sigma}^{-1}\mathbf{B}^{-1}\mathbf{Y}
\end{aligned}
$$

$$
= \mathbf{Y}'(\mathbf{B}\boldsymbol{\Sigma}\mathbf{B}')^{-1}\mathbf{Y} = \mathbf{Y}'\mathbf{Y}, \text{ by Equation (A.1).}
$$

Hence by Equation (A.2), as the d elements of $\mathbf{Y}$ are i.i.d. N(0, 1) random variables, $Z \sim \chi_d^2$.

A.4: Markov chains

The song of blackbirds can be broken down into a number of different types. When one type is finished, another begins until the song is finished, and we can describe the resulting string of types, developing over time, as a Markov chain, with a transition matrix $\mathbf{Q}$. The (i, j)th element of $\mathbf{Q}$, q_{ij}, is the probability that the ith type is followed by the jth type. The matrix $\mathbf{Q}$ is said to be *stochastic* as its row sums are unity: $\sum_j q_{ij} = 1$. The stochastic property is readily verified for the mover-stayer model of Exercise 5.13, which is a Markov chain.

We can appreciate that in a Markov chain the probability rules determining the future development of the system depend only on the current state of the system, and are not influenced by the past. An important class of Markov chains is one in which there is only a finite number of states. Suppose the distribution of state occupancy at time n is given by the probability row vector $\mathbf{p}^{(n)}$, then by considering all the different possibilities for changing states, we can see that the distribution of state occupancy at time $(n + 1)$ is given by:

$$
\mathbf{p}^{(n+1)} = \mathbf{p}^{(n)}\mathbf{Q}.
$$

The *equilibrium* distribution, $\boldsymbol{\mu}$, if it exists, satisfies the equation

$$
\boldsymbol{\mu} = \boldsymbol{\mu}\mathbf{Q}.
$$

A Markov chain is said to be *irreducible* if it is possible to move from any one of its states to any other one, though not necessarily in one step. A Markov chain is said to be *aperiodic* if there is not a positive integer such that, for example, one may only return to a state, having left it, at times which are multiples of that integer. A necessary, but insufficient, condition for a Markov chain to be periodic is that the diagonal terms of $\mathbf{Q}$ are all zero, i.e., $q_{ii} = 0$ for all i. If a finite Markov chain is irreducible and aperiodic, then it is also said to be *ergodic*. Ergodic Markov chains have a unique equilibrium distribution $\boldsymbol{\mu}$. The jth element of $\boldsymbol{\mu}$, μ_j, is the probability that the system is in state j after a long period of time. We see from the above that $\boldsymbol{\mu} = \lim_{n \to \infty} \mathbf{p}^{(n)}$. The probability μ_j is the reciprocal of the *mean recurrence time* of the jth state, which is the mean number of steps it takes to return to that state once it is left.

Markov chains are useful and important stochastic processes, and they have a well-developed theory (Cox and Miller, 1965, Chapter 3) which extends also to cover the case of an infinite number of states; applications to the alternations, of wet and dry days, and types of bird song and rhesus monkey behaviours are given respectively by Cox and Miller (1965, p.78) and Morgan (1976). It is interesting to see, from the work of Chapter 7, how Markov chains have become so important for evaluating the behaviour of complex Bayesian models.

Whilst Markov chains were originally defined as having a discrete state space, the terminology now extends also to the case of a continuous state space — see Meyn and Tweedie (1993). However all Markov chains share the concept of discrete time. The output from the Metropolis-Hastings method is a Markov chain.

Appendix B: Computing

In this appendix we describe aspects of version 7 of MATLAB which will help to explain the operation of the MATLAB examples in this book. We shall make brief comparisons with R, and also outline the facilities of WinBUGS. A very large amount of helpful material can be found on the World Wide Web.

B.1: MATLAB

The name MATLAB is short for *matrix laboratory*. As the name suggests, MATLAB is an interactive computer program which works with rectangular matrices rather than just single numbers. In addition it naturally calculates a wide range of standard mathematical functions such as the beta function, and possesses good features for numerical integration, several robust numerical optimisation routines, powerful matrix handling functions and good graphical facilities. It is therefore a valuable tool for statistical modelling. Its detailed facilities are described in a MATLAB User's Guide such as Hanselman and Littlefield (2005) and much useful material and advice are given by Hahn (1997), Hunt et al. (2006), and Lindfield and Penny (1995). A large number of books make use of MATLAB, and a list, as well as a wealth of other material, are to be found on the World Wide Web at: http://www.mathworks.com. A student version of MATLAB provides a useful reduced version of the program, at a reduced price compared with the full version. Readers who wish to run MATLAB programs will need to consult a MATLAB User's Guide for full details of commands and examples of their use. MATLAB has a *help* system which will answer most questions; this can be accessed by clicking on the help icon, or directly typing *help* 'subject' in the command window (see below). The facilities of MATLAB are greatly extended by means of a number of toolboxes; for example there are toolboxes for *statistics*, *optimisation*, *symbolic algebra*, which provide the facilities of MAPLE, and *control theory*. Much useful information can be found at http://www.mathworks.matlabcentral/.

B.1.1: m-files and functions

Once MATLAB is entered, by typing matlab on a computer containing MATLAB, then MATLAB commands can be executed. The default MATLAB worktop is divided into three parts. The *Command* window is where one types commands to be executed, and where results are displayed. The *Command History* window keeps a record of the commands entered in the command window, and it is possible to re-run commands by suitably copying from the command history window. The third window may be used to display either the contents of the *Current Directory* or the variables in the current *Workspace*, and one can move easily between the two alternatives. It is possible to change the values of variables directly in the workspace window, using an array editor. It is also easy and convenient to plot selected variables directly from the workspace window. The name of the current directory is displayed at the top of the MATLAB screen, and it is easily changed by suitably clicking at that part of the screen.

It is often convenient for sequences of MATLAB commands to be stored on file in the order in which they are to be executed. MATLAB can execute such sequences of commands that are stored in disk files. These files are called m-files, as they have file names like *betageo.m*. Typically, we create m-files using a suitable editor, and then, when in MATLAB, if we type the name of the m-file (but without the '.m' suffix) then the appropriate commands are executed. m-files may be 'script' files, which just contain a sequence of commands, or 'function' files, when the file's first line is a function statement. An example of a function can be found at the end of this section. A function file evaluates the file function at a particular value, determined by the function call — e.g., geolik(0.5). This is useful for maximising a likelihood, for example, with the function determining the value of the likelihood at a particular point. It is sensible for the function name to be the same as the file name, but without the '.m' suffix. For short programs one can easily construct them within MATLAB, by clicking on 'File,' then 'New,' and then 'M-FILE.' This activates the MATLAB editor, and when the program is completed, then it can be suitably filed.

An attractive feature of MATLAB is that we can write our own m-files and then use them whenever we want. For instance, an m-file written in Chapter 6 for the computer simulation of random-variables with a gamma distribution is located in a file: *gamsim.m* (the name *gamma* cannot be used as it is reserved by MATLAB for the gamma function). We make use of this function in the program of Figure 7.2. Functions can return more than one variable, and functions may also operate on other functions. An example of this is provided by the *grad* and *hessian* functions of Figure 3.3. This is done by making use of the feval command, as illustrated in Figure 3.3. The global command is used to specify variables which are to be used by several m-files. Variables in function files which are not declared as global are not available outside the function. A simple example of this is seen in the *grad* function used in Figure 3.3(a); the variable t which is defined by *grad* is a local variable, whose value is not returned after the function call. Thus MATLAB functions use their own individual workspaces. Very simple functions can be included directly within programs using the inline command. In addition, MATLAB has useful facilities for debugging programs and also for improving their efficiency. Computing error messages can often seem incredible or even impossible. For example, this can occur if a script file uses the same variables as a program which calls the script file. When programs that are run from files fail, their names are highlighted in the command window. Clicking on the highlighted name then opens an editor screen with a cursor conveniently positioned at the line with the error.

When one is running a MATLAB program from a script file, by typing its name in the command window, editing is easily accomplished by having the program displayed in a current editor window. If several different programs are current, then they can all be open in the editor, with their names conveniently listed at the foot of the window, allowing the user to determine which to edit by simply clicking on the relevant program name.

B.1.2: Some useful MATLAB features

Here we list a number of features which are useful for constructing and running MATLAB programs. Note that use of the "↑" key on the computer keyboard within MATLAB produces the last command line, which is a very convenient facility, and saves typing.

The diary command is essential for producing hard-copy output. The command diary filename results in all subsequent keyboard input and most of the resulting output to be written to the file with the name *filename*. This file is an ASCII file, i.e., a standard text file, which may be edited and printed. In order to suspend *diary*, we type diary off.

Note that graphics appear in separate windows. If more than one figure window is being used, the command figure(n) causes the nth figure window to become the current figure window. Any figure window can contain several plots, by using the subplot command. It is possible to edit displayed figures, for instance to add comments and headings, and it is also easy to save figures for display in talks or documents, for example as postscript files. Greek letters may be included.

MATLAB programs run faster if the *zeros* function is used initially to set aside storage for a matrix which is going to have elements generated one-at-a-time or a row or a column at a time. This is done in Figure 3.3, for example.

The clear command clears the workspace, but does not clear global variables; clear global does this. Variables may also be cleared from the workspace window.

If we wish to save the material in the current workspace, this can be done using the save command, which places the information in the file: *matlab.mat*. The load command will then restore the information at a later date.

It is especially important for statisticians to know how to enter data into a MATLAB computer program. Small sets of data are easily entered directly into vectors or matrices within MATLAB. An example of this is provided by the commands later in this Appendix. For large data matrices it is better, and more elegant, to form a separate data file using an editor. If the data are to enter the matrix with the name *emu*, for example, then the data file should be called *emu.dat*. The command load emu.dat in MATLAB then automatically reads the data from that file into the matrix *emu*.

MATLAB is most efficient when the practice of looping, which may be familiar from computing languages such as BASIC and FORTRAN, is not used.

Consider, for example, the program used to produce Figure 2.1. In order to form the log-likelihood, we need

$$s1 = \sum_{j=1}^{r} n_j \quad \text{and} \quad s2 = \sum_{j=1}^{r+1}(j-1)n_j .$$

To obtain these values using a loop, we proceed as follows:

```
data1 = [29 16 17 4 3 9 4 5 1 1 1 3 7]; % data 1 is a row vector
tic
s1 = 0 ; s2 = 0;
   for i = 1:13
   s1 = s1+ data1(i);
   s2 = s2 + (i-1) * data1(i);
   end
s1
s2
toc
```

However, without using a loop we use the commands:

```
data1 = [29 16 17 4 3 9 4 5 1 1 1 3 7]; % data1 is a row vector
tic
s1 = sum (data1);
int = 0:12; s2 = int * data1';
s1
s2
toc
```

In the second of the two examples above, efficiency is achieved from realising that the variable $s2$ is simply obtained from multiplying two vectors, one of which, *int*, is the row vector, $[0\ 1\ 2\ \ldots\ 12]$.

Here we use tic and toc to time the operations between them. In order to obtain a reliable measure of the time taken by MATLAB commands it is best to exclude comments (which follow '%' signs placed at the start of lines) from the operations between tic and toc.

Note that if we do not place a ';' at the end of an instruction then the results of that instruction are displayed as output on the screen. This is useful for de-bugging programs that do not work, when we need to follow calculations through, line-by-line.

The MATLAB elementwise operations for matrices: .* ./ and .̂ are very useful and important. Thus, for example, if $\mathbf{x}$ and $\mathbf{y}$ are vectors of the same dimension, k, then $\mathbf{x}.*\mathbf{y}$ is a vector with elements, $(x_1y_1,\ x_2y_2,\ldots,\ x_ky_k)$.

Note that if $\mathbf{A}$ and $\mathbf{B}$ are matrices with appropriate dimensions for the operations to take place, then $\mathbf{A}\backslash\mathbf{B}$ results in $\mathbf{A}^{-1}\mathbf{B}$, while $\mathbf{A}/\mathbf{B}$ produces $\mathbf{A}\,\mathbf{B}^{-1}$; $\mathbf{A}\,\hat{}\,n$ provides the nth power of $\mathbf{A}$, but for small values of n it is more efficient to write out in full components of products, for example, $\mathbf{B} = \mathbf{A}*\mathbf{A}$, and $\mathbf{C} = \mathbf{A}*\mathbf{B}$. In order to obtain the transpose of $\mathbf{A}$ we write $\mathbf{A}'$.

Individual elements of matrices are obtained straightforwardly as one would expect, for example as $\mathbf{A}(i,j)$. In addition, entire rows and columns can be obtained by using ':' in place of the corresponding subscript. For example, $\mathbf{A}(5,:)$ is the fifth row of $\mathbf{A}$ and $\mathbf{A}(:,2)$ is the second column. Submatrices can also easily be obtained from suitable use of ':', for example, $\mathbf{A}(4:7,\ 2:5)$ is the matrix resulting as a submatrix of A with entries from rows 4 to 7 and columns 2 to 5.

Logical and relational operations can result in very simple code. The illustration given below (provided by Ted Catchpole) is a function for obtaining a random sample from a multinomial distribution.

The expression, `sum (m < cump(j))` compares each of the uniform, U(0,1), variates in `rand(N,1)` with `cump(j)` and records how many are `< cump(j)`.

```
function x=multsim(N,theta)
%
% MULTSIM simulates a multinomial distribution
%    'N' is the index of the distribution.
%    'theta' is the vector containing the cell probabilities.
%    Code due to Ted Catchpole.
%_____
if sum(theta) < 1
   theta=[theta,1-sum(theta)];     % the procedure can cope with
end                                % the cell probabilities summing to
n=length(theta);                   % less than unity
cump=cumsum(theta);                % 'cump' holds the crucial cumulative
m=rand(N,1);                       % sums of probabilities
cumx=zeros(1,n); for j=1:n
   cumx(j)=sum(m<cump(j));         % uses the relational operator, '<' :
end                                % creates a vector of 0/1s : entry = 1
x=[cumx(1),diff(cumx)];            % iff corresponding entry of m < cump(j)
                                   % ref: Hunt et al.,(2006), p 87.
```

B.2: R

The computing system R presents an extensive range of statistical programs and high quality graphics; the package is written by Ihaka and Gentleman (1996). R is open-source and free, and may be obtained from http://CRAN.R-project.org/. R and S-PLUS are very similar, and have developed from S. See for instance, Venables and Ripley (2002), Kruase and Olson (2000) and Murrell (2006). R is now widely used, and illustrative applications in statistics can be found in Bowman and Azzalini (1997), Rizzo (2008), Kneib and Petzoldt (2007), Crawley (2005, 2007), and Faraway (2005). The book by Albert (2007) is just one in a series entitled *Use R!*. The similarity of R to MATLAB can be appreciated from comparing the following R program with the MATLAB program in Figure 6.3(b). Once one is familiar with both systems then it is not difficult to translate programs from one to the other.

```
poissim <- function(lambda, n) { # # POISSIM simulates n independent
Po(lambda) random variables #
        y <- rep(0, n)
        e1 <- exp( - lambda)
        z <- runif(n)
        for(i in 1:n) {
                k <- 0
                u <- z[i]
                while(u > e1) {
                        u <- u * runif(1)
                        k <- k + 1
                }
                y[i] <- k
        }
        y
}
```

All the MATLAB programs in the book are also available in R and available on the book Web site: http://www.crcpress.com/e_products/downloads/down load.asp?cat_no=C6668. There is a wealth of material available on the World Wide Web. It should be noted that its default matrix operations are all element-wise. For optimisation it provides a wide range of options, including simulated annealing.

B.3: WinBUGS

The computer package WinBUGS provides an extremely simple and easy to use way of performing a wide range of basic Bayesian analyses; see Lunn et al. (2000). It is currently available for free download from http://www.mrc-bsu.c am.ac.uk/bugs/, while an open source version can be found at: http://maths tat.helsinki.fi/openbugs/. Illustrative code for the geometric model of Section 2.2, applied to the smokers data of Table 1.1, is given below

```
model {
#We specify the prior distribution; here uniform
p~dbeta(1,1)
#We specify the multinomial likelihood
m[1:13]~dmulti(q[1:13],n)
n<- sum(m[])
#We specify the probabilities for geometric model
for(i in 1:12){
q[i]<-p*pow((1-p),(i-1))
}
q[13]<-pow((1-p),12)
}
#We give the smokers data
list(m=c(29, 16, 17,4,3,9,4,5,1,1,1,3,7))
#We initialise p
list(p=0.5)
```

An interesting feature is that one does not need to specify a likelihood explicitly, only its components. As in R, $<-$ is used to assign values to parameters, and **pow** is used to power up terms. The **dmulti** term specifies that a multinomial distribution is to be used, and the **dbeta** term gives p a uniform prior distribution. Thus this code may be compared with Figure 7.9 of Chapter 7. It is simple to modify the code to replace the geometric distribution by a beta-geometric distribution, or this can be achieved by giving p a beta distribution with parameters α, β, and then assigning prior distributions to those parameters.

The WinBUGS package comes with a range of very helpful worked examples, for instance for fitting logit and survival models and the beta-binomial model, all encountered in this book. It is possible to interact with R and MATLAB, through respective use of R2WinBUGS (Sturtz et al., 2005) and `http://www.cs.ubc.ca/ murphuk/Software/MATBUGS/matbugs.html`.

At the time of writing, a *Jump* extension to WinBUGS (Lunn et al., 2006) implements a particular form of RJMCMC. The WinBUGS package allows for testing for convergence (Brooks and Gelman, 1998) of the MCMC iterations, and also provides the Deviance Information Criterion, DIC, for use in model-selection (Spiegelhalter et al., 2004). Critical discussion of the use of the DIC is provided by Celeux et al. (2006) and Gimenez et al. (2008). The books by Congdon (2003, 2006) provide additional applications of WinBUGS.

Appendix C: Kernel Density Estimation

If a random variable X has pdf $f(x)$, then

$$f(x) = \lim_{h \to 0} \frac{1}{2h} P(x - h < X < x + h).$$

This motivates the following naive estimator $\hat{f}(x)$ of $f(x)$ when we have a random sample, $\{X_1 \ldots, X_n\}$:

$$\hat{f}(x) = \frac{1}{2nh} \{\text{No. of } \{X_i\} \in (x - h, \ x + h)\} \qquad \text{(C.1)}$$

$$\text{for some } h > 0.$$

If we define $w(x) = \begin{cases} \frac{1}{2} & \text{if } |x| < 1 \\ 0 & \text{otherwise,} \end{cases}$ then (C.1) is equivalent to:

$$\hat{f}(x) = \frac{1}{nh} \sum_{i=1}^{n} w\left(\frac{x - X_i}{h}\right). \qquad \text{(C.2)}$$

This expression is generalised by the kernel density estimate:

$$\hat{f}(x) = \frac{1}{nh} \sum_{i=1}^{n} K\left(\frac{x - X_i}{h}\right),$$

where $K(x)$ is a non-negative function such that $\int_{-\infty}^{\infty} K(x)dx = 1$. Usually $K(x)$ is taken as a symmetric pdf. An illustration of kernel density estimation is given in Figure C.1. The smoothing parameter h is called the window width, or bandwidth.

The Epanechnikov kernel (not shown here) is optimal in a particular sense (Silverman, 1982, p.42). However Wand and Jones (1995, p.31) concluded that there is little difference in performance between alternative unimodal pdfs used for the kernel, $K(x)$. Much research has been devoted to the question of how to choose h — Wand and Jones (1995, Chapter 3). There is no single rule which will work well in all situations, and we would normally expect to try several values of h in order to appreciate the results from different degrees of smoothing, as in Figure C1. An attractively simple choice (Silverman, 1986, p.48) is to set

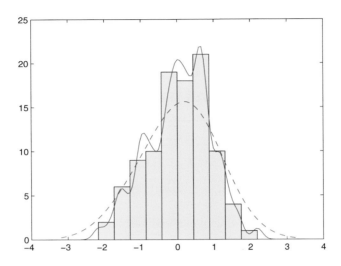

Figure C.1 *Kernel density estimation. A histogram of* 100 $N(0,1)$ *random variates and two kernel density estimates,* —: $h = 2.35 * \hat{\sigma}/n^{\frac{1}{5}}$; —— : $h = 0.9 * \hat{\sigma}/n^{\frac{1}{5}}$; *where n is the sample-size (here $n = 100$) and $\hat{\sigma}$ is an estimate of the sample standard deviation.*

$$h = 0.9 \, An^{-\frac{1}{5}},$$

where n is the sample size and $A = \min$ (standard deviation, interquartile range /1.34). A more complex approach evaluates the asymptotic mean integrated squared error (AMISE) of the estimator of the underlying density, which was encountered in Exercise 6.31. This leads on to what are called 'plug-in' estimates of bandwidth (Wand and Jones, 1995, p.71). A two-stage plug-in estimate of the bandwidth which performs well is described by Shaether and Jones (1991); computerised routines for this procedure exist in both MATLAB and R, and are readily available from the World Wide Web.

We give below simple illustrative MATLAB programs for taking a sample of data, $\{x_i\}$, and plotting a histogram and kernel density estimate for one value of the smoothing parameter h.

First, we have to obtain a sample of size 100 from the $N(0,1)$ distribution, and a histogram of the sample, and we do that by means of the commands:

```
x = randn (100, 1);
hist (x);
hold ;%this retains the graphics window
```

Next we run the set of commands in the file: `kerplot.m`, shown below. This program uses the function *kernel*, which in turn uses the function *delta*, both of which are displayed.

(a)
```
% Program to plot a kernel density estimate for a sample in 'data'.
% 'k1' determines smoothing used in the kernel.
% Calls the function KERNEL.
%--------------------------------------------------------------------
k1=1;
n1=length(data);
range=max(data)-min(data);
mn=min(data)-range/4;mx=max(data)+range/4;
                        % extends the range of 'x-values'
range=range*1.5;        % for kernel plotting

y=zeros(1,n1); f=zeros(1,n1);
for i=1:n1                  % sets up the grid of y-values
  y(i)=mn+i*range/n1;       % and forms the kernel estimate
  f(i)=kernel(y(i),data,k1); % at each point, ready for
end                        % plotting

[n,loca]=hist(data);       % we scale before we
width=diff(loca);          % plot histogram and kernel
scale=sum(n)*width(1);     % estimates on the same graph.
plot(y,f*scale,'-')        % We do this by scaling up the
                           % kernel estimate by the area of
                           % the histogram, in 'scale'.



function z=kernel(y,data,k1)
%
% KERNEL Calculates a kernel density estimate for a sample in 'data'.
%    'y' is the value at which we want the kernel estimate.
%    'k1' can be chosen to vary the amount of smoothing;
%    Calls the function DELTA.
%--------------------------------------------------------------------
n=length(data);
h=k1*std(data)/n^0.2;      % 'h' is the window-width. A particular
z=0;                       % dependence on std() and n is used here
for i=1:n
  z=z+delta((y-data(i))/h); % delta is the function which specifies
end                        % the form of kernel to be used
z=z/(n*h);
```

```
function y=delta(u)
%
% DELTA calculates a normal kernel
%----------------------------------
y=(exp(-u*u/2))/sqrt(2*pi);
```

We conclude this appendix with a small number of further comments. Naive use of kernel methods will result in problems when the range of a random variable is constrained by boundaries. For example the random variable may be non-negative, as in Exercise 6.18. This is discussed by Silverman (1986, Section 2.10). One solution is to reflect the random sample about and beyond the boundary, form a standard density estimate and then double the result for the correct range and ignore the result beyond the boundary; in fact the reflection should only need to be done for a fraction of the sample near the boundary.

Rather than use a fixed window-width, it may be sensible to use an adaptive approach in which the window-width adopted reflects the local density of the sample — see Silverman (1986, Section 5.3). Kernel density estimates may also be obtained from multivariate data (see for example, Wand and Jones, 1994, Chapter 4). An interesting bi-variate application is to the estimation of animal home ranges (Worton, 1989; ESRI, 1999).

Solutions and Comments for Selected Exercises

Chapter 1

1.3 A machine may switch between a normal mode of operation, when it makes no mistakes, and an aberrant mode, when it makes mistakes according to a Poisson distribution. Ridout et al. (1998) distinguish between structural zeros, which are inevitable, and sampling zeros, which occur by chance.

1.4 One might expect ϕ to vary with age and time. Such a model has been developed for herons by North and Morgan (1979). In order to allocate probabilities to cell entries, we need to include in the model probabilities of dead birds being found and reported.

1.6 Suppose r female birds choose their mate from a pool consisting of their previous mates and n additional males. A test statistic for mate fidelity is the number of females that choose their previous mate. In fact, $r + m$ females are choosing mates, and not all birds will find mates.

1.7 For the last 3 months of 2006, the actual figures were, in order: 15, 3, 7. For the first 6 months of 2007, the numbers of papers submitted were, in order: 17, 11, 8, 7, 13, 13.

1.8 The figure of $1/3$ is comparable to the estimates in Section 1.1. Additional estimates of the probability of conception in a month are given in Sections 2.2 and 4.2 (Tables 4.1 and 4.2). The book *Inconceivable* by Elton (1999) gives a figure of $1/5$, following in-vitro fertilisation. *The Independent* of 3 July 1999 reported research that indicated that men become less fertile as they age. Thus it would be useful to include factors such as age as covariates. Maternity wards have limited capacity, and are not designed for features such as a millenium effect.

1.9 Quail build their nests on the ground, so they may leave the nest at an early stage after hatching. If hatching was not synchronised the mother bird would find it harder to protect the young from predation. The same feature is identified in Nile crocodile *Crocodylus niloticus* by Vergne and Mathevon (2008).

Chapter 2

2.2 The data of Table 1.1 were collected in a more controlled manner than those of this question, which were obtained from maternity wards. Possibly as a result of this there is pronounced 'heaping' of responses into 6 months and 12 months. One way of describing this by means of a stochastic model is given in Example 4.2.

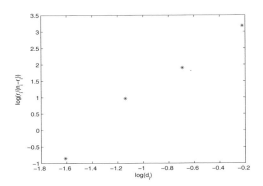

Figure S.1 *Plot of $\log(r_i/(n_i - r_i))$ versus $\log(d_i)$ for male four beetle data.*

2.3 First, note that in models of this kind it is usual to take log (dose), but not essential. It is sensible if there is natural mortality, and Abbott's formula is to be used — see Section 5.6.1. The graphs needed here are plots of r_i/n_i vs log (d_i). In order to judge whether a logistic model is appropriate, we might plot $\log(r_i/(n_i - r_i))$ vs log d_i, and consider whether the plot looks like a straight line. For the males, for example, we obtain Figure S.1. For only four data points, there is little to be said. This could be a more useful diagnostic procedure when many more doses are involved.

The likelihood is equal to:

$$L(\alpha_1, \alpha_2) = \prod_{i=1}^{4} \binom{n_i}{r_i} P(d_i)^{r_i} (1 - P(d_i))^{n_i - r_i},$$

where we focus on just one sex at a time, and r_i die out of n_i exposed at dose d_i, and

$$P(d_i) = 1/[1 + \exp\{-(\alpha_1 + \alpha_2 \log d_i)\}].$$

Thus the likelihood is the product of four binomial distributions, which give the binomial response at each dose separately. We obtain the following maximum-likelihood point estimates:

	α_1	α_2	ED_{50}
Males	4.2704	3.1380	0.2564
Females	2.5618	2.5816	0.3707

The ED_{50} is the dose which we would expect to kill 50% of individuals. We therefore see that the males are more susceptible than the females — it needs less to kill 50%. This could be due to a known difference in the size of male and female flour beetles.

2.4 Suppose the number of encounters with a host is $Po(\lambda)$, and that at the first encounter an egg is always laid, with probability 1, while at all subsequent encounters an egg is laid with probability δ. The total number of eggs laid has probability generating function

$$
\begin{aligned}
Q(z) &= (1 - \delta + \delta z)^{-1}[z \exp\{\lambda\delta(z-1)\} + e^{-\lambda}(1-\delta)(1-z)] \\
&= \sum_{i=0}^{\infty} z^i p_i.
\end{aligned}
$$

$$
p_o = e^{-\lambda},
$$

$$
p_r = \delta^{r-1}(\delta-1)^{-r} e^{-\lambda\delta} \sum_{i=r}^{\infty} \frac{\lambda^i(\delta-1)^i}{i!}, \quad \text{for } r = 1, 2, \ldots.
$$

The likelihood is then $\prod_{r=0}^{5} p_r^{n_r}$, where $n_0 = 19$, $n_1 = 169$, etc.

2.6 In one possible model the probability of confusing j for i, denoted q_{ij}, is given by:

$$
\begin{aligned}
q_{ij} &= (1-s_i)p_j, \quad \text{for } i \neq j \\
&= s_i + (1-s_i)p_i, \quad \text{for } i = j.
\end{aligned}
$$

For the example, this provides a saturated model, with six parameters and six degrees of freedom. The maximum-likelihood estimates are then the result of equating cell values to expected values and solving the equations. There is more discussion of this model in Exercise 5.13. Morgan (1974) discusses alternative possible models for acoustic confusion.

2.7
$$
pr(X = k|p) = \binom{n}{k} p^k (1-p)^{n-k}, \quad k = 0, 1, \ldots, n.
$$

$$
f(p) = \frac{p^{\alpha-1}(1-p)^{\beta-1}}{B(\alpha, \beta)}
$$

$$
pr(X = k) = \int_0^1 pr(X = k|p) f(p) dp
$$

$$
= \frac{\binom{n}{k} \int_0^1 p^{k+\alpha-1}(1-p)^{n-k+\beta-1} dp}{B(\alpha, \beta)}
$$

$$
= \binom{n}{k} \frac{B(\alpha+k, \, n-k+\beta)}{B(\alpha, \beta)}, \quad k = 0, 1, \ldots, n.
$$

Hence
$$
pr(X = k) = \frac{n!}{(n-k)!k!} \frac{\Gamma(\alpha+k)\Gamma(n-k+\beta)\Gamma(\alpha+\beta)}{\Gamma(n+\alpha+\beta)\Gamma(\alpha)\Gamma(\beta)}
$$

$$
= \frac{n!(\alpha+k-1)\ldots\alpha(n-k+\beta-1)\ldots\beta}{(n+\alpha+\beta-1)\ldots(\alpha+\beta)}.
$$

At the first drawing, the denominator of the probability of getting a red ball is: $(r + b)$.

At the ith drawing, it is: $(r + b + i - 1)$.

If k red balls are drawn in all, then the numerators of the probabilities must contain the terms: $r, (r+1), \ldots (r+k-1)$. The remaining numerators are: $b, (b+1), \ldots (n-k+b-1)$. Thus we see that when n balls are drawn from the urn, the probability of k red balls is beta-binomial (equating $b = \beta$, $r = \alpha$).

2.9 $\ell(\lambda) = C - \lambda A + B \log \lambda$

$$\frac{dl}{d\lambda} = -A + \frac{B}{\lambda}; \quad \hat{\lambda} = B/A.$$

$$\frac{d^2 l}{d\lambda^2} = -\frac{B}{\lambda^2} < 0, \quad \text{since } B \text{ is positive.}$$

2.10 $f(p|X > j) = Pr(X > j|p)f(p)/Pr(X > j)$

$$\propto (1-p)^j f(p),$$

as $Pr(X > j)$ does not involve p.

Hence $f(p|X > j) \propto (1-p)^{j+\beta-1} p^{\alpha-1}$,

i.e., $f(p|X > j) \sim Be(\alpha, \beta + j)$.

But $\alpha = \mu/\theta$; $\beta = (1-\mu)/\theta$.

Hence $f(p|X > j) \sim Be\left(\dfrac{\mu}{\theta}, \; j + \dfrac{1-\mu}{\theta}\right)$.

$$\mathbb{E}[X - j|X > j, p] = \frac{1}{p}.$$

$$\mathbb{E}[X - j|X > j] = \int_0^1 \frac{p^{\alpha-2}(1-p)^{j+\beta-1}}{B(\alpha-1, j+\beta)} \cdot \frac{B(\alpha-1, j+\beta)}{B(\alpha, j+\beta)} dp$$

$$= \frac{B(\alpha-1, j+\beta)}{B(\alpha, j+\beta)} = \frac{\Gamma(\alpha-1)\Gamma(j+\beta)\Gamma(\alpha+\beta+j)}{\Gamma(\alpha+j+\beta-1)\Gamma(\alpha)\Gamma(j+\beta)}$$

$$= (\alpha+\beta+j-1)/(\alpha-1)$$

$$= 1 + \frac{\beta+j}{\alpha-1} = \left(\frac{1-\theta+j\theta}{\mu-\theta}\right).$$

Note that integrating with respect to $f(p|X > j)$ is valid as long as we take $\alpha > 1$, which is necessary for a finite expectation.

2.11 These expressions for mean and variance provide a convenient way of evaluating the mean and variance of the beta-geometric distribution without using its probability function directly. Conditional on p, $pr(X = k|p) = (1-p)^{k-1}p$, for $k \geq 1$. Thus X has probability generating function (pgf),

$$
\begin{aligned}
G(z) \quad &= \quad \mathbb{E}[z^X|p] = \sum_{k=1}^{\infty} z^k (1-p)^{k-1} p \\
&= \quad pz/\{1 - z(1-p)\} \\
\log G(z) &= \quad \log p + \log z - \log(1 - z(1-p)) \qquad (a) \\
\frac{1}{G}\frac{dG}{dz} &= \quad \frac{1}{z} + \frac{(1-p)}{\{1 - z(1-p)\}}
\end{aligned}
$$

$$
-\frac{1}{G^2}\left(\frac{dG}{dz}\right)^2 + \frac{1}{G}\frac{d^2G}{dz^2} = -\frac{1}{z^2} + \frac{(1-p)^2}{\{1 - z(1-p)\}^2}. \qquad (b)
$$

We now set $z = 1$ in (a) and (b) and make use of the facts that $G(1) = 1$, $G'(1) = \mathbb{E}[X|p]$ and $G''(1) = \mathbb{E}[X(X-1)|p]$.

From (a), $\mathbb{E}[X|p] = 1 + \dfrac{1-p}{p} = \dfrac{1}{p}$.

From (b), $-\dfrac{1}{p^2} + \mathbb{E}[X(X-1)|p] = -1 + \dfrac{(1-p)^2}{p^2}$, so that

$$
\begin{aligned}
\mathrm{Var}\,(X|p) \quad &= \quad -1 + \frac{1}{p^2} + \frac{(1-p)^2}{p^2} + \frac{1}{p} - \frac{1}{p^2} \\
&= \quad (1-p)/p^2 \quad \text{(after elementary algebra).}
\end{aligned}
$$

Hence,

$$
\begin{aligned}
\mathbb{E}[X] \quad &= \quad \int_0^1 \frac{1}{p}\frac{p^{\alpha-1}(1-p)^{\beta-1}\,dp}{B(\alpha,\beta)} = \frac{B(\alpha-1,\,\beta)}{B(\alpha,\,\beta)} \\[2mm]
&= \quad \frac{\Gamma(\alpha-1)\Gamma(\beta)}{\Gamma(\alpha+\beta-1)}\frac{\Gamma(\alpha+\beta)}{\Gamma(\alpha)\Gamma(\beta)} = \left(\frac{\alpha+\beta-1}{\alpha-1}\right) \\[2mm]
&= \quad \left(\frac{1-\theta}{\mu-\theta}\right), \quad \text{where} \quad \mu = \frac{\alpha}{\alpha+\beta} \quad \text{and} \quad \theta = \frac{1}{\alpha+\beta}.
\end{aligned}
$$

$$
\begin{aligned}
\mathrm{Var}(X) &= \mathbb{E}_p\left[\mathrm{Var}\ (X|p)\right] + \mathrm{Var}_p\left[\mathbb{E}\ [X|p]\right]\\
&= \int_0^1 \frac{(1-p)}{p^2}\ \frac{p^{\alpha-1}(1-p)^{\beta-1}}{B(\alpha,\beta)} dp +\\
&\quad \int \frac{1}{p^2}\ \frac{p^{\alpha-1}(1-p)^{\beta-1}}{B(\alpha,\beta)}\ dp - \left(\frac{1-\theta}{\mu-\theta}\right)^2\\
&= 2\mathbb{E}\left(\frac{1}{p^2}\right) - \mathbb{E}\left(\frac{1}{p}\right) - \left(\frac{1-\theta}{\mu-\theta}\right)^2.
\end{aligned}
$$

$$
\begin{aligned}
\text{Now}\quad \mathbb{E}\left(\frac{1}{p^2}\right) &= \frac{B(\alpha-2,\beta)}{B(\alpha,\beta)} = \frac{\Gamma(\alpha-2)\Gamma(\beta)}{\Gamma(\alpha+\beta-2)}\ \frac{\Gamma(\alpha+\beta)}{\Gamma(\alpha)\Gamma(\beta)}\\
&= \frac{(\alpha+\beta-1)(\alpha+\beta-2)}{(\alpha-1)(\alpha-2)} = \frac{(1-\theta)(1-2\theta)}{(\mu-\theta)(\mu-2\theta)}
\end{aligned}
$$

$$
\begin{aligned}
\text{Hence,}\quad \mathrm{Var}\ (X) &= \frac{2(1-\theta)(1-2\theta)}{(\mu-\theta)(\mu-2\theta)} - \left(\frac{1-\theta}{\mu-\theta}\right) - \frac{(1-\theta)^2}{(\mu-\theta)^2}\\
&= \frac{(1-\theta)\mu(1-\mu)}{(\mu-\theta)^2(\mu-2\theta)}\ .
\end{aligned}
$$

Note that we can calculate $\mathbb{E}[X|p]$, $\mathrm{Var}\ (X|p)$ directly, without using a pgf. For example:

$$
\begin{aligned}
\mathbb{E}[X|p] &= \sum_{k=1}^{\infty} (1-p)^{k-1} pk = p\left(\frac{d}{dx}\sum_{k=1}^{\infty} x^k\right)\Bigg|_{x=(1-p)}\\
&= p\frac{d}{dx}\left(\frac{x}{1-x}\right)\Bigg|_{x=1-p}\\
&= \frac{p}{(1-(1-p)^2} = \frac{1}{p}.
\end{aligned}
$$

The pgf approach is more elegant.

2.13 $$pr(X = k|\lambda) = \frac{e^{-\lambda}\lambda^k}{k!}, \quad k \geq 0;$$

$$f(\lambda) = \frac{e^{-\frac{\lambda}{\beta}}\lambda^{\alpha-1}\beta^{-\alpha}}{\Gamma(\alpha)} \quad \text{for} \quad \lambda \geq 0. \qquad (*)$$

$$pr(X = k) = \int_0^\infty pr(X = k|\lambda)f(\lambda)d\lambda$$

$$= \int_0^\infty \frac{e^{-\lambda}\lambda^k}{k!} \cdot \frac{e^{-\frac{\lambda}{\beta}}\lambda^{\alpha-1}\beta^{-\alpha}}{\Gamma(\alpha)}d\lambda$$

$$= \frac{\beta^{-\lambda}}{k!\Gamma(\alpha)} \int_0^\infty e^{-\lambda(1+\frac{1}{\beta})}\lambda^{k+\alpha-1}d\lambda. \qquad (\dagger)$$

But, from $(*)$, we know that $\int_0^\infty e^{-ax}x^{b-1}dx = \Gamma(b)a^{-b}$, as the integral of a gamma pdf over its range is 1. If we now use this result in $(\dagger)$, we obtain

$$pr(X = k) = \frac{\beta^{-\alpha}}{k!\Gamma(\alpha)} \Gamma(k+\alpha) \left(1 + \frac{1}{\beta}\right)^{-(k+\alpha)}$$

$$= \frac{(k+\alpha-1)!}{k!(\alpha-1)!} \frac{\beta^k}{(1+\beta)^k} \frac{1}{(1+\beta)^\alpha}, \quad k \geq 0,$$

since we are told that α is a positive integer, that is the probability function of a negative-binomial random variable.

2.14 See Exercise 5.22.

2.17 $L(\theta) = \prod_{i=1}^n \left(\frac{u_i}{\theta}\right) = \theta^{-n} \left(\prod_{i=1}^n u_i\right)$ for $0 \leq u_i \leq \theta$, for all i.

We can see that the maximum-likelihood estimator, $\hat{\theta}$, is given by $\hat{\theta} = \max_i(u_i)$.

Chapter 3

3.5 This is a model from signal-detection theory — see Grey and Morgan (1972) and cf. Morgan (1998). The motivation is that there is a latent response which is compared to the cut-offs, z_1 and z_2. The cdf for the latent response is $F(x)$ if only noise is present, and it is $F(bx - a)$ if signal is present. For the example, there are four parameters and four degrees of freedom. Thus the model is saturated (cf. solution to Exercise 2.6) and the fitted values equal the observed values. Comparing a latent variable to cut-offs provides a method for analysing general contingency tables with ordered categories (McCullagh, 1980).

3.10 $(\hat{a}, \hat{b}) = (-14.8084, 0.2492)$

3.12 $(\hat{\alpha}, \hat{\beta}) = (-0.2671, 1.9946)$

3.13 For Exercise 3.10:   0.00001332, –0.00005056
      For Exercise 3.12:   0.0001, 0.0064

3.14 The results are, respectively, 0.2241, 0.3317. The MATLAB function *geo* is shown below.

3.15 If the random variable X has an $Ex(\lambda)$ distribution then

$$P(X > \tau) = \int_{\tau}^{\infty} \lambda e^{-\lambda x} dx = [e^{-\lambda x}]_{\infty}^{\tau} = e^{-\lambda \tau}.$$

In this example the likelihood is made up of a product of pdfs, corresponding to the exactly observed values, $x_1, \ldots, x_m$, and a product of $(n - m)$ probabilities each of which is: $e^{-\lambda \tau}$. Thus,

$$L(\lambda) = \left(\prod_{i=1}^{m} \lambda e^{-\lambda x_i} \right) e^{-(n-m)\tau \lambda}$$

$$= \lambda^m \exp \left[-\lambda \left\{ \sum_{i=1}^{m} x_i + (n - m)\tau \right\} \right],$$

$$\ell(\lambda) = \log_e L(\lambda) = m \log \lambda - \lambda \left\{ \sum_{i=1}^{m} x_i + (n - m)\tau \right\}.$$

Therefore $\dfrac{dl}{d\lambda} = \dfrac{m}{\lambda} - \left\{ \sum_{i=1}^{m} x_i + (n - m)\tau \right\}$, so that from setting $\dfrac{dl}{d\lambda} = 0$,

we get the maximum-likelihood estimate: $\hat{\lambda} = m \left/ \left\{ \sum_{i=1}^{m} x_i + (n - m)\tau \right\} \right.$.

Note that $\dfrac{d^2 l}{d\lambda^2} < 0$, confirming that we have a maximum.

Note also the similarity with $\hat{p}$ in Equation (2.1). The geometric and exponential distributions are both waiting-time distributions, one in discrete time and one in continuous time.

3.18 The maximum-likelihood estimate is $[0.7285, 0.7182]$.

From running the MATLAB program of Figure 3.7, we get the iterations:

$$0.7422 \quad 0.7223 \quad 0.7231 \quad 0.7282 \quad 0.7285 \quad 0.7285$$
$$\rightarrow \hspace{10cm} \rightarrow$$
$$0.4417 \quad 0.6059 \quad 0.6980 \quad 0.7173 \quad 0.7182 \quad 0.7182$$

It is quite easy to make the Newton-Raphson procedure diverge. For example, starts such as $[10, 10]$, $[2, 0.3]$ result in divergence. The implication here is that we need to take care when choosing the start for iterative methods. How we do this is typically problem-dependent.

3.22 See Palmer et al. (2008) for detailed discussion of alternative possibilities, including gamma, inverse Gaussian and tempered stable.

Chapter 4

4.1(a) The MATLAB program *preference* fits a digit-preference model with four parameters: $[\mu, \theta, \alpha, \beta]$. As we can see from the code, the optimisation takes place with respect to, for example, $m1$, where $\mu = 1/(1 + \exp(m1))$.

We know that this is one way of ensuring that μ does not stray outside the range $[0, 1]$. In fact this may not be an appropriate transformation for θ, but in this case it works well. If we start from: $[0.1, 0.2, 0.3, 0.4]$, then we obtain the results:

$$g * 10^3 =$$

$$\left.\begin{array}{r} -0.0688 \\ -0.0217 \\ -0.5886 \\ 0.0721 \end{array}\right\} \begin{array}{l} \text{approximation to the} \\ \text{gradient vector at max;} \\ - \text{ seems suitably small} \end{array}$$

$$\left.\begin{array}{r} 1.4762 \\ 1.1633 \\ 3.1153 \\ 62.0470 \end{array}\right\} \begin{array}{l} \text{eigen-values of Hessian} \\ \text{approximation at max; } - \\ \text{all positive as required} \end{array}$$

$$\left\{\begin{array}{|c|cccc} \hline 0.7821 & 0.5958 & 0 & 0 & 0 \\ 0.2602 & -0.0408 & 0.8348 & 0 & 0 \\ 0.9546 & 0.0331 & -0.0285 & 0.1849 & 0 \\ 2.4935 & -0.2379 & -0.0920 & 0.6867 & 0.8877 \\ \hline \end{array}\right.$$

Note for instance that the third of these is $\widehat{m1}$, not $\widehat{\mu}$. We get, for example,

$$\widehat{\mu} = 1/(1 + e^{\widehat{m1}}) = 0.278,$$

$$\widehat{\theta} = 1/(1 + e^{2 \cdot 4935}) = 0.076.$$

An interesting feature of the analysis is the high standard errors for the parameters that measure the confusion, but they do not suggest that we can set $\alpha = \beta = 0$. They do suggest that we can take $\alpha = \beta$. This constraint may be imposed by using the MATLAB constrained optimisation function: *fmincon*.

(b) It is pleasing to note the good correspondence of $\widehat{\mu}$- values for the non-smokers and the other contraceptive users.

4.2   We use *fmax* on the two programs, the first of which uses, $p = 1/(1 + e^{\alpha + \beta x})$, and the second uses $p = 1/(1 + e^{\beta(x - \mu)})$. The outputs we get are:

| α | 14.8084 | 1.2881 | | μ | 59.4310 | 0.5285 | 0 |
| β | -0.2492 | -0.9948 | 0.0214 | β | -0.2492 | -0.0922 | 0.0214 |

Note that $\widehat{\mu} = -\widehat{\alpha}/\widehat{\beta}$, and that $\widehat{\beta}$ and its standard error remain unchanged.

4.5 This is done in the code of Figure 4.4.

4.7 Take for illustration the pill users. Here, the sample size is $n = 1274$, $\log_e n = 7.15$. Recall that

$$\text{AIC} = -2\log L + 2d; \quad \text{BIC} = -2\log L + d\log_e n$$

Thus we obtain

Model	p	$-2\log L$	AIC	BIC
1	2	5252	5256	5266
2	4	5096	5104	5125
3	3	5118	5124	5139
4	5	5082	5092	5118

Here we see that the BIC selects model 4, just as the AIC did.

4.8 The Aranda-Ordaz model for quantal assay data:

$$P(d) = 1 - (1 + \lambda e^{\alpha + \beta d})^{-\frac{1}{\lambda}},$$

$\lambda = 1$ gives the logit model.

The score test statistic is, from Equation (4.7),

$$\mathbf{U}(\boldsymbol{\phi}_0, \widehat{\boldsymbol{\psi}}_0)'\{\mathbf{J}_{11} - \mathbf{J}_{12}\mathbf{J}_{22}^{-1}\mathbf{J}_{12}'\}^{-1}\mathbf{U}(\boldsymbol{\phi}_0, \widehat{\boldsymbol{\psi}}_0).$$

Here $\mathbf{U}(\boldsymbol{\phi}_0, \widehat{\boldsymbol{\psi}}_0)$ is the vector of scores for $\boldsymbol{\phi}$ alone, evaluated when $\boldsymbol{\phi} = \boldsymbol{\phi}_0$, and $\boldsymbol{\psi} = \widehat{\boldsymbol{\psi}}_0$, which is the maximum-likelihood estimate of $\boldsymbol{\psi}$ when we fix $\boldsymbol{\phi} = \boldsymbol{\phi}_0$.

In our example, we have the parameters (λ, α, β) and $\phi_0 = \lambda = 1$.

Thus for $\mathbf{U}$, all we need is $\frac{\partial \ell}{\partial \lambda}$ evaluated when $\lambda = 1$, and $\widehat{\alpha}$, $\widehat{\beta}$, the last two being the logit (the case $\lambda = 1$) maximum-likelihood estimates.

We have:

$$L \propto \prod_i P(d_i)^{r_i}(1 - P(d_i))^{n_i - r_i}$$

$$\ell = \text{const} + \sum_i \{r_i \log P(d_i) + (n_i - r_i)\log(1 - P(d_i))\}$$

$$\frac{\partial \ell}{\partial \lambda} = \sum_i \left\{ \frac{1}{P(d_i)} r_i \frac{\partial P}{\partial \lambda} - \frac{(n_i - r_i)}{1 - P(d_i)} \frac{\partial P}{\partial \lambda} \right\}. \qquad (a)$$

Note that $\log(1 - P) = -\frac{1}{\lambda}\log(1 + \lambda e^{\alpha + \beta d})$, leading after some algebra, to:

$$\frac{\partial P}{\partial \lambda} = (1 - P)\{\log(1 - P) + P\}. \qquad (b)$$

Thus we substitute $\widehat{P}(d_i)$ for $P(d_i)$ in (a) and (b), where $\widehat{P}(d_i)$ is the fitted probability of response to dose d_i under the logit model.

We find ultimately that

$$\frac{\partial \ell}{\partial \lambda} = \sum_i \left(\frac{r_i - n_i \widehat{P}_i}{\widehat{P}_i} \right) \{\widehat{P}_i + \log(1 - \widehat{P}_i)\},$$

where $\widehat{P}_i \equiv \widehat{P}(d_i)$.

Regarding the elements of $\mathbf{J}$, we already have the second-order derivatives for the parameters α and β, from fitting the logit model. We also need the terms, $\mathbb{E}\left[\frac{\partial^2 \ell}{\partial \lambda^2}\right]$, $\mathbb{E}\left[\frac{\partial^2 \ell}{\partial \lambda \partial \alpha}\right]$, $\mathbb{E}\left[\frac{\partial^2 \ell}{\partial \lambda \partial \beta}\right]$. Take, for example, $\frac{\partial^2 \ell}{\partial \lambda^2}$:

$$\frac{\partial^2 \ell}{\partial \lambda^2} = \sum_i \left\{ \frac{-r_i}{P(d_i)^2} \left(\frac{\partial P}{\partial \lambda}\right)^2 - \frac{(n_i - r_i)}{(1 - P(d_i))^2} \left(\frac{\partial P}{\partial \lambda}\right)^2 \right\}$$

since $\mathbb{E}(r_i) = n_i P(d_i)$. We obtain:

$$-\mathbb{E}\left[\frac{\partial^2 \ell}{\partial \lambda^2}\right] = \sum_i \frac{n_i(1 - \widehat{P}_i)}{\widehat{P}_i} \{\widehat{P}_i + \log(1 - \widehat{P}_i)\}^2$$

and there are similar results for $\mathbb{E}\left[\frac{\partial^2 \ell}{\partial \lambda \partial \beta}\right]$ and $\mathbb{E}\left[\frac{\partial^2 \ell}{\partial \lambda \partial \alpha}\right]$. This may seem like a lot of work, but once it has been done then it may be applied routinely to any set of quantal response data.

4.13   See the solution to Exercise 2.17.

4.15(a) $\begin{aligned} L(\lambda; x_1, x_2) &= (a\lambda)e^{-a\lambda x_1}(b\lambda)e^{-b\lambda x_2} \\ &\propto \lambda^2 \exp\{-\lambda(ax_1 + bx_2)\} \\ \ell(\lambda; x_1, x_2) &= 2\log\lambda - \lambda(ax_1 + bx_2) \end{aligned}$

$\begin{aligned} \frac{dl}{d\lambda} &= \frac{2}{\lambda} - ax_1 - bx_2, & \frac{d^2 l}{d\lambda^2} &= -\frac{2}{\lambda^2} < 0. \\ \hat{\lambda} &= 2/(ax_1 + bx_2), & \lambda^{-1} &= (ax_1 + bx_2)/2. \end{aligned}$

(b) $\mathbb{E}[\widehat{\lambda^{-1}}] = \frac{1}{2}\left(\frac{a}{a\lambda} + \frac{b}{b\lambda}\right) = \frac{1}{\lambda}.$

$\mathrm{Var}(\widehat{\lambda^{-1}}) = \frac{a^2}{4}\frac{1}{(a^2\lambda^2)} + \frac{b^2}{4}\frac{1}{b^2\lambda^2} = \frac{1}{2\lambda^2}.$

$$\mathbb{E}\left[\frac{ab(X_1 + X_2)}{(a+b)}\right] = \frac{1}{\lambda}, \qquad \text{Var}\left[\frac{ab(X_1 + X_2)}{a+b}\right] = \frac{1}{\lambda^2}\left\{\frac{b^2 + a^2}{(a+b)^2}\right\};$$

$$\mathbb{E}\left[\frac{ab(X_1 - X_2)}{(b-a)}\right] = \frac{1}{\lambda}, \qquad \text{Var}\left[\frac{ab(X_1 - X_2)}{(b-a)}\right] = \frac{1}{\lambda^2}\left\{\frac{b^2 + a^2}{(b-a)^2}\right\}.$$

Note that $\dfrac{a^2 + b^2}{(a+b)^2} - \dfrac{1}{2} = \dfrac{(a-b)^2}{2(a+b)^2} > 0.$

Also, $\dfrac{a^2 + b^2}{(a-b)^2} - \dfrac{1}{2} = \dfrac{(a+b)^2}{2(a-b)^2} > 0,$

and so in order of decreasing variance, the estimators are:

$$\frac{ab(X_1 - X_2)}{(b-1)}, \quad \frac{ab(X_1 + X_2)}{(a+b)}, \quad (aX_1 + bX_2)/2.$$

4.16
$$L(\theta; \mathbf{x}) = \prod_i \frac{1}{x!}\exp(\theta x_i - e^\theta)$$
$$\propto \exp(n\theta\bar{x} - ne^\theta)$$
$$\ell(\theta; \mathbf{x}) = n\theta\bar{x} - ne^\theta,$$
$$\frac{dl}{d\theta} = n\bar{x} - ne^\theta; \quad \frac{d^2 l}{d\theta^2} = -ne^\theta < 0.$$

The maximum-likelihood estimator is $\hat{\theta} = \log \bar{x}$.

This comes as no surprise, since if we set $\lambda = e^\theta$, we see that $pr(X = k)$ is Poisson, with $\hat{\lambda} = \bar{x}$.

4.34 In the notation for the Weibull distribution given in Appendix A, we are testing $\kappa = 1$. Under the exponential hypothesis, in an obvious notation, the appropriate leading term in the inverse of the observed information matrix has the form
$$v = (I_{\kappa\kappa} - I_{\kappa\rho}^2/I_{\rho\rho})^{-1}.$$

The score test statistic is then $u^2 v$.

4.36
$$\frac{d^2\ell}{dp^2} = -\frac{\sum_1^r n_i}{p^2} - \frac{\sum_1^{r+1}(i-1)n_i}{(1-p)^2} < 0.$$

$\mathbb{E}\left[\dfrac{d^2\ell}{dp^2}\right]$ requires terms like $\mathbb{E}[n_i]$.

We have $\mathbb{E}[n_i] = np(1-p)^{i-1}$, for $i = 1, \ldots, r$, and $\mathbb{E}[n_{r+1}] = n(1-p)^r$. Simple algebra then gives the required result.

Hence, the asymptotic variance of the maximum likelihood estimate $\hat{p}$ is
$$\text{Var}(\hat{p}) = \frac{p^2 q}{n(1-q^r)},$$

where $q = 1 - p$. Also, $\text{Var}(\tilde{p}) = pq/n$. Hence $\text{Var}(\hat{p})/\text{Var}(\tilde{p}) = p/(1-q^r) <$

1, as expected.

Chapter 5

5.4 We have
$$
\begin{aligned}
f(x) &= \kappa\rho(\rho x)^{\kappa-1}\,\exp\{-(\rho x)^{\kappa}\}, \text{ for } x \ge 0 \\
y &= x^{\kappa};\ \frac{dy}{dx} = \kappa x^{\kappa-1} \\
f(y) &= \rho^{\kappa}\,\exp(-\rho^{\kappa}y) \text{ for } y \ge 0. \\
\text{Thus} &\qquad f(y) \sim \text{Ex}(\rho^{\kappa}). \\
\text{If } z &= \rho^{\kappa}x^{\kappa},\quad f(z) \sim \text{Ex}(1).
\end{aligned}
$$

5.6 Set $f(y; \psi, \phi) = \dfrac{\psi}{\phi}\,\dfrac{y^{\psi-1}}{\phi^{\psi-1}}\,\exp\left\{-\left(\dfrac{y}{\phi}\right)^{\phi}\right\}.$

We shall re-parameterise, from (ψ, ϕ) to (ψ, λ).

Let $\quad l = \log f$:

$$
l = \log\psi + (\psi-1)\log y - \psi\log\phi - \left(\frac{y}{\phi}\right)^{\psi}
$$

$$
\frac{\partial l}{\partial \phi} = -\frac{\psi}{\phi} + \frac{\psi y^{\psi}}{\phi^{\psi+1}},
$$

$$
\frac{\partial^2 l}{\partial \phi^2} = \frac{\psi}{\phi^2} - \frac{\psi(\psi+1)y^{\psi}}{\phi^{\psi+2}}.
$$

From Exercise 5.4, we know that $\left(\dfrac{Y}{\phi}\right)^{\psi}$ has the exponential distribution, of mean 1.

Hence, $\quad \mathbb{E}\left[\dfrac{\partial^2 l}{\partial \phi^2}\right] = \dfrac{\psi}{\phi^2}\{1 - \psi - 1\} = -\left(\dfrac{\psi}{\phi}\right)^2.$

Also, $\quad \dfrac{\partial^2 l}{\partial \phi \partial \psi} = -\dfrac{1}{\phi} + \dfrac{y^{\psi}}{\phi^{\psi+1}} + \dfrac{\psi}{\phi}\left(\dfrac{y}{\phi}\right)^{\psi}\log\left(\dfrac{y}{\phi}\right).$

But when $X \sim e^{-x}$, $\mathbb{E}[X\log X] = 1 - \gamma$, where γ is Euler's constant ($\gamma \approx 0.577215$).

Hence, $\quad \mathbb{E}\left[\dfrac{\partial^2 l}{\partial \phi \partial \psi}\right] = -\dfrac{1}{\phi} + \dfrac{1}{\phi} + \dfrac{1}{\phi}(1-\gamma) = \dfrac{(1-\gamma)}{\phi}.$

Hence, Equation (5.6) is here:

$$\frac{d\phi}{d\psi}\left(\frac{\psi}{\phi}\right)^2 = \frac{(1-\gamma)}{\phi}.$$

$$\frac{d\phi}{\phi} = (1-\gamma)\frac{d\psi}{\psi^2},$$

$$\log\phi = \left(\frac{\gamma-1}{\psi}\right) + \log\lambda, \quad \text{say, so that}$$

$$\phi = \lambda\exp\left(\frac{\gamma-1}{\psi}\right),$$

$$\lambda = \phi\exp\left(\frac{1-\gamma}{\psi}\right).$$

Note that the survivor function for the Weibull distribution can now be written as $S(y) = \exp\left\{-\left(\frac{y}{\lambda}\right)^\psi e^{1-\gamma}\right\}.$

5.7 The function *logit* is provided in Exercise 3.10. *fmax* (*'logit'*, [1,2]) produces the result:

Parameter	Estimate	Std. errors + correlation	
α	14·8084	1·2881	0
β	−0·2492	−0.9948	0·0214

Here, the model is:

$$P(d) = \frac{1}{1 + e^{-(\alpha+\beta d)}}.$$

If we change the parameterisation to give:

$$P(d) = \frac{1}{1 + e^{-\beta(d-\mu)}},$$

we obtain:

Parameter	Estimate	Std. errors + correlation	
μ	59·4310	0·5285	0
β	−0·2492	−0.0922	0·0214

We note that $\widehat{\mu} = -\widehat{\alpha}/\widehat{\beta}$, a standard result for maximum-likelihood estimators.

We note that $\widehat{\beta}$ has not changed and it also has the same standard error. We note that $\text{Corr}(\widehat{\mu}, \widehat{\beta}) << \text{Corr}(\widehat{\alpha}, \widehat{\beta})$. Now

$$\text{Corr}(\widehat{\beta}, \widehat{\alpha}) = \text{Cov}(\widehat{\alpha}, \widehat{\beta})\{se(\widehat{\alpha})se(\widehat{\beta})\}.$$

Hence $\text{Cov}(\widehat{\alpha}, \widehat{\beta}) = -0.9948 \times 1.2881 \times 0.0214 = -0.02742$. Hence the variance-covariance matrix for $(\widehat{\alpha}, \widehat{\beta})$ is:

$$\mathbf{S} = \left[\begin{array}{cc} 1{\cdot}6592 & -0{\cdot}02742 \\ -0{\cdot}02742 & 0{\cdot}00045796 \end{array} \right].$$

By the δ-method,

$$\text{Var}(\widehat{\mu}) \approx \left(-\frac{1}{\widehat{\beta}}, \frac{\widehat{\alpha}}{\widehat{\beta}^2} \right) \mathbf{S} \left(\begin{array}{c} -\frac{1}{\widehat{\beta}} \\ \frac{\widehat{\alpha}}{\widehat{\beta}^2} \end{array} \right),$$

where

$$\left(-\frac{1}{\widehat{\beta}^2}, \frac{\widehat{\alpha}}{\widehat{\beta}^2} \right) = (4{\cdot}01284, 238{\cdot}4581).$$

This gives $\text{Var}(\widehat{\mu}) \approx 0{\cdot}2787$.

From the direct analysis above, we get $\text{Var}(\widehat{\mu}) = 0{\cdot}5285^2 = 0{\cdot}2793$.

Thus in this instance the two approaches produce results that agree well.

5.12 *Poismix* is a program to fit a mixture of two Poisson distributions:

$$pr(i) = \pi \frac{e^{-\theta_1}\theta_1^i}{i!} + (1 - \pi)\frac{e^{-\theta_2}\theta_2^i}{i!},$$

using maximum-likelihood. It does this by using the EM algorithm. The end result is:

$$\widehat{\pi} = 0{\cdot}3608, \ \widehat{\theta}_1 = 1{\cdot}2577, \ \widehat{\theta}_2 = 2{\cdot}6645.$$

This agrees well with the solution given in the question, which has been obtained by maximising the likelihood directly, using *fminsearch*. Note the (slow) speed of convergence. Note also the dramatic improvement of fit obtained by moving from a single Poisson to a mixture of two Poissons.

5.16 This is a question designed to extend our use of the EM algorithm for fitting an exponential model when there are censored data present.

The Weibull extends the exponential model, with the addition of a single extra parameter. However it is hard to implement the EM algorithm here for two reasons: we cannot form $\mathbb{E}[X|X > y]$ easily. This can be done in MATLAB by using the numerical integration facilities of *quad*, or *quad8*. Even then, however, the maximisation stage is still not explicit, as we saw in Example 5.2.

Let us now consider the gamma pdf, as an alternative extension of the exponential pdf:

$$f(x) = \frac{e^{-\lambda x}\lambda^n x^{n-1}}{\Gamma(n)}, \quad \text{for } x \geq 0.$$

We note that $n = 1$ returns us to the exponential case. As an illustration, we shall take $n = 2$ and show how to form $\mathbb{E}[X|X > y]$ in this instance:

$$\mathbb{E}[X|X > y] = \frac{\int_y^\infty xe^{-\lambda x}\lambda^2 x dx}{\lambda^2 \int_y^\infty e^{-\lambda x} x dx}.$$

Let $I = \lambda^2 \int_y^\infty xe^{-\lambda x} dx$. We integrate by parts:

$$
\begin{aligned}
I = -\lambda \int_y^\infty xd(e^{-\lambda x}) &= -\lambda\left[xe^{-\lambda x}\right]_y^\infty + \lambda \int_y^\infty e^{-\lambda x} dx \\
&= e^{-\lambda y}(1 + \lambda y)
\end{aligned}
$$

(checks when $y = 0$ and when $y = \infty$).

Let $J = \lambda^2 \int_y^\infty x^2 e^{-\lambda x} dx$. We can show that $J = \lambda y^2 e^{-\lambda y} + 2I/\lambda$, and after algebra, we obtain $J = \frac{e^{-\lambda y}}{\lambda}(\lambda^2 y^2 + 2\lambda y + 2)$.

Hence,

$$\mathbb{E}[X|X > y] = \frac{(\lambda^2 y^2 + 2\lambda y + 2)}{\lambda(1 + \lambda y)} = \frac{2}{\lambda} + \frac{\lambda y^2}{(1 + \lambda y)}. \qquad (*)$$

We can check that when $y = 0$, we get $2/\lambda$, which is $\mathbb{E}[X]$, as it should be.

This completes the expectation stage – we replace each censored value y_i by the result of $(*)$, after substituting y_i for y and using the current estimate of λ. We then treat the values resulting from $(*)$ as if they were uncensored and carry out the maximisation as if there were no censored values. This is easily done, as you can readily verify, and so the iteration proceeds, until convergence. Note finally that the adjustment of $(*)$ is not as simple as in the exponential case: $\left(y + \frac{1}{\lambda}\right)$. This is because the gamma distribution has 'memory,' which we know is not true of the exponential distribution – see Appendix A.

5.22 We note that four of the data entries are incompatible with the assumption that parent branches produce at most two offspring branches. This is rare (Ridout et al., 1999) and these cases are discussed separately in the source paper. For example, take $m = 3$; the possible values for X are: 0, 1, 2, ..., 6, with probabilities $\{q_i, \quad i = 0, 1, \ldots, 6\}$, say. If the corresponding output frequencies are $\{r_i, \quad i = 0, 1, \ldots, 6\}$, then we have the multinomial likelihood:

$$L(\pi, \rho) \propto \prod_{i=0}^6 q_i^{r_i}.$$

For illustration, we continue to consider the case $m = 3$, and we list below the possible outcomes and corresponding values of X (the outcome is the number of offspring branches at each of the three parent branches).

Outcome $\{u, v, w\}$	X	Underlying frequency	Probability	Observed frequency	Probability
$\{0,0,0\}$	0	r_0	p_0^3	r_0	p_0^3
$\{0,0,1\}$	1	r_1	$3p_0^2 p_1$	r_1	$3p_0^2 p_1$
$\{0,0,2\}$	2	s_2	$3p_0^2 p_2$		
$\{0,1,1\}$	2	t_2	$3p_0 p_1^2$	r_2	$3p_0^2 p_2 + 3p_0 p_1^2$
$\{0,1,2\}$	3	s_3	$6p_0 p_1 p_2$		
$\{1,1,1\}$	3	t_3	p_1^3	r_3	$6p_0 p_1 p_2 + p_1^3$
$\{1,1,2\}$	4	s_4	$3p_1^2 p_2$		
$\{0,2,2\}$	4	t_4	$3p_0 p_2^2$	r_4	$3p_1^2 p_2 + 3p_0 p_2^2$
$\{1,2,2\}$	5	r_5	$3p_1 p_2^2$	r_5	$3p_1 p_2^2$
$\{2,2,2\}$	6	r_6	p_2^3	r_6	p_2^3

If the underlying frequencies were known then the likelihood is trinomial in form, with p_1 raised to the power:

$$\begin{aligned} \nu_1 &= r_1 + 2t_2 + s_3 + 3t_3 + 2s_4 + r_5, \text{ and } p_2 \text{ raised to the power:} \\ \upsilon_2 &= s_2 + s_3 + s_4 + 2t_4 + 2r_5 + 3r_6. \end{aligned}$$

In this case, the maximum-likelihood estimate of π would be:

$$\widehat{\pi} = (\nu_1 + 2\nu_2)/(6n),$$

where n is the number of inflorescences. Thus

$$\begin{aligned} \widehat{\pi} &= \{r_1 + 2(s_2 + t_2) + 3(s_3 + t_3) + 4(s_4 + t_4) + 5r_5 + 6r_6)\}/(6n) \\ &= (r_1 + 2r_2 + 3r_3 + 4r_4 + 5r_5 + 6r_6)/(6n) \\ &= \bar{X}/6. \end{aligned}$$

We see that $\widehat{\pi}$ depends only on the $\{r_i\}$, and not on the unobserved frequencies t_i and s_i $(i = 2, 3, 4)$. Thus if we estimate parameters by the EM algorithm, replacing t_i and s_i by their expectations at each iteration, then at each iteration we have $\widehat{\pi} = \bar{X}/6$, which must therefore be the maximum-likelihood estimate of π. Additional models are considered by Cole et al. (2003, 2005).

5.23 The negative binomial distribution can be expressed in slightly different ways. It is not hard to see how the expression of this question can be cast in the form given in Appendix A.

5.30 We have
$$\mathbf{x}' = \mathbf{Ax} + \mathbf{Bg}$$
As $\mathbf{x}(0) = \mathbf{0}, \quad s\tilde{\mathbf{x}} = \mathbf{A}\tilde{\mathbf{x}} + \mathbf{B}\tilde{\mathbf{g}},$ and
$$\tilde{y} = \mathbf{c}'(s\mathbf{I} - \mathbf{A})^{-1}\mathbf{B}\tilde{\mathbf{g}}, \text{ and we can write}$$
$$\tilde{y} = Q(s)\tilde{g}, \text{ where}$$
$$Q(s) = \mathbf{c}'(s\mathbf{I} - \mathbf{A})^{-1}\mathbf{B}.$$

In our case,

$$Q(s) = \theta_1\theta_3/\{s^3 + (\Sigma\theta_i)s^2 + \theta_1\theta_3 + \theta_1\theta_4 + \theta_1\theta_5 + \theta_2\theta_4 + \theta_2\theta_5 + \theta_3\theta_5)s + \theta_1\theta_3\theta_5\},$$

which we may write as:

$$\kappa_1/(s^3 + \kappa_2 s^2 + \kappa_3 s + \kappa_4).$$

To seek estimable parameters, we need to solve:

$$\boldsymbol{\alpha}' \frac{\partial y(t)}{\partial\boldsymbol{\theta}} = 0.$$

This is equivalent to $\boldsymbol{\alpha}' \dfrac{\partial\tilde{y}}{\partial\boldsymbol{\theta}} = 0$, which is equivalent to $\boldsymbol{\alpha}' \dfrac{\partial Q(s)}{\partial\boldsymbol{\theta}} = 0$. By

the chain rule, this becomes: $\boldsymbol{\alpha}' \dfrac{\partial\boldsymbol{\kappa}}{\partial\boldsymbol{\theta}} \dfrac{\partial Q(s)}{\partial\boldsymbol{\kappa}} = 0.$

As this equation must hold for any s, then the effective derivative matrix for this exercise is:

$$\mathbf{D} = \frac{\partial\boldsymbol{\kappa}}{\partial\boldsymbol{\theta}} = \begin{pmatrix} \theta_3 & 1 & \theta_3 + \theta_4 + \theta_5 & \theta_3\theta_5 \\ 0 & 1 & \theta_4 + \theta_5 & 0 \\ \theta_1 & 1 & \theta_1 + \theta_5 & \theta_1\theta_5 \\ 0 & 1 & \theta_1 + \theta_2 & 0 \\ 0 & 1 & \theta_1 + \theta_2 + \theta_3 & \theta_1\theta_3 \end{pmatrix}.$$

The solution to $\boldsymbol{\alpha}'\mathbf{D} = \mathbf{0}$ tells us that θ_5 is estimable, as are: $\theta_1\theta_3$, $\theta_1 + \theta_2 + \theta_3 + \theta_4$ and $(\theta_1 + \theta_2 - \theta_5)\theta_4$ — see Section 5.3.3 and Catchpole et al. (1998).

Chapter 6

6.1 Let $\{E_i, \ i = 1, 2, \ldots\}$ be a sequence of independent $Ex(\lambda)$ random variables. Suppose $S_0 = 0$ and $S_k = \displaystyle\sum_{i=1}^{k} E_i$, for $k = 1, 2, \ldots$. From the Poisson process results of Appendix A, we know that the random variable K defined by:

$$S_K \leq 1 < S_{K+1} \quad \text{is} \quad Po(\lambda).$$

We know that if $\{U_i, i = 1, 2, \ldots\}$ is a sequence of independent $U(0, 1)$ random variables, then we obtain $\{E_i\}$ from setting $E_i = -\dfrac{1}{\lambda}\log(U_i)$. Hence

$$S_k = -\frac{1}{\lambda}\log\left(\prod_{i=1}^{k} U_i\right), \text{ and we can see that the comparison } S_k > 1 \text{ is}$$

equivalent to $\prod_{i=1}^{k} U_i < e^{-\lambda}$. It is this check which is directly coded in the MATLAB program of Figure 6.3(b).

6.2 For the Weibull distribution, the pdf has the form:

$$f(x) = \kappa\rho(\rho x)^{\kappa-1}\exp\{-(\rho x)^{\kappa},\} \quad \text{for } x \geq 0.$$

Thus the cdf is given by:

$$F(x) = 1 - \exp\{-(\rho x)^{\kappa}\},$$

leading to the simulation algorithm:

$$X = \frac{1}{\kappa}\{-\log(U)\}^{\frac{1}{\rho}},$$

where $U \sim U(0,1)$. (We use U, rather than $1-U$.)

In the logistic case, $F(x) = \dfrac{1}{1+e^{-(\alpha+\beta x)}}$, leading to the simulation algorithm:

$X = \dfrac{1}{\beta}\{\log\left(\dfrac{U}{1-U}\right) - \alpha\}$, where $U \sim U(0,1)$.

6.3 The half-normal pdf is given by:

$$f(x) = \sqrt{\frac{2}{\pi}}\, e^{-\frac{x^2}{2}} \quad \text{for } x \geq 0.$$

To find the multiple, k, of an $Ex(1)$ pdf to form an envelope, we set

$$ke^{-x} = \sqrt{\frac{2}{\pi}}\, e^{-\frac{x^2}{2}}.$$

This results in a quadratic equation in x. The envelope will just touch $f(x)$ if the quadratic has equal roots, which is when $k = \sqrt{\dfrac{2e}{\pi}}$.

The rejection method then works as follows:
We set $X = -\log U_1$, where $U_1 \sim U(0,1)$.
We set $Y = kU_2 e^{-X}$, where U_2 is independent $U(0,1)$.

Thus, $Y = kU_1U_2$. We accept X if and only if $Y < \sqrt{\dfrac{2}{\pi}}e^{-X^2/2}$, that is

$kU_1U_2 < \sqrt{\dfrac{2}{\pi}}e^{-\frac{X^2}{2}}$,
or $U_1U_2 < \exp\{-(1+X^2)/2\}$.

This is equivalent to: $\log U_1 + \log U_2 < -\dfrac{(1+X^2)}{2}$,

$$\therefore \quad 2X_1 + 2X_2 > 1 + X_1^2,$$
$$\therefore \quad (X_1-1)^2 < 2X_2.$$

This rejection method is illustrated in Figure S.2.

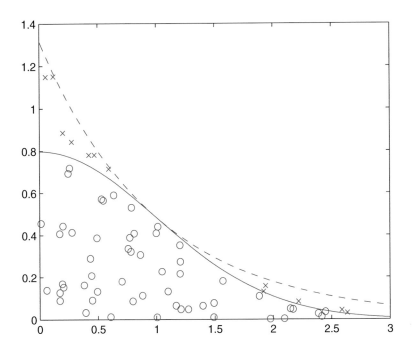

Figure S.2 *Illustration of the rejection method for simulating half-normal random variates, using an exponential envelope. The points denoted by crosses are rejected, and realisations of the required random variable are the x-values of the points denoted by circles.*

6.4 Let $z = f(x)/h(x)$, where $f(x) = \dfrac{x^{r-1}e^{-x}}{\Gamma(r)}$, for $x \geq 0$, and $h(x) = e^{-\frac{x}{r}}/r$ for $x \geq 0$. Then $\log z = (r-1)\log x - x + \dfrac{x}{r} + \log(r/\Gamma(r))$.

It is easily shown that $\log z$ is maximised by setting $x = r$, making use of the fact that $r > 1$. When $x = r$, $z = \dfrac{r^r e^{1-r}}{\Gamma(r)}$.

Thus $f(x) \leq kh(x)$, where $k = r^r e^{1-r}/\Gamma(r)$.

It is readily verified that this approach also produces k for Exercise 6.3.

6.6 We can write $\quad \Phi(x) = \displaystyle\int_{-\infty}^{x} \dfrac{k\phi(y)}{f(y)}\,\dfrac{f(y)}{k}\,dy,$

where k is chosen so that $f(y)/k$ is a pdf over the range $(-\infty, x)$. Hence,

$$f(y)/k = \frac{\pi \exp(-\pi y/\sqrt{3})\{1 + \exp(-\pi x/\sqrt{3})\}}{\sqrt{3}\{1 + \exp(-\pi y/\sqrt{3})\}^2}, \quad -\infty < y < x. \quad (*).$$

If Y is a random variable with the pdf of (*), then

$$\Phi(x) = \{1 + \exp(-\pi x/\sqrt{3})\}^{-1} \mathbb{E}[\phi(Y)/f(Y)],$$

and we see that we can estimate $\Phi(x)$ by:

$$\hat{\Phi}(x) = \frac{1}{n}\{1 + \exp(-\pi x/\sqrt{3})\}^{-1} \sum_{i=1}^{n} \phi(Y_i)/f(Y_i),$$

when $\{Y_i, \quad i = 1, 2, \ldots, n\}$ is a random sample from the pdf of (*). The $\{Y_i\}$ are readily obtained by the inversion method. Morgan (1984, pp.170, 184) provides examples of $\hat{\Phi}(x)$ for example values of x and n.

6.19 We wish to simulate from the pdf $f(x)$, and the enveloping function is $g(x) = kh(x)$, where $h(x)$ is also a pdf.

$$\text{Probability of rejection} = \frac{\int\{g(x) - f(x)\}dx}{\int g(x)dx} = 1 - \frac{1}{k}.$$

Chapter 7

7.1
$$\pi(\theta) = \frac{e^{-\frac{1}{2}(\theta - \mu_0)^2/\sigma_0^2}}{\sigma_0\sqrt{2\pi}}$$

$$L(\theta, \sigma^2) = \prod_{i=1}^{n} \frac{e^{-\frac{1}{2}(x_i - \theta)^2/\sigma^2}}{\sigma\sqrt{2\pi}}$$

$$\pi(\theta|\mathbf{x}) \propto \exp\left[\frac{1}{2}\left\{\sum_{i=1}^{n}\frac{(x_i - \theta)^2}{\sigma^2} + \frac{(\theta - \mu_0)^2}{\sigma_0^2}\right\}\right].$$

In brackets { } we get:

$$\theta^2\left(\frac{1}{\sigma_0^2} + \frac{n}{\sigma^2}\right) - 2\theta\left(\frac{n\bar{x}}{\sigma^2} + \frac{\mu_0}{\sigma_0^2}\right) + \frac{\sum_{i=1}^{n}x_i^2}{\sigma^2} + \frac{\mu_0^2}{\sigma_0^2}$$

$$= \left\{\theta - \frac{\left(\frac{n\bar{x}}{\sigma^2} + \frac{\mu_0}{\sigma_0^2}\right)}{\left(\frac{n}{\sigma^2} + \frac{1}{\sigma_0^2}\right)}\right\}^2\left(\frac{n}{\sigma^2} + \frac{1}{\sigma_0^2}\right) + \frac{\sum_{i=1}^{n}x_i^2}{\sigma^2} + \frac{\mu_0^2}{\sigma_0^2} - \frac{\left(\frac{n\bar{x}}{\sigma^2} + \frac{\mu_0}{\sigma_0^2}\right)^2}{\left(\frac{n}{\sigma^2} + \frac{1}{\sigma_0^2}\right)}.$$

Hence, $\pi(\theta|\mathbf{x}) \propto \exp\left\{-\frac{1}{2}(\theta - \mu_n)^2/\sigma_n^2\right\},$

where $\mu_n = \left(\frac{n\bar{x}}{\sigma^2} + \frac{\mu_0}{\sigma_0^2}\right)\Big/\left(\frac{n}{\sigma^2} + \frac{1}{\sigma_0^2}\right)$

and $\sigma_n^{-2} = n\sigma^{-2} + \sigma_0^{-2}.$

Thus $\pi(\theta|\mathbf{x}) \sim N(\mu_n, \sigma_n^2).$

7.2 $\pi(\theta|r)$ $\propto$ $\theta^r \, (1-\theta)^{n-r} \, \theta^{a-1} \, (1-\theta)^{b-1}$
                  $=$ $\theta^{a+r-1}(1-\theta)^{n-r+b-1}$,
        $\therefore \pi(\theta|r)$ $\sim$ $Be(a+r, \; n-r+b)$.

7.3 $f(y|n)$ $=$ $f(x,y|n)/pr(x|y, \, n)$
               $\propto$ $y^{x+\alpha-1}(1-y)^{n-x+\beta-1}/\{y^x(1-y)^{n-x}\}$
               $\sim$ $Be(\alpha, \beta)$.

7.4 $s\mathbf{Q}$ $=$ $s\mathbf{Q}_y \mathbf{Q}_x$

$$= \quad (p_1+p_3, \quad p_2+p_4) \begin{bmatrix} \dfrac{p_1}{p_1+p_3} & \dfrac{p_3}{p_1+p_3} \\[2ex] \dfrac{p_2}{p_2+p_4} & \dfrac{p_4}{p_2+p_4} \end{bmatrix} \begin{bmatrix} \dfrac{p_1}{p_1+p_2} & \dfrac{p_2}{p_1+p_2} \\[2ex] \dfrac{p_3}{p_3+p_4} & \dfrac{p_4}{p_3+p_4} \end{bmatrix}$$

$$= \quad (p_1+p_2, \quad p_3+p_4) \begin{bmatrix} \dfrac{p_1}{p_1+p_2} & \dfrac{p_2}{p_1+p_2} \\[2ex] \dfrac{p_3}{p_3+p_4} & \dfrac{p_4}{p_3+p_4} \end{bmatrix}$$

$$= \quad (p_1+p_3, \quad p_2+p_4).$$

7.5 Models are considered in Example 7.2. Possibilities, designated by the parameter sets, are:

$$(\phi_1, \phi_a, \lambda);$$
$$(\phi_1, \phi_a, \{\lambda_i\});$$
$$(\{\phi_{1,i}\}, \phi_a, \{\lambda_i\});$$
$$(\{\phi_{1i}\}, \phi_a, \lambda);$$
$$(\phi_1, \phi_2, \ldots, \phi_{12}, \lambda).$$

The recovery data are typically sparse, and so additional models can be expected to produce flat likelihood surfaces. A conservative approach would be to set all prior model probabilities to be equal. As in Figure 7.6, beta prior distributions for model probabilities are a natural choice. A conservative approach is to use uniform distributions.

7.7 For Equation (7.14), $\pi_i p_{ij}$ $=$ $\pi_i q_{ij} \; \pi_j/\pi_i$, if $\pi_j \leq \pi_i$,
                            $=$ $\pi_i q_{ij}$,   if $\pi_j \geq \pi_i$.

Thus if $\pi_j \leq \pi_i$, $\pi_i p_{ij}$ $=$ $\pi_j q_{ij}$
                          $=$ $\pi_j q_{ji}$, by symmetry,
                          $=$ $\pi_j p_{ji}$, as $\alpha_{ji} = 1$.

Additionally, if $\pi_i \geq \pi_j$,
                 $\pi_i p_{ij}$ $=$ $\pi_i q_{ij} = \pi_i q_{ji}$, by symmetry,
                              $=$ $\pi_j p_{ji}$, as $\alpha_{ji} = \pi_i/\pi_j$.

For Equation (7.10), in outline, write $p(\mathbf{a}, \mathbf{b}) = pr(\mathbf{X}^{(i+1)} = \mathbf{b}|\mathbf{X} = \mathbf{a})$. Then, $\pi(\mathbf{a})p(\mathbf{a}, \mathbf{b}) = \pi(\mathbf{a})q(\mathbf{a}, \mathbf{b})\alpha(\mathbf{a}, \mathbf{b})$.

Suppose $\quad \pi(\mathbf{b})q(\mathbf{b},\mathbf{a}) \quad < \quad \pi(\mathbf{a})q(\mathbf{a},\mathbf{b}),$

then $\qquad\qquad \alpha(\mathbf{b},\mathbf{a}) \;=\; 1, \text{ and } \alpha(\mathbf{a},\mathbf{b}) = \dfrac{\pi(\mathbf{b})q(\mathbf{b},\mathbf{a})}{\pi(\mathbf{a})q(\mathbf{a},\mathbf{b})}. \qquad (*)$

$$
\begin{aligned}
\text{Now,} \quad \pi(\mathbf{b})p(\mathbf{b},\mathbf{a}) &= \pi(\mathbf{b})q(\mathbf{b},\mathbf{a})\alpha(\mathbf{b},\mathbf{a}). \\
\text{Hence,} \quad \pi(\mathbf{b})p(\mathbf{b},\mathbf{a}) &= \pi(\mathbf{b})q(\mathbf{b},\mathbf{a}) \\
&= \pi(\mathbf{a})q(\mathbf{a},\mathbf{b})\alpha(\mathbf{a},\mathbf{b}), \quad \text{from} \quad (*) \\
&= \pi(\mathbf{a})p(\mathbf{a},\mathbf{b}).
\end{aligned}
$$

If $\pi(\mathbf{b})q(\mathbf{b},\mathbf{a}) \geq \pi(\mathbf{a})q(\mathbf{a},\mathbf{b})$, then reversibility follows from a similar argument to that above.

7.8 In this example, for a symmetric function $q(\mathbf{x},\mathbf{y})$,

$$
\alpha(\mathbf{x},\mathbf{y}) = \min\left[\frac{\exp\{-\frac{1}{2}(\mathbf{y}-\boldsymbol{\mu})'\boldsymbol{\Sigma}^{-1}(\mathbf{y}-\boldsymbol{\mu})\}}{\exp\{-\frac{1}{2}(\mathbf{x}-\boldsymbol{\mu})'\boldsymbol{\Sigma}^{-1}(\mathbf{x}-\boldsymbol{\mu})\}},1\right],
$$

where we wish to simulate from $N_2(\boldsymbol{\mu},\boldsymbol{\Sigma})$.

Chib and Greenberg (1995) provide a range of functions for $q(\mathbf{x},\mathbf{y})$, such as:

(i) $\mathbf{y}=\mathbf{x}+\mathbf{z}$, where $z_i \sim U(-\delta_i,\delta_i)$, $i=1,2$, and z_1 and z_2 are independent. They let $\delta_1 = 0.75$ and $\delta_2 = 1$.

(ii) $\mathbf{y} = \mathbf{x}+\mathbf{z}$, where $\mathbf{z} \sim N_2(0,\mathbf{D})$ and $\mathbf{D}$ is the diagonal matrix with diagonal elements, 0.6, 0.4.

7.9 Tanner (1993, p.138) sets

$$
\pi(\boldsymbol{\theta}|\mathbf{x}) \propto (2+\theta)^{125}(1-\theta)^{38}\theta^{34}.
$$

The normalising constant is not needed. Additionally, he sets

$$
q(a,b)=1, \quad \text{i.e., } b \sim U(0,1).
$$

7.10 The elements of $\mathbf{Q}$ have to be chosen so that the acceptance probability of Equation (7.10) results in $\{p_{ij}\}$ which correspond to Equation (7.13). For example,

$$
\text{set} \qquad \frac{q_{ji}}{q_{ij}} = \frac{\pi_i}{(\pi_i+\pi_j)}, \qquad \text{to correspond to}
$$
$$
\pi_i q_{ij} > \pi_j q_{ji}.
$$

7.14 From Tables 7.2 and 7.3, the model-averaged estimate of ϕ_n is given by:

$$
\bar{\phi}_n = 0.205 \times 0.561 + 0.795 \times 0.609 = 0.599.
$$

The model-averaged estimate of ϕ_n^2 is given by:

$$
\overline{\phi_n^2} = 0.205 \times (0.561^2 + 0.025^2) + 0.795 \times (0.609^2 + 0.031^2) = 0.3603.
$$

Hence the model-averaged estimate of the variance of ϕ_n is 0.0018.

7.15 Both X and Y are bounded random variables. The use of Figure 7.7(b) makes it easy to ensure that both random variables remain in range.

7.16 When the prior distribution is proper and flat, the posterior distribution is proportional to the likelihood.

7.17 The Metropolis-Hastings method involves two stages, viz. the selection of a candidate transition, and then the acceptance or rejection of making the move. The probabilities of both of these stages depend only on the current state of the system, and so the Metropolis-Hastings method results in a Markov chain. Although the transition probabilities in a Markov chain only depend on the current state of the chain, we can envisage a situation in which successive transitions all occur with high probability, resulting in correlations between state occupancies several steps apart.

7.19 The SD values are indeed calculated as customary standard deviations. The posterior distributions are not classical sampling distributions. There is an interesting disparity between p-values and goodness of fit. A model with many parameters can explore regions of the parameter space with small likelihood, and so will tend to have low posterior probability, $P(M|x)$. The Bayesian p-value does not penalise overparameterisation. For more detailed discussion, see Brooks et al. (2000a). When there is little information on parameters in the likelihood, as is true of a parameter like ϕ_{12}, the posterior distribution is dominated by the prior. In this case, we obtain a mean close to 0.5 and a variance close to $\frac{1}{12}$, the value for a $U(0,1)$ random variable.

7.20 The AVSS data set appears to be different from the others, which more clearly indicate a 'best' model in terms of posterior model probability. The difference between mean probabilities of response for the main components is greatest for the data sets E2 and AVSS, when there is greatest uncertainty regarding the best model. It is striking how SDs can increase as a result of model averaging. This reflects the variation introduced by including different models.

7.33 This is done in King et al. (2009), who obtain model probabilities of 0.217 and 0.783 for models C/C and $C2/C$, when uniform prior distributions are used, in very good agreement with those of Table 7.3.

Chapter 8

8.3 Although the data are multinomial, we may analyse them using a Poisson distribution. Suppose the number in the jth row and kth column is denoted by the random variable X_{jk}; then a model for independence of the two categories has:

$$\log\ (\mathbb{E}[X_{jk}]) = \mu + \alpha_j + \beta_k.$$

8.4
$$L(\alpha, \beta; \mathbf{y}) = \prod_{i=1}^{n} \frac{1}{y_i!} (\alpha + \beta x_i)^{y_i} e^{-(\alpha + \beta x_i)}$$

$$\propto \left\{ \prod_{i=1}^{n} (\alpha + \beta x_i)^{y_i} \right\} e^{-(n\alpha + \beta \Sigma x_i)},$$

$$\ell(\alpha, \beta; \mathbf{y}) = \text{constant} + \sum_{i=1}^{n} y_i \log (\alpha + \beta x_i) - n\alpha - n\beta \bar{x}.$$

$$\frac{\partial \ell}{\partial \alpha} = \sum_{i=1}^{n} \left(\frac{y_i}{\alpha + \beta x_i} \right) - n, \qquad \frac{\partial \ell}{\partial \beta} = \sum_{i=1}^{n} \frac{y_i x_i}{(\alpha + \beta x_i)} - n\bar{x}.$$

For the data, we get:

$$\frac{1}{\hat{\alpha}} + \frac{12}{\hat{\alpha} + \hat{\beta}} + \frac{46}{\hat{\alpha} + 2\hat{\beta}} = 6, \qquad \frac{12}{\hat{\alpha} + \hat{\beta}} + \frac{92}{\hat{\alpha} + 2\hat{\beta}} = 6.$$

Hence, $\qquad \dfrac{1}{\hat{\alpha}} = \dfrac{46}{\hat{\alpha} + 2\hat{\beta}}; \quad 2\hat{\beta} = 45\hat{\alpha}.$

Substitute for $\hat{\beta}$ and solve to get $\hat{\alpha} = 0.4184$; $\hat{\beta} = 9.4154$.

8.5 The number of deaths is increasing with time. As we have counts, a natural model is to assume a Poisson distribution, with mean μ_i corresponding to the ith quarter, and to set

$$\mu_i = i^{\theta}.$$

As $\log(\mu_i) = \theta \log i$, we have a GLM.

8.6 Suppose there are n sites. The expected count at the ith site is given by:

$$\mu_{it} = \exp(\hat{\alpha}_i + \hat{\beta}_t).$$

Thus for these sites, for year t, we get an expected total count of:

$$\sum_{i=1}^{n} \mu_{it} = e^{\hat{\beta}_t} \sum_{i=1}^{n} e^{\hat{\alpha}_i}.$$

Fewster et al. (1999) define the abundance index for year t as: $e^{(\hat{\beta}_t - \hat{\beta}_1)}$. One might plot $(\hat{\beta}_t - \hat{\beta}_1)$ against $\log t$.

8.7
$$pr(Y = j) = \frac{e^{-\lambda} \lambda^j}{j!(1 - e^{-\lambda})}$$
$$= \exp \left\{ j \log(\lambda) - \lambda - \log(1 - e^{-\lambda}) - \log(j!) \right\}$$

The distribution is therefore a member of the exponential family. It is the canonical form, and the natural parameter is $\log(\lambda)$.

8.10 It is sometimes the case for certain diseases that mortality reaches a peak, after which it declines. In such a case one would require a hazard function which also reaches a peak. For the log-logistic model,

$$F(t) = \frac{1}{1 + t^{-\phi}e^{-\theta}}$$

$$f(t) = \frac{\phi t^{-(\phi+1)}e^{-\theta}}{(1 + t^{-\phi}e^{-\theta})^2},$$

and so the hazard function is:

$$h(t) = f(t)/(1 - F(t))$$

$$= \frac{\phi}{t(1 + t^{-\phi}e^{-\theta})^2}.$$

This function of t has a maximum at $t = \{(\phi - 1)e^{-\theta}\}^{\frac{1}{\phi}}$.

For any two individuals sharing the same parameter ϕ, the hazard ratio $\to 1$ as $t \to \infty$. We see that

$$\log\left\{\frac{F_1(t)}{1 - F_1(t)}\right\} - \log\left\{\frac{F_2(t)}{1 - F_2(t)}\right\} = \boldsymbol{\beta}'(\mathbf{x}_1 - \mathbf{x}_2),$$

which is independent of t, so that we have a proportional odds model. Bennett and Whitehead (1981) provide GLIM macros. We see that we might set $\beta_2 = 0$.

8.13 We see from Example 8.1 that for the Poisson case the natural parameter is $\log(\theta)$, and from Example 8.2, for the binomial case the natural parameter is $\log\{\theta/(1 - \theta)\}$.

8.15 From Equation (8.4), in the canonical case,

$$\log(f(w;\theta)) = b(\theta) + c(\theta) + d(w).$$

The only functional difference compared with the expression in Section 5.3.2 is that there the canonical parameter is taken as proportional to θ. A more general function is obtained by simply reparameterising.

8.22 The results of Figure 8.1 suggest that parallel regression curves may be used. Anderson and Catchpole (1998) found it useful to break the results down by year.

Bibliography

Adby, P.R. and Dempster, M.A.H. (1974) *Introduction to Optimization Methods*. Chapman & Hall, London.

Aitkin, M., Anderson, D., Francis, B. and Hinde, J. (1989) *Statistical Modelling in GLIM*. Clarendon Press, Oxford.

Albert, J. (2007) *Bayesian Computation with R*. Springer, New York.

Albert, J.H. and Chib, S. (1993) Bayesian analysis of binary and polychotomous response data. *Journal of the American Statistical Association*, **88**, 422, 669–679.

Aldrich, J. (1997) R.A. Fisher and the making of maximum likelihood 1912–1922. *Statistical Science*, **12**, 162–176.

Altham, P.M.E. (1978) Two generalizations of the binomial distribution. *Applied Statistics*, **27**, 162–167.

Altham, P.M.E. (1984) Improving the precision of estimation by fitting a model. *Journal of the Royal Statistical Society, Series B*, **46**, 1, 118–119.

Anderson, B.R. and Catchpole, E.A. (1998) Comparing the performance of NSW and Victorian students in a first-year mathematics course. *Australian & New Zealand Journal of Statistics*, **40**, 901–918.

Anderson, D.A. and Aitkin, M.A. (1985) Variance component models with binary response interviewer variability. *Journal of the Royal Statistical Society, Series B*, **47**, 203–210.

Anderson, D. and Burnham, K.P. (1998) *Model Selection and Inference: a Practical Information Theoretic Approach*. Springer-Verlag, New York.

Aranda-Ordaz, F.J. (1981) On two families of transformations to additivity for binary response data. *Biometrika*, **68**, 357–364. Correction: *Biometrika*, **70**, 303.

Aranda-Ordaz, F.J. (1983) An extension of the proportional hazards model for grouped data. *Biometrics*, **39**, 109–118.

Atkinson, A.C. (1982) *Plots, Transformations and Regression*. Clarendon Press, Oxford.

Azalini, A. (1996) *Statistical Inference Based on the Likelihood*. Chapman & Hall, London.

Baird, D.D. and Wilcox, A.J. (1985) Cigarette smoking associated with delayed conception. *Journal of the American Medical Association*, **253**, 2979–2983.

Balding, D.J. and Donnelly, P. (1995) Inference in forensic identification. *Journal of the Royal Statistical Society, Series A*, **158**, 21–54.

Ball, F. (1995) Laplace transform based inference for single ion channel data incorporating time interval omission. *21st European Meeting of Statisticians*, Aarhus.

Barker, A.A. (1965) Monte Carlo calculations of radial distribution functions for a proton-electron plasma. *Australian Journal of Physics*, **18**, 119–133.

Barnard, G.A. (1963) Discussion of Professor Bartlett's paper. *Journal of the Royal Statistical Society, Series B*, **25**, 294.

Barndorff-Nielsen, O.E. and Cox, D.R. (1994) *Asymptotic techniques for use in statistics*. Chapman & Hall, London.

Barry, S.C., Brooks, S.P., Catchpole, E.A., and Morgan, B.J.T. (2003) The analysis of ring-recovery data using random effects. *Biometrics*, **59**, 54–65.

Bayarri, M.J. and Berger, J.O. (1998) Quantifying surprise in the data and model verification. In: *Bayesian Statistics VI*. Eds. J. Bernardo, J. Berger, A.P. Dawid, and A.F.M. Smith, Oxford University Press, Oxford.

Bennett, S. (1983) Log-logistic regression models for survival data. *Applied Statistics*, **32**, 2, 165–171.

Bennett, S. and Whitehead, J. (1981) Fitting logistic and log-logistic regression models to censored data using GLIM. *GLIM Newsletter*, **4**, 12–19. Correction, **5**, 3.

Besag, J., Green, P., Higdon, D. and Mengersen, K. (1995) Bayesian computation and stochastic systems. *Statistical Science*, **10**, 3–66.

Besbeas, P.T. (1999) Parameter estimation based on empirical transforms. Unpublished Ph.D. thesis, University of Kent, England.

Besbeas, P. and Morgan, B.J.T. (2001) Integrated squared error estimation of Cauchy parameters. *Statistics and Probability Letters*, **55**, 397–401.

Besbeas, P. and Morgan, B.J.T. (2003) Integrated squared error estimation of normal mixtures. *Computational Statistics and Data Analysis*, **44**, 517–526.

Besbeas, P. and Morgan, B.J.T. (2004) Efficient and robust estimation for the one-sided stable distribution of index $1/2$. *Statistics and Probability Letters*, **66**, 251–257.

Besbeas, P. and Morgan, B.J.T. (2008) Improved estimation of the stable laws. *Statistics and Computing*, **18**, 219–231.

Bishop, Y.M.M., Fienberg, S.E. and Holland, P.W. (1975) *Discrete Multivariate Analysis: Theory and Practice*. MIT Press, Cambridge, Massachusetts.

Bliss, C.L. (1938) The determination of dosage-mortality curves from small numbers. *Quarterly Journal of Pharmacology*, **11**, 192–216.

Bohachevsky, I., Johnson, M.E. and Stein, M.L. (1986) Generalized simulated annealing for function optimization. *Technometrics*, **28**, 209–217.

Borchers, D.L., Buckland, S.T. and Zucchini, W. (2002) *Estimating Animal Abundance*. Springer-Verlag, London.

Bowman, A.W. and Azzalini, A. (1997) *Applied Smoothing Techniques for Data Analysis: The Kernel Approach with S-Plus Illustrations*. Clarendon Press, Oxford.

Box, G.E.P. and Muller, M.E. (1958) A note on the generation of random normal deviates. *Annals of Mathematical Statistics*, **29**, 610–611.

Breslow, N.E. and Clayton, D.G. (1993) Approximate inference in generalized linear mixed models. *Journal of the American Statistical Association*, **88**, 9–25.

Brooks, R.J., James, W.H. and Gray, E. (1991) Modelling sub-binomial variation in the frequency of sex combinations in litters of pigs. *Biometrics*, **47**, 2, 403–418.

Brooks, S.P. (1995) AS298: A hybrid optimisation algorithm. *Applied Statistics*, **44**, 4, 530–533.

Brooks, S.P. (1998) Markov chain Monte Carlo method and its application. *The Statistician*, **47**, 69–100.

Brooks, S.P. (2001) On Bayesian analyses and finite mixture models for proportions. *Statistics and Computing*, **11**, 179–190.

Brooks, S.P. (2003) Bayesian computation: a statistical revolution. *Transactions of the Royal Society, Series A*, **361**, 2681–2697.

Brooks, S.P., Catchpole, E.A. and Morgan, B.J.T. (2000a) Bayesian animal survival estimation. *Statistical Science*, **15**, 357–376.

Brooks, S.P., Catchpole, E.A., Morgan, B.J.T. and Barry, S. (2000b) On the Bayesian analysis of ring-recovery data. *Biometrics*, **56**, 951–956.

Brooks, S.P., Friel, N. and King, R. (2003) Classical model selection via simulated annealing. *Journal of the Royal Statistical Society, Series B*, **65**, 503–520.

Brooks, S.P. and Gelman, A. (1998) General methods for monitoring convergence of iterative simulations. *Journal of Computational and Graphical Statistics*, **7**, 434–455.

Brooks, S.P., Giudici, P. and Roberts, G.O. (2003) Efficient construction of reversible jump MCMC proposal distributions (with discussion). *Journal of the Royal Statistical Society, Series B*, **65**, 3–55.

Brooks, S.P. and Morgan, B.J.T. (1994) Automatic starting point selection for function optimisation. *Statistics and Computing*, **4**, 173–177.

Brooks, S.P. and Morgan, B.J.T. (1995) Optimisation using simulated annealing. *The Statistician*, **44**, 2, 241–257.

Brooks, S.P., Morgan, B.J.T., Ridout, M.S. and Pack, S.E. (1997) Finite mixture models for proportions. *Biometrics*, **53**, 1097–1115.

Brooks, S.P. and Roberts, G.O. (1999) Diagnosing convergence of Markov chain Monte Carlo algorithms. *Statistics and Computing*, **8**, 319–335.

Brown, C.C. (1982) On a goodness of fit test for the logistic model based on score statistics. *Communications in Statistics — Theory and Methods*, **11**(10), 1087–1105.

Brownie, C., Anderson, D.R., Burnham, K.P. and Robson, D.S. (1985). Statistical inference from band recovery data — a handbook. Technical report, US Dept. of Interior, Fish and Wildlife Service.

Buck, C.E., Litton, C.D. and Stephens, D.A. (1993). Detecting a change in the shape of a prehistoric corbelled tomb. *The Statistician*, **42**, 483–490.

Buckland, S.T., Burnham, K.P. and Augustin, N.H. (1997) Model selection: an integral part of inference. *Biometrics*, **53**, 2, 603–618.

Bunday, B.D. (1984) *Basic Optimisation Methods*. Edward Arnold, London.

Burnham, K.P. and Anderson, D.R. (2002) *Model Selection and Multimodel Inference*, Springer-Verlag, New York.

Burnham, K.P., Anderson, D.R. and White, G.C. (1995) Selection among open population capture-recapture models when capture probabilities are heterogeneous. *Journal of Applied Statistics*, **22**, 5 & 6, 611–624.

Buse, A. (1982) The likelihood ratio, Wald and Lagrange multiplier tests: an expository note. *The American Statistician*, **36**, 3, 153–157.

Campbell, E.P. (1992) Robustness of estimation based on empirical transforms. Unpublished Ph.D. thesis, University of Kent, England.

Campbell, E.P. (1993) Influence for empirical transforms. *Communications in Statistics — Theory and Methods*, **22**, 2491–2502.

Carlin, B.P. and Louis, T.A. (1996). *Bayes and Empirical Bayes Methods for Data Analysis*. Chapman & Hall, London.

Carroll, J.D. and Green, P.E. (1997) *Mathematical Tools for Applied Multivariate Analysis* (revised edition with contributions by A. Chaturvedi). Academic Press, San Diego, California.

Casella, G. and George, E.I. (1992) Explaining the Gibbs sampler. *The American Statistician*, **46**, 3, 167–174.

Catchpole, E.A. (1995) MATLAB: an environment for analyzing ring-recovery and recapture data. *Journal of Applied Statistics*, **22**, 801–816.

Catchpole, E.A., Freeman, S.N. and Morgan, B.J.T. (1993) On boundary estimation in ring recovery models and the effect of adding recapture information, pp 215-228. In: *Marked Individuals in the Study of Bird Populations*. Eds. J.-D. Lebreton and P.M. North. Birkhauser-Verlag, Basel.

Catchpole, E.A., Freeman, S.N., Morgan, B.J.T. and Harris, M.P. (1998a) Integrated recovery/recapture data analysis. *Biometrics*, **54**, 1, 33–46.

Catchpole, E.A. and Morgan, B.J.T. (1991) A note on Seber's model for ring-recovery data. *Biometrika*, **78**, 917–919.

Catchpole, E.A. and Morgan, B.J.T. (1994) Boundary estimation in ring recovery models. *Journal of the Royal Statistical Society, Series B*, **56**, 2, 385–391.

Catchpole, E.A. and Morgan, B.J.T. (1996) Model selection in ring-recovery models using score tests. *Biometrics*, **52**, 664–672.

Catchpole, E.A. and Morgan, B.J.T. (1997) Detecting parameter redundancy. *Biometrika*, **84**, 187–196.

Catchpole, E.A. and Morgan, B.J.T. (2000) Structural constants of parameter redundancy. Technical Report, University of Kent, UKC/IMS/00/06.

Catchpole, E.A., Morgan, B.J.T. and Boucher, M. (1997) Score tests. *The Ring*, **19**, 179–184.

Catchpole, E.A., Morgan, B.J.T., Coulson, T.N., Freeman, S.N. and Albon, S.D. (2000) Factors influencing Soay sheep survival. *Applied Statistics*, **49**, 453–472.

Catchpole, E.A., Morgan, B.J.T. and Freeman, S.N. (1998b) Estimation in parameter-redundant models. *Biometrika*, **85**, 2, 462–468.

Catchpole, E.A., Morgan, B.J.T. and O'Dowd, D. (1987) Fitting models to data describing the distribution of insects over flower heads of *Helichrysum bracteatum*. *Biometrics*, **43**, 4, 767–782.

Catchpole, E.A., Morgan, B.J.T. and Viallefont, A. (2002) Solving problems in parameter redundancy using computer algebra. *Journal of Applied Statistics*, **29**, 1–4, 625–636.

Celeux, G., Forbesy, F., Robert, C.P. and Titterington, D.M. (2006) Deviance information criteria for missing data models. *Bayesian Analysis*, **1**, 651–674.

Celeux, G. and Diebolt, J. (1985) The SEM algorithm: A probabilistic teacher algorithm derived from the EM algorithm for the mixture problem. *Computational Statistics Quarterly*, **2**, 73–82.

Chambers, J.M. and Hastie, T.J. (1993) *Statistical Models in S*. Chapman & Hall, London.

Chatfield, C. (1984) *The Analysis of Time Series: An Introduction* (3rd edition). Chapman & Hall, London.

Chatfield, C. (1995) Model uncertainty, data mining and statistical inference. *Journal of the Royal Statistical Society, Series A*, **158**, 3, 413–466.

Cheng, R.C.H. and Traylor, L. (1995) Non-regular maximum-likelihood problems. *Journal of the Royal Statistical Society, Series B*, **57**, 1, 3–44.

Chernoff, H. (1954) On the distribution of the likelihood ratio. *Annals of Mathematical Statistics*, **25**, 573–578.

Chernoff, H. and Lander, E. (1995) Asymptotic distribution of the likelihood ratio test that a mixture of two binomials is a single binomial. *Journal of Statistical Planning and Inference*, **43**, 19–40.

Chib, S. and Greenberg, E. (1995) Understanding the Metropolis-Hastings algorithm. *The American Statistician*, **49**, 4, 327–335.

Chu, C.-K. and Marron, J.S. (1991) Choosing a kernel regression estimator. *Statistical Science*, **6**, 4, 404–436.

Clarke, F.R. (1957) Constant-ratio rule for confusion matrices in speech communication. *Journal of the Acoustical Society of America*, **30**, 715–720.

Clarke, G.M. and Cooke, D. (2004) *A Basic Course in Statistics* (5th edition). Edward Arnold, London.

Clements, R.R. (1989) *Mathematical Modelling: A Case Study Approach*. Cambridge University Press, Cambridge.

Cleveland, W.S., Grosse, E. and Shyu, W.M. (1992) Local regression models. Chapter 8. In: *Statistical Models in S*, Eds. J.M. Chambers and T.J. Hastie. Wadsworth, Pacific Grove, California.

Cole, D.J. and Morgan, B.J.T. (2007) Parameter redundancy with covariates. University of Kent Technical Report, UKC/IMS/07/007.

Cole, D.J., Morgan, B.J.T. and Ridout, M.S. (2003) Generalized linear mixed models for strawberry inflorescence data. *Statistical Modelling*, **3**, 273–290.

Cole, D.J., Morgan, B.J.T. and Ridout, M.S. (2005) Models for strawberry inflorescence data. *JABES*, **10**, 4, 411–423.

Cole, D.J., Ridout, M.S., Morgan, B.J.T., Byrne, L.J. and Tuite, M.F. (2004)

Estimating the number of prions in yeast cells. *Mathematical Medicine and Biology*, **21**, 369–395.

Congdon, P. (2003) *Applied Bayesian Modelling*. Wiley, Chichester.

Congdon, P. (2006) *Bayesian Statistical Modelling* (2nd edition). Wiley, New York.

Cook, D. (1986) Assessment of local influence. *Journal of the Royal Statistical Society, Series B*, **48**, 133–169.

Cook, R.D. and Weisberg, S. (1982) *Residuals and Influence in Regression*. Chapman & Hall, London.

Copas, J.B. (1975) On the unimodality of the likelihood for the Cauchy distribution. *Biometrika*, **62**, 701–707.

Copas, J.B. and Hilton, F.J. (1990) Record linkage: statistical models for matching computer records. *Journal of the Royal Statistical Society, Series A*, **153**, 3, 287–320.

Cormack, R.M. (1970) Statistical appendix to Fordham's paper. *Journal of Animal Ecology*, **39**, 24–27.

Cormack, R.M. (1992) Interval estimation for mark-recapture studies of closed population. *Biometrics*, **48**, 2, 567–576.

Cormack, R. (1994) Unification of mark-recapture analyses by loglinear modelling, pp. 19–32. In: *Proceedings of the conference on statistics in ecology and environmental monitoring*. Eds. D.J. Fletcher and B.F.J. Manly, University of Otago Press, Dunedin.

Cox, D.R. (1972) Regression models and life tables. *Journal of the Royal Statistical Society, Series B*, **34**, 187–220.

Cox, D.R. and Hinkley, D.V. (1974) *Theoretical Statistics*. Chapman & Hall, London.

Cox, D.R. and Miller, H.D. (1965) *The Theory of Stochastic Processes*. Methuen, London.

Cox, D.R. and Reid, N. (1987) Parameter orthogonality and approximate conditional inference. *Journal of the Royal Statistical Society, Series B*, **49**, 1, 1–39.

Crawley, M.J. (2005) *Statistics: An Introduction Using R*. Wiley, New York.

Crawley, M.J. (2007) *The R Book*. Wiley, New York.

Cressie, N. (1991) *Statistics for Spatial Data*. Wiley, New York.

Currie, I.D. (1995) Maximum likelihood estimation and Mathematica. *Applied Statistics*, **44**, 3, 379–394.

Dagpunar, J. (1988) *Principles of Random Variate Generation*. Clarendon Press, Oxford.

Dagpunar, J.S. (2007) *Simulation and Monte Carlo with applications in finance and MCMC*. Wiley, Chichester.

Davis, T.A. and Sigmon, K. (2005) *MATLAB Primer* (7th edition). Chapman & Hall/CRC, Boca Raton, Florida.

Davison, A.C. (2003) *Statistical Models*. Cambridge University Press, Cambridge.

Davison, A.C. and Hinkley, D.V. (1997) *Bootstrap Methods and Their Application*. Cambridge University Press, Cambridge.

Davison, A.C., Hinkley, D.V. and Schechtman, E. (1987) Efficient bootstrap simulation. *Biometrika*, **74**, 555–566.

de Groot, M.H. (1986) *Probability and Statistics* (2nd edition). Addison-Wesley, Reading, Massachusetts.

Dempster, A.P., Laird, N.M. and Rubin, D.B. (1977) Maximum likelihood from incomplete data via the EM algorithm. *Journal of the Royal Statistical Society, Series B*, **39**, 1, 1–38.

Devroye, L. (1986) *Non-uniform Random Variate Generation*. Springer-Verlag, New York.

Diebolt, J. and Ip, E.H.S. (1996) Stochastic EM method and application, pp. 259–273. In: *Markov Chain Monte Carlo in Practice*. Eds. W.R. Gilks, S. Richardson and D.J. Spiegelhalter, Chapman & Hall, London.

Diggle, P.J. (1983) *Statistical Analysis of Spatial Point Patterns*. Academic Press, London.

Diggle, P.J. and Gratton, R.J. (1984) Monte Carlo methods of inference for implicit statistical models. *Journal of the Royal Statistical Society, Series B*, **46**, 2, 193–227.

Dobson, A.J. (2002) *An Introduction to Generalized Linear Models* (2nd edition). Chapman & Hall/CRC, London.

Draper, D. (1995) Assessment and propagation of model uncertainty. *Journal of the Royal Statistical Society, Series B*, **57**, 1, 45–98.

Edwards, A.W.F. (1972) *Likelihood*. Cambridge University Press, Cambridge (expanded edition, 1992, Johns Hopkins University Press, Baltimore).

Efron, B. (1979) Bootstrap methods: another look at the jackknife. *Annals of Statistics*, **7**, 1–26.

Efron, B. and Hinkley, D.V. (1978) Assessing the accuracy of the maximum likelihood estimator. *Biometrika*, **65**, 457–488.

Efron, B. and Thisted, R., (1976) Estimating the number of unseen species: How many words did Shakespeare know? *Biometrika*, **63**, 3, 435–448.

Efron, B. and Tibshirani, R. (1993) *An Introduction to the Bootstrap*. Chapman & Hall, New York.

Elton, B. (1999) *Inconceivable*. Bantam, London.

Engel, B. (1997) Extending generalized linear models with random effects and components of dispersion. Ph.D. thesis, Agricultural University of Wageningen, Holland.

Engel, B. and Buist, W. (1996) Analysis of a generalized linear mixed model: a case study and simulation results. *Biometrical Journal*, **38**, 1, 61–80.

Engel, B. and Keen, A. (1994) A simple approach for the analysis of generalized linear mixed models. *Statistica Neerlandica*, **48**, 1–22.

ESRI (1999) ArcView GIS 3.2, in: ESRI, Inc., Redlands, CA.

Everitt, B.S. (1977) *The Analysis of Contingency Tables*. Chapman & Hall, London.

Everitt, B.S. (1987) *Introduction to Optimization Methods and their Application in Statistics.* Chapman & Hall, London.

Everitt, B.S. and Hand, D.J. (1981) *Finite Mixture Distributions.* Chapman & Hall, London.

Faraway, J.R. (2005) *Extending the Linear Model with R. Generalized Linear, Mixed Effects and Nonparametric Regression Models.* Chapman & Hall/CRC, London.

Farewell, V.T. (1982) The use of mixture models for the analysis of survival data with long-term survivors. *Biometrics,* **38**, 1041–1046.

Feng, Z. and McCulloch, C.E. (1992) Statistical inference using maximum likelihood estimation and the generalized likelihood ratio when the true parameter is on the boundary of the parameter space. *Statistics and Probability Letters,* **13**, 325–332.

Fewster, R.M., Buckland, S.T., Siriwardena, G.M., Baillie, S.R. and Wilson, J.D. (1999) Analysis of population trend for farmland birds using generalized additive models. *Ecology,* **81**, 1970–1984.

Fewster, R.M. and Jupp, P.E. (2008) Inference on population size in binomial detectability models. Submitted for publication, *Biometrika.*

Fienberg, S.E. (1981) *The Analysis of Cross-Classified Categorical Data* (2nd edition). MIT Press, Cambridge, Massachusetts.

Firth, D. (1993) Bias reduction of maximum likelihood estimates. *Biometrika,* **80**, 27–38.

Fisher, R.A. (1925) Theory of statistical estimation. *Proceedings of the Cambridge Philosophical Society,* **22**, 700–725.

Forsythe, G.F., Malcolm, M.A. and Moler, C.B. (1976) *Computer Methods for Mathematical Computations.* Prentice Hall, New Jersey.

Fouskakis, D. and Draper, D. (2002) Stochastic optimization: a review. *International Statistical Review,* **70**, 3, 315–350.

Freeman, S.N. and Morgan, B.J.T. (1990) Studies in the analysis of ring-recovery data. *The Ring,* **13**, 271–287.

Freeman, S.N. and Morgan, B.J.T. (1992) A modelling strategy for recovery data from birds ringed as nestlings. *Biometrics,* **48**, 1, 217–236.

Gad Attay, A.M. (1999) Fitting selection models to longitudinal data with dropout using the stochastic EM algorithm. Unpublished Ph.D. thesis, University of Kent, England.

Gamerman, D. (1997). *Markov Chain Monte Carlo: Stochastic Simulation for Bayesian Inference.* Chapman & Hall, London.

Gamerman, D. and Lopes, H.F. (2006) *Markov Chain Monte Carlo. Stochastic Simulation for Bayesian Inference* (2nd edition). Chapman & Hall/CRC, Boca Raton, Florida.

Garren, S.T., Smith, R.L. and Piegorsch, W.W. (2000) On a likelihood-based goodness-of-fit test of the beta-binomial model. *Biometrics,* **56**, 948–950.

Gelfand, A.E. and Smith, A.F.M. (1990) Sampling-based approaches to calculating marginal densities. *Journal of the American Statistical Association,* **85**, 398–409.

Gelman, A., Carlin, J.B., Stern, H.S. and Rubin, D.B. (2004) *Bayesian Data Analysis* (2nd edition). Chapman & Hall/CRC, Boca Raton, Florida.

Geman, S. and Geman, D. (1984) Stochastic relaxation, Gibbs distributions and the Bayesian restoration of images. *IEEE Transactions on Pattern Analysis and Machine Intelligence*, **6**, 721–741.

Gerson, M. (1975) The technique and uses of probability plotting. *The Statistician*, **24**, 4, 235–257.

Geyer, C.J. and Thompson, E.A. (1992) Constrained Monte Carlo maximum likelihood for dependent data. *Journal of the Royal Statistical Society, Series B*, **54**, 3, 657–700.

Gilks, W.R. and Wild, P. (1992) Adaptive rejection sampling for the Gibbs sampler. *Computing Science and Statistics*, **24**, 439–448.

Gill, P.E., Murray, W. and Wright, M.H. (1981) *Practical Optimization.* Academic Press, London.

Giltinan, D.M. Capizzi, T.P. and Malani, H. (1988) Diagnostic tests for similar action of two compounds. *Applied Statistics*, **37**, 1, 39–50.

Gimenez, O., Bonner, S., King, R., Parker, R.A., Brooks, S.P., Jamieson, L.E., Grosbois, V., Morgan, B.J.T. and Thomas, L. (2008) WinBUGS for population ecologists: Bayesian modeling using Markov chain Monte Carlo methods. pp 885–918. In: Modelling Demographic Processes in Marked Populations: Series: *Environmental and Ecological Statistics*, **3**. Eds. D.L. Thomson, E.G. Cooch and M.J.Conroy.

Gimenez, O., Choquet, R. and Lebreton, J-D. (2003) Parameter redundancy in multistate capture-recapture models. *Biometrical Journal*, **45**, 704–722.

Gimenez, O., Crainiceanu, C., Barbraud, C., Jenouvrier, S. and Morgan, B.J.T. (2006) Semiparametric regression in capture-recapture modeling. *Biometrics*, **62**, 691–698.

Giordano, F.R. and Weir, M.D. (1985) *A First Course in Mathematical Modelling.* Brooks/Cole Pub.Co., California.

Givens, G.H. and Hoeting, J.A. (2005) *Computational Statistics.* Wiley, New Jersey.

Goldman, N. (1993) Statistical tests of models of DNA substitution. *Journal of Molecular Evolution*, **36**, 182–198.

Grace, A. (1994) *Optimization Toolbox, for Use with* MATLAB: *User's Guide.* The Maths. Works., Inc., Natick, Massachusetts.

Green, P.J. (1984) Iteratively reweighted least squares for maximum likelihood estimation, and some robust and resistant alternatives. *Journal of the Royal Statistical Society, Series B*, **46**, 149–192.

Green, P.J. (1995). Reversible jump Markov chain Monte Carlo computation and Bayesian model determination. *Biometrika*, **82**, 711–732.

Green, P.J. and Silverman, B.W. (1994) *Nonparametric Regression and Generalized Linear Models.* Chapman & Hall, London.

Grey, D.R. and Morgan, B.J.T. (1972) Some aspects of ROC curve-fitting: normal and logistic models. *Journal of Mathematical Psychology*, **9**, 128–139.

Grieve, A.P. (1996) On likelihood and Bayesian methods for interval estimation of the LD_{50}, pp. 87–100. In: *Statistics in Toxicology*, Ed. B.J.T. Morgan, Clarendon Press, Oxford.

Griffiths, D.A. (1977) Avoidance-modified generalised distributions and their application to studies of super-parasitism. *Biometrics*, **33**, 1, 103–112.

Grimmett, G.R. and Stirzaker, D.R. (1992) *Probability and Random Processes*. Clarendon Press, Oxford.

Gross, D. and Harris, C.M. (1974) *Fundamentals of Queueing Theory*. Wiley, Toronto.

Hahn, B.D.(1997) *Essential* MATLAB *for Scientists and Engineers*. Arnold, London.

Hanselman, D. and Littlefield, B. (2005) *Mastering MATLAB 7*, Pearson/Prentice Hall, Upper Saddle River, New Jersey.

Harlap, S. and Baras, H. (1984) Conception-waits in fertile women after stopping oral contraceptives. *International Journal of Fertility*, **29**, 73–80.

Hasselblad, V. (1969) Estimation of finite mixtures of distributions from the exponential family. *Journal of the American Statistical Association*, **64**, 1459–1471.

Hastie, T.J. and Tibshirani, R.J. (1990) *Generalized Additive Models*. Chapman & Hall, London.

Hastings, W.K. (1970) Monte Carlo sampling methods using Markov chains and their applications. *Biometrika*, **57**, 97–109.

Healy, M.J.R. (1986) *GLIM: An Introduction*. Clarendon Press, Oxford.

Hewlett, P.S. (1974) Time for dosage to death in beetles *Tribolium cataneum*, treated with pyrethrin or DDT, and its bearing on dose-mortality relations. *Journal of Stored Product Research*, **10**, 27–41.

Hinkley, D.V. (1988) Bootstrap methods. *Journal of the Royal Statistical Society, Series B*, **50**, 3, 321–337.

Hjorth, J.S.U. (1994) *Computer Intensive Statistical Methods*. Chapman & Hall, London.

Hope, A.C.A. (1968) A simplified Monte Carlo significance test procedure. *Journal of the Royal Statistical Society, Series B*, **30**, 582–598.

Horner, F.R. (1853) Liberation of the chick from the shell, pp. 186–188. In: *The Poultry Book* , Eds. W. Wingfield and G.W. Johnson. William S. Orr, London.

Hunt, B.R., Lipsman, R.L. and Rosenberg, J.M. with Coombes, K.R., Osborn, J.E. and Stuck, G.J. (2006) *A Guide to* MATLAB *for Beginners and Experienced Users*. Cambridge University Press, Cambridge.

Ihaka, R. and Gentleman, R. (1996) R: a language for data analysis and graphics. *Journal of Computational and Graphical Statistics*, **5**, 299–314.

Ip, E.H.S. (1994) A stochastic EM estimator in the presence of missing data: Theory and applications. Technical report. Division of Biostatistics, Stanford University, Stanford, California.

Jansen, J. (1993) Generalized linear mixed models and their application in plant breeding research. Ph.D. dissertation. Agricultural University, Wageningen, Holland.

Jarrett, R.G. and Morgan, B.J.T. (1984) A critical reappraisal of the patch-gap model. *Biometrics*, **40**, 203–207.

Jensen, T.K., Hjollund, N.H.I., Henriksen, T.B., Scheike, T., Kolstad, H., Giwercman, A., Ernst, E., Bonde, J.P., Skakkebaek, N.E. and Olsen, J. (1998) Does moderate alcohol consumption affect fertility? Follow up study among couples planning first pregnancy. *British Medical Journal*, **317**, 505–510.

Jolliffe, I.T. (2002) *Principal Component Analysis* (2nd edition). Springer-Verlag, New York.

Jones, P.W. and Smith, P. (2001) *Stochastic Processes*, Arnold, London.

Jørgensen, M., Keiding, N. and Skakkebaek, N.E. (1991) Estimation of spermarche from longitudinal spermaturia data. *Biometrics*, **47**, 1, 177–194.

Kadane, J.B. and Wolfson, L.J. (1998). Experiences in elicitation. *The Statistician*, **47**, 3–20.

Kahn, H. (1956) *Application of Monte Carlo*. Rand Corp., Santa Monica, California.

Kalbfleisch, J.D., Lawless, J.F. and Vollmer, W.M. (1983) Estimation in Markov models from aggregate data. *Biometrics*, **39**, 4, 907–919.

Karlin, S. and Taylor, H.M. (1975) *A First Course in Stochastic Processes* (2nd edition). Academic Press, New York.

Kass, R.E. and Raftery, A.E. (1995) Bayes Factors. *Journal of the American Statistical Association*, **90**, 773–795.

Kay, R. and Little, S. (1986) Assessing the fit of the logistic model: a case study of children with the haemolytic uraemic syndrome. *Applied Statistics*, **35**, 16–30.

Kelly, F.P. (1979) *Reversibility and Stochastic Networks*. Wiley, Chichester.

Kendall, D.G. and Kendall, W. (1980) Alignments in two-dimensional random sets of points. *Advances in Applied Probability*, **12**, 380–424.

Kennedy, W.J. and Gentle, J.E. (1980) *Statistical Computing*, Marcel Dekker, New York.

Kinderman, A.J. and Monahan, J.F. (1977) Computer generation of random variables using the ratio of normal deviates. *Association for Computing Machinery Transactions on Mathematical Software*, **3**, 257–260.

King, R., Brooks, S.P., Mazzetta, C., Freeman, S.N. and Morgan, B.J.T. (2008) Identifying and diagnosing population declines: A Bayesian assessment of lapwings in the UK. *Applied Statistics*, **57**, 609–632.

King, R. and Brooks, S.P. and Morgan, B.J.T. and Coulson, T. (2006) Bayesian sheep survival, *Biometrics*, **62**, 211–220.

King, R., Morgan, B.J.T., Gimenez, O. and Brooks, S.P. (2009) *Bayesian Analysis for Population Ecology*. In preparation.

Kneib, T. and Petzoldt, T. (2007) Introduction to the special volume on "Ecology and Ecological Modelling in R." *Journal of Statistical Software*, **22**, 1–7.

Krause, A. and Olson, M. (2000) *The Basics of S and S-Plus*. Springer, New York.

Kvasnicka, V. and Pospichal, J. (1997) A hybrid of simplex method and simulated annealing. *Chemometrics and Intelligent Laboratory Systems*, **39**, 161–173.

Lai, K.-L. and Crassidis, J.L. (2008) Extensions of the first and second complex-step derivative approximations. *Journal of Computational and Applied Mathematics*, **219**, 276–293.

Laurence, A.F. and Morgan, B.J.T. (1987) Selection of the transformation variable in the Laplace transform method of estimation. *Australian Journal of Statistics*, **29**, 2, 113–127.

Laurence, A.F. and Morgan, B.J.T. (1989) Observations on a stochastic model for quantal assay data. *Biometrics*, **45**, 733–744.

Lawless, J.R. (2002) *Statistical Models and Methods for Lifetime Data* (2nd edition). Wiley, New York.

Lebreton, J-D., Burnham, K.P., Clobert, J. and Anderson, D.R. (1992) Modeling survival and testing biological hypotheses using marked animals: a unified approach with case studies. *Ecological Monographs*, **62**, 67–118.

Lebreton, J-D., Morgan, B.J.T., Pradel, R. and Freeman, S.N. (1995) A simultaneous survival rate analysis of dead recovery and live recapture data. *Biometrics*, **51**, 1418–1428.

Lebreton, J.-D. and North, P.M. (Eds.) (1993) *Marked Individuals in the Study of Bird Populations*. Birkhäuser Verlag, Basel.

Liang, K.Y. and Zeger, S.L. (1986) Longitudinal data analysis using generalized linear models. *Biometrika*, **73**, 13–22.

Lindfield, G. and Penny, J. (1995) *Numerical Methods Using* MATLAB. Ellis Horwood, London.

Link, W.A., Cam, E., Nichols, J.D. and Cooch, E. G. (2002) Of BUGS and birds: Markov chain Monte Carlo for hierarchical modeling in wildlife research. *Journal of Wildlife Management*, **66**, 277–291.

Lloyd, C.J. (1999) *Statistical Analysis of Categorical Data*. Wiley, New York.

Louis, T.A. (1982) Finding observed information using the EM algorithm. *Journal of the Royal Statistical Society, Series B*, **44**, 98–130.

Lunn , D.J., Best, N. and Whittaker, J. (2005) Generic reversible jump MCMC using graphical models. Technical Report, Imperial College Department of Epidemiology and Public Health, *TR EPH-2005-01*.

Lunn, D.J., Thomas, A., Best, N. and Spiegelhalter, D. (2000) WinBUGS — a Bayesian modelling framework: concepts, structure and extensibility. *Statistics and Computing*, **10**, 325–337.

MacKenzie, D. I., Nichols, J.D., Royle, J.A., Pollock, K.H., Bailey, L.L. and Hines, J.E. (2006) *Occupancy Estimation and Modelling*. Academic Press, Amsterdam.

McCullagh, P. (1980) Regression models for ordinal data. *Journal of the Royal Statistical Society, Series B*, **42**, 109–142.

McCullagh, P. and Nelder, J.A. (1989) *Generalized Linear Models* (2nd edition). Chapman & Hall, London.

McGilchrist, C.A. (1994) Estimation in generalized mixed models. *Journal of the Royal Statistical Society, Series B*, **56**, 61–69.

McLachlan, G.J. (1987) On bootstrapping the likelihood ratio test statistic for the number of components in a normal mixture. *Applied Statistics*, **36**, 318–324.

McLachlan, G.J. and Basford, K.E. (1988) *Mixture Models: Inference and Applications to Clustering*. Marcel Dekker, New York.

McLachlan, G.J. and Krishnan, T. (1997) *The EM Algorithm and Extensions*. Wiley, New York.

McLeish, D.L. and Tosh, D.H. (1990) Sequential designs in bioassay. *Biometrics*, **46**, 103–116.

Maindonald, J. (1984) *Statistical Computation*. Wiley, New York.

Marquardt, D.W. (1980) You should standardize the predictor variables in your regression models. *Journal of the American Statistical Association*, **75**, 369, 87–91.

Marriott, F.H.C. (1979) Barnard's Monte Carlo tests: How many simulations? *Applied Statistics*, **28**, 1, 75–77.

Marsaglia, G. and Bray, T.A. (1964) A convenient method for generating normal variables. *SIAM Review*, **6**, 260–264.

Martinez, W.L. and Martinez, A. R. (2008) *Computational Statistics Handbook with* MATLAB (2nd edition). Chapman & Hall/CRC, Boca Raton, Florida.

Meng, X.-L. and van Dyk, D. (1997) The EM algorithm — an old folk-song sung to a fast new tune. *Journal of the Royal Statistical Society, Series B*, **59**, 3, 511–568.

Mesterton-Gibbons, M. (1989) *A Concrete Approach to Mathematical Modelling*. Addison-Wesley, California.

Metropolis, N., Rosenbluth, A.W., Rosenbluth, M.N., Teller, A.H. and Teller, E. (1953) Equations of state calculations by fast computing machines. *Journal of Chemical Physics*, **21**, 1087–1091.

Meyn, S.P. and Tweedie, R.L. (1993) *Markov Chains and Stochastic Stability*. Springer-Verlag, Berlin.

Milicer, H. and Szotka, F. (1966) Age at menarche in Warsaw girls in 1965. *Human Biology*, **38**, 199–203.

Millar, R.B. (2004) Sensitivity of Bayes estimators to hyper-parameters with an application to maximum yield from fisheries. *Biometrics*, **60**, 536–542.

Molenberghs, G. and Verbeke, G. (2005) *The Generalized Linear Mixed Model (GLMM)*. Springer, New York.

Moran, P.A.P. (1971) Maximum likelihood estimators in non-standard conditions. *Proceedings of the Cambridge Philosophical Society*, **70**, 441–450.

Morgan, B.J.T. (1974) On Luce's Choice Axiom. *Journal of Mathematical Psychology*, **11**, 107–123.

Morgan, B.J.T. (1976) Stochastic models of grouping changes. *Advances in Applied Probability*, **8**, 30–57.

Morgan, B.J.T. (1978) A simple comparison of Newton-Raphson and the Method-of-Scoring. *International Journal of Mathematical Education in Science and Technology*, **9**, 3, 343–348.

Morgan, B.J.T. (1982) Modelling polyspermy. *Biometrics*, **38**, 885–898.

Morgan, B.J.T. (1984) *Elements of Simulation*, Chapman & Hall, London.

Morgan, B.J.T. (1992) *Analysis of Quantal Response Data*. Chapman & Hall, London.

Morgan, B.J.T. (1993) Expected size distributions in models of group dynamics. *Journal of Applied Probability*, **30**, 1–16.

Morgan, B.J.T. (1998) Quantal Response. pp. 3618–3625. In: *Encyclopedia of Biostatistics*. Eds. P. Armitage and R. Colton, Wiley, Chichester, **5**.

Morgan, B.J.T. and Freeman, S.N. (1989) A model with first-year variation for ring-recovery data. *Biometrics*, **45**, 4, 1087–1102.

Morgan, B.J.T. and North, P.M. (1985) A model for avian lung ventilation, and the effect of accelerating stimulation in Japanese quail embryos. *Biometrics*, **41**, 215–226.

Morgan, B.J.T., Pack, S.E. and Smith, D.M. (1989) QUAD: A computer package for the analysis of QUantal Assay Data. *Computer Methods and Programs in Biomedicine*, **30**, 265–278.

Morgan, B.J.T., Palmer, K.J. and Ridout, M.S. (2007) Negative score test statistic. *The American Statistician*, **61**, 4, 285–288,

Morgan, B.J.T., Revell, D.J. and Freeman, S.N. (2007) Simplified likelihoods for site occupancy models. *Biometrics*, **63**, 618–621.

Morgan, B.J.T. and Ridout, M.S. (2008a) A new mixture model for recapture heterogeneity. *Applied Statistics*, **57**, 4, 433–446.

Morgan, B.J.T. and Ridout, M.S. (2008b) Estimating N: a robust approach to recapture heterogeneity. pp. 1071–1082. In: Modelling Demographic Processes in Marked Populations: Series: *Environmental and Ecological Statistics*, **3**. Eds. D.L. Thomson, E.G. Cooch and M.J.Conroy.

Morgan, B.J.T., Ridout, M.S. and Ruddock, L.W. (2003) Models for yeast prions. *Biometrics*, **59**, 562–569.

Morgan, B.J.T. and Thomson, D. (Eds.) (2002) Statistical analysis of data from marked bird populations, special issue of *Journal of Applied Statistics*, **29**, Nos. 1–4.

Morgan, B.J.T. and Titterington, D.M. (1977) A comparison of iterative methods for obtaining maximum-likelihood estimates in contingency tables with a missing diagonal. *Biometrika*, **64**, 265–269.

Morgan, B.J.T. and Watts, S.A. (1980) On modelling microbial infections. *Biometrics*, **36**, 2, 317–321.

Morisita, M. (1971) Measuring of habitat value by environmental density method, pp. 379–401. In: *Statistical Ecology* Eds. G.P. Patil, E.C. Pielou and W.E.Waters. Pennsylvania State University Press, University Park.

Morrison, D.F. (1976) *Multivariate Statistical Methods*. McGraw-Hill, Kogaku-sha, Tokyo.

Murrell, P. (2006) *R Graphics*. Chapman & Hall/CRC, Boca Raton, Florida.

NAG (2006) Numerical Algorithms Group Library Manual, Mark 21, Oxford: Numerical Algorithms Group Ltd.

Nash, J.C. and Walker-Smith, M. (1987) *Nonlinear Parameter Estimation: An Integrated System in BASIC.* Marcel Dekker Inc., New York.

Nelder, J.A. and Mead, R. (1965) A simplex method for function minimization. *The Computer Journal,* **7**, 308–313.

Nelder, J.A. and Wedderburn, R.W.M. (1972) Generalized linear models. *Journal of the Royal Statistical Society, Series A,* **135**, 370–384.

Newcomb, S. (1886) A generalized theory of the combination of observations so as to obtain the best result. *American Journal of Mathematics,* **8**, 343–366.

North, P.M. and Morgan, B.J.T. (1979) Modelling heron survival using weather data. *Biometrics,* **35**, 667–681.

Oakes, D. (1999) Direct calculation of the information matrix via the EM algorithm. *Journal of the Royal Statistical Society, Series B,* **61**, 2, 479–482.

O'Hagan, A. (1998) Eliciting expert beliefs in substantial practical applications. *The Statistician,* **47**, 1, 21–36.

O'Hagan, A., Buck, C.E., Daneshkhah, A., Eiser, J.E., Garthwaite, P.H., Jenkinson, D.J., Oakley, J.E. and Rakow, T. (2006). *Uncertain Judgements: Eliciting Expert Probabilities.* Wiley, Chichester.

Osman, I.H. (1993) Metastrategy simulated annealing and tabu search algorithms for the vehicle routing problem. *Annals of Operational Research,* **41**, 3, 421–451.

Pack, S.E. (1986) The analysis of proportions from toxicological experiments. Unpublished Ph.D. thesis, University of Kent, England.

Pack, P., Jolliffe, I.T. and Morgan, B.J.T. (1988) Influential observations in principal component analysis: a case study. *Journal of Applied Statistics,* **15**, 1, 37–50.

Pack, S.E. and Morgan, B.J.T. (1990). A mixture model for interval-censored time-to-response quantal assay data. *Biometrics,* **46**, 3, 749–758.

Palmer, K.J., Ridout, M.S. and Morgan, B.J.T. (2008) Modelling cell generation times using the tempered stable distribution. *Applied Statistics,* 57, 4, 379–397.

Pannekoek, J. and van Strien, A. (1996) TRIM (TRends and Indices for Monitoring data). Research paper 9634, Statistics Netherlands, Voorburg, The Netherlands.

Paradis, E. (2006) *Analysis of Phylogenetics and Evolution with R.* Springer, New York.

Parzen, E. (1962) On estimation of a probability density function and mode. *Annals of Mathematical Statistics,* **33**, 1065–1076.

Patterson, H.D. and Thompson, R. (1971) Recovery of inter-block information when block sizes are unequal. *Biometrika,* **58**, 545–554.

Paul, S.R. (2005) Testing goodness of fit of the geometric distribution: an application to human fecundability data. *Journal of Modern Applied Statistical Methods*, **4**, 425–433.

Pawitan, Y. (2001) *In All Likelihood: Statistical Modelling and Inference using Likelihood.* Oxford University Press, Oxford.

Peskun, P.H. (1973) Optimum Monte Carlo sampling using Markov chains. *Biometrika*, **60**, 607–612.

Pielou, E.C. (1963a) The distribution of diseased trees with respect to healthy ones in a patchily infected forest. *Biometrics*, **19**, 450-459.

Pielou, E.C. (1963b) Runs of healthy and diseased trees in transects through an infected forest. *Biometrics*, **19**, 603–614.

Pierce, D.A., Stewart, W.H. and Kopecky, K.J. (1979) Distribution-free regression analysis of grouped survival data. *Biometrics*, **35**, 785–793.

Pledger, S. and Bullen, L. (1998) Tests for mate and nest fidelity in birds with application to little blue penguins (*Eudyptula minor*). *Biometrics*, **54**, 61–66.

Pregibon, D. (1981) Logistic regression diagnostics. *Annals of Statistics*, **9**, 705–724.

Pregibon, D. (1982) Score tests in GLIM with applications, pp 87–97. In: *GLIM82: Proceedings of the International Conference on Generalised Linear Models.* Ed. R. Gilchrist, Springer-Verlag, New York.

Prentice, R.L. (1973) Exponential survivals with censoring and explanatory variables. *Biometrika*, **60**, 279–288.

Presley, R. and Baker, P.F. (1970) Kinetics of fertilization in the sea-urchin: a comparison of methods. *Journal of Experimental Biology*, **52**, 455–468.

Price, K.L. and Seaman, J.W. (2006) Bayesian modeling of retrospective time-to-pregnancy data with digit preference bias. *Mathematical and Computer Modelling*, **43**, 1424–1433.

Propp, J.G. and Wilson, D.B. (1996) Exact sampling with coupled Markov Chains and applications to statistical mechanics. *Random Structures and Algorithms*, **9**, 223–252.

Quandt, R.E. and Ramsey, J.B. (1978) Estimating mixtures of normal distributions and switching regressions. *Journal of the American Statistical Association*, **73**, 730–738.

Racine, A., Grieve, A.P., Fluhler, H. and Smith, A.F.M. (1986) Bayesian methods in practice: experiences in the pharmaceutical industry. *Applied Statistics*, **35**, 2, 93–150.

Rai, S.N. and Matthews, D.E. (1993) Improving the EM algorithm. *Biometrics*, **49**, 587–591.

Rao, C.R. (1973) *Linear Statistical Inference and its Applications* (2nd edition). Wiley, New York.

Ratkowski, D.A. (1983) *Nonlinear Regression Modeling.* Marcel Dekker, New York.

Ratkowski, D.A. (1988) *Handbook of Nonlinear Regression Models.* Marcel Dekker, New York.

Reinsch, C. (1967) Smoothing by spline functions. *Numerical Mathematics*, **10**, 177–183.

Richardson, S. and Green, P.J. (1997) On Bayesian analysis of mixtures with an unknown number of components. *Journal of the Royal Statistical Society, Series B*, **59**, 731–792.

Ridout, M.S. (1999) Memory in coal tits: an alternative model. *Biometrics*, **55**, 2, 660–662.

Ridout, M.S. (2009) Statistical applications of the complex-step method of numerical differentiation. To appear, *The American Statistician*.

Ridout, M.S., Cole, D.J., Morgan, B.J.T., Byrne, L.J. and Tuite, M.F. (2006) New approximations to the Malthusian parameter. *Biometrics*, **62**, 1216–1223.

Ridout, M.S. and Demétrio, C.G.B. (1992) Generalized linear models for positive count data. *Revista de Matemática e Estatística*, **10**, 139–148.

Ridout, M., Demétrio, C.G.B. and Hinde, J. (1998) Models for count data with many zeros. Proceedings of the XIXth International Biometric Conference, Cape Town. Invited Papers, pp. 179–192.

Ridout, M.S., Faddy, M.J. and Solomon, M.G. (2006) Modelling the effects of repellent chemicals on foraging bees. *Applied Statistics*, **55**, 1, 63–75.

Ridout, M.S. and Morgan, B.J.T. (1991) Modelling digit preference in fecundability studies. *Biometrics*, **47**, 1423–1433.

Ridout, M.S., Morgan, B.J.T. and Taylor, D.R. (1999) Modelling variability in the branching structure of strawberry inflorescenses. *Applied Statistics*, **48**, 2, 185–196.

Ripley, B.D. (1987) *Stochastic Simulation*. Wiley, New York.

Rizzo, M.L. (2008) *Statistical Computing with R*. Chapman & Hall/CRC, Boca Raton, Florida.

Roberts, G.O., Gelman, A. and Gilks, W.R. (1994) Weak convergence and optimal scaling of random walk Metropolis algorithms. Technical Report, University of Cambridge.

Roberts, G., Martyn, A.L., Dobson, A.J. and McCarthy, W.H. (1981) Tumour thickness and histological type in malignant melanoma in New South Wales, Australia, 1970–1976. *Pathology*, **13**, 763–770.

Ross, G.J.S. (1990) *Nonlinear Estimation*. Springer-Verlag, New York.

Rothschild, Lord and Swann, M.M. (1950) The fertilization reaction in the sea-urchin. The effect of nicotine. *Journal of Experimental Biology*, **27**, 400–406.

Royston, J.P. (1982) Basal body temperature, ovulation and the risk to conception, with special reference to lifetimes of sperm and egg. *Biometrics*, **38**, 397–406.

Rubinstein, R.Y. and Kroese, D. P. (2008) *Simulation and the Monte Carlo Method*, (2nd edition). Wiley, New York.

Sartwell, P.E. (1950) The distribution of incubation periods of infectious diseases. *American Journal of Hygiene*, **51**, 310–318.

Schall, R. (1991) Estimation in generalized linear models with random effects. *Biometrika*, **78**, 719–728.

Seber, G.A.F. (1971) Estimating age-specific survival rates for birds from bird-band returns when the reporting rate is constant. *Biometrika*, **58**, 491–497.

Seber, G.A.F. and Wild, C.J. (1989) *Nonlinear Regression*. Wiley, New York.

Self, S.G. and Liang, K.-Y. (1987) Asymptotic properties of maximum likelihood estimators and likelihood ratio tests under nonstandard conditions. *Journal of the American Statistical Association*, **82**, 605–610.

Sheather, S.J. (2004) Density estimation. *Statistical Science*, **19**, 588–597.

Sheather, S.J. and Jones, M.C. (1991) A reliable data-based band-width selection method for kernel density estimation. *Journal of the Royal Statistical Society, Series B*, **53**, 683–690.

Silverman, B.W. (1986) *Density Estimation for Statistics and Data Analysis*. Chapman & Hall, London.

Silverman, B.W. and Young, G.A. (1987) The bootstrap: to smooth or not to smooth? *Biometrika*, **74**, 469–479.

Silvey, S.D. (1975) *Statistical Inference*. Chapman & Hall, London.

Simmonds, F.J. (1956) Superparasitism by *Spalangia drosophilae* Ashm. *Bulletin of Entomological Research*, **47**, 361–376.

Smith, R.L. (1989) A survey of nonregular problems, pp 353-372. In: *Proceedings of the International Statistical Institute Conference, 47th Session*, Paris.

Soubeyrand, S., Held, L., Höle and Sache, I. (2008) Modelling the spread in space and time of an airbourne plant disease. *Applied Statistics*, **57**, 3, 253–272.

Spendley, W., Hext, G.R. and Himsworth, F.R. (1962) Sequential applications of simplex designs in optimization and evolutionary operation. *Technometrics*, **4**, 441–461.

Spiegelhalter, D.J., Thomas, A. and Best, N.G. (1996). Computation on Bayesian graphical models. In: *Bayesian Statistics 5*. Eds. J.M. Bernardo, A.F.M. Smith, A.P. Dawid and J.O. Berger, Oxford University Press, Oxford.

Spiegelhalter, D.J., Thomas, A., Best, N.G. and Gilks, W.R. (1996). *BUGS: Bayesian Inference using Gibbs Sampling. Version 0.50*. MRC Biostatistics Unit, Cambridge.

Stigler, S.M. (2007) The epic story of maximum likelihood. *Statistical Science* **22**, 598–620.

Student (1908) The probable error of a mean. **VI**, 1.

Tanner, M.A. (1993) *Tools for Statistical Inference*. (2nd edition). Springer-Verlag, New York.

Terrill, P.K. (1997) Statistical models in the assessment of biological control of insects. Unpublished Ph.D. thesis, University of Kent, Canterbury.

Terrill, P.K., Morgan, B.J.T. and Fenlon, J.S. (1998) Stochastic models for biological control. In: *Proceedings of XIXth International Biometric Conference*, Cape Town, pp. 215–225.

Thompson, J.R. (1989) *Empirical Model Building*. Wiley, New York.

Thompson, J.R. and Tapia, R.A. (1990) *Nonparametric Function Estimation, Modeling, and Simulation.* Siam, Philadelphia.

Tierney, L. (1994). Markov chains for exploring posterior distributions. *Annals of Statistics*, **22**, 1701–1762.

Titterington, D.M., Smith, A.F.M. and Makov, U.E. (1985) *Statistical Analysis of Finite Mixture Distributions.* Wiley, London.

van den Broek, J. (1995) A score test for zero inflation in a Poisson distribution. *Biometrics*, **51**, 738–743.

Venables, W.N. and Ripley, B.D. (2002) *Modern Applied Statistics with S-plus, fourth edition.* Springer-Verlag, New York.

Venzon, D.J. and Moolgavkar, S.H. (1988) A method for computing profile-likelihood-based confidence intervals. *Applied Statistics*, **37**, 1, 87–94.

van Duijn, M.A.J. (1993) Mixed Models for Repeated Count Data. Ph.D. Dissertation. University of Groningen. DSWO Press, Leiden, Holland.

Vergne, A.L. and Mathevon, N. (2008) Crocodile egg sounds signal hatching time. *Current Biology*, **18**, R513–R514.

Vince, M.A. (1968) Effect of rate of stimulation on hatching time in Japanese quail. *British Poultry Science*, **9**, 87–91.

Vince, M.A. (1979) Effects of accelerating the stimulation on different indices of development in Japanese quail embryos. *The Journal of Experimental Zoology*, **208**, 201–212.

Vince, M.A. and Cheng, R.C.H. (1970) Effects of stimulation on the duration of lung ventilation in quail fetuses. *The Journal of Experimental Zoology*, **175**, 477–486.

Vuong, Q.H. (1989) Likelihood ratio tests for model selection and non-nested hypotheses. *Econometrica*, **57**, 307–333.

Wakefield, J.K., Gelfand, A.E. and Smith, A.F.M. (1991) Efficient computation of random variates via the ratio-of-uniforms method. *Statistics and Computing*, **1**, 129–133.

Wand, M.P. and Jones, M.C. (1995) *Kernel Smoothing.* Chapman & Hall, London.

Wang, P., Puterman, M.L., Cockburn, I. and Le, N. (1996) Mixed Poisson regression models with covariate dependent rates. *Biometrics*, **52**, 381–400.

Wei, G.C.G. and Tanner, M.A. (1990) A Monte Carlo implementation of the EM algorithm and the poor man's data augmentation algorithm. *Journal of the American Statistical Association*, **85**, 699–704.

Weinberg, C.R. and Gladen, B.C. 1986) The beta-geometric distribution applied to comparative fecundability studies. *Biometrics*, **42**, 547–560.

Weisberg, S. (1980) *Applied Linear Regression.* Wiley, New York.

West, M. (1993) Approximating posterior distributions by mixtures. *Journal of the Royal Statistical Society, Series B*, **55**, 409–422.

Wetherill, G.B. (1981) *Intermediate Statistical Methods.* Chapman & Hall, London.

White, R.E. (2007) *Elements of Matrix Modeling and Computing with* MAT-LAB. Chapman & Hall/CRC, Boca Raton, Florida.

Williams, D.A. (1970) Discussion of Atkinson, A.C., A Method for discriminating between models. *Journal of the Royal Statistical Society, Series B*, **32**, 323–353.

Williams, D.A. (1975) The analysis of binary responses from toxicological experiments involving reproduction and teratogenicity. *Biometrics*, **31**, 949–952.

Williams, D.A. (1986) Interval estimation of the median lethal dose. *Biometrics*, **42**, 3, 641–646.

Williams, D.A. (1988) Reader reaction: estimation bias using the beta-binomial distribution in teratology. *Biometrics*, **44**, 1, 305–307.

Wolfinger, R. (1993) Laplace's approximation for nonlinear mixed models. *Biometrika*, **80**, 791–795.

Wolfinger, R. and O'Connell, M. (1993) Generalized linear mixed models: a pseudo-likelihood approach. *Journal of Statistical Computation and Simulation*, **48**, 233–243.

Wolynetz, M.S. (1979) Maximum likelihood estimation from confined and censored normal data. *Applied Statistics*, **28**, 185–195.

Wood, S.N. (2006) *Generalized Additive Models. An Introduction with R.* Chapman & Hall/CRC, London.

Worton, B.J. (1989) Kernel methods for estimating the utilization distribution in home-range studies. *Ecology*, **70**, 164–168.

Wu, C.F.J. (1988) Optimal design for percentile estimation of a quantal response curve. *Optimal Design and Analysis of Experiments*, pp. 213–223. In: Eds. T. Dodge, V.V. Fedorov and H.P. Wynn, Elsevier, Amsterdam.

Yao, Q. and Morgan, B.J.T. (1999) Empirical transform estimation for indexed stochastic models. *Journal of the Royal Statistical Society, Series B*, **61**, 1, 127–141.

Young, G.A. (1990) Alternative smoothed bootstraps. *Journal of the Royal Statistical Society, Series B*, **52**, 477–484.

Young, G.A. (1994) Bootstrap: more than a stab in the dark? *Statistical Science*, **9**, 3, 382–415.

Young, P.J., Morgan, B.J.T., Sonksen, P., Till, S. and Williams, C. (1995) Using a mixture model to predict the occurrence of diabetic retinopathy. *Statistics in Medicine*, **14**, 2599–2608.

Zeger, S.L. and Karim, M.R. (1991) Generalized linear models with random effects; a Gibbs sampling approach. *Journal of the American Statistical Association*, **86**, 413, 79–86.

Zeger, S.L., Liang, K.Y. and Albert, P.S. (1988) Models for longitudinal data: a generalized estimating equation approach. *Biometrics*, **44**, 1049–1060.

Index

Abbott's formula, 137, 156
ABO blood group, 164
acceptance function
  in Metropolis-Hastings, 213
acclimatisation, 115
accuracy of simulated annealing, 67
acf, 209
acoustic confusion, 289
adaptive kernel density estimation, 286
adaptive recursive Newton-Cotes, 175
adaptive recursive Simpson's rule, 175
age
  effect of on fertility, 287
age of menarche, 111, 159
AIC, 96
AIDS, 3, 257
Akaike information criterion, 96
AMISE criterion for choosing
  smoothing parameter, 197, 284
animal abundance, 119
animal home range estimation, 286
annealing, 62
  schedule, 63
annual mortality, 93
annual survival probabilities, 5
ant-guard, 8
  hypothesis, 37
ant-lions, 9, 41, 193
ants, 9
aperiodic Markov chain, 272
apparent error rate, 107
approximate orthogonality, 54
approximating derivatives, 47
Aranda-Ordaz model, 109, 244, 261
  score test for, 105, 296
array editor in MATLAB, 276
asymptotic distribution of
  maximum-likelihood estimators, 78,
   79
asymptotic equivalence of tests, 103

asymptotic mean integrated squared
  error, 284
asymptotic multivariate normal
  distribution, 78
Australian Defence Force Academy, 248
autocorrelation function, 209
avian recovery data, 177

backfitting, 251
bacteria, 121
balanced simulation
  use in bootstrap, 189
bandwidth, 283
BASIC program for simplex method, 59
Bayes
  Thomas, 199
Bayes factor, 221
Bayes' Theorem, 199
Bayesian information criterion, 96
Bayesian p-values, 210
  for dipper models, 223
bees, 13
beliefs
  quantifying, 199
Bernouilli trials, 37
beta distribution, 22
beta random variables
  as ratio of gamma variables, 171
beta-binomial distribution, 24, 191, 269
  for dead fetuses in mouse litters, 70
  for litter effects, 245
  induced correlation, 245
  urn derivation, 37
beta-binomial simulation, 196
beta-geometric distribution, 22, 194,
  269
  alternative expressions for, 40
  determining moments using the pgf,
   291
  mean for, 290

of derivative matrix, 134
rug plot, 250

saddlepoints
in Bohachevsky surface, 60
sampling distribution of likelihood-ratio
test-statistic, 80
sampling distributions, 180
saturated model, 41, 106, 241, 289
in signal-detection, 293
scale parameter
in Cauchy distribution, 51
score test
in GLMs, 241
avoiding the algebra for GLMs, 241
example of negative test statistic, 103
for comparing models in
capture-recapture, 105
illustrations of use, 103
of Weibull vs exponential, 121
use in generalised linear models, 105
use of expected Hessian, 79
scores vector, 77
expectation of, 77
script files in MATLAB, 276
sea urchins, 3
search methods
deterministic, 48
secondary fertilisations in polyspermy,
29, 111
seed for initiating a sequence of
pseudo-random variables, 170
seed knowledge
use in repeating simulations, 177
SEM, 144, 164, 168
sensitivity to prior, 211, 236
sequential quadratic programming
algorithm, 61, 135
sex difference
in mortality of flour beetles, 97
shags, 177
Shakespeare, 8
shape parameter
for beta distribution, 23
short-term survivors, 115
signal-detection, 293
simplex
definition of, 59
simplex method, 58

BASIC program for, 59
simplification
arising from method-of-scoring, 55
Simpson's rule, 175
simulated annealing, 62, 196
accuracy of, 67
applied to Cauchy log-likelihood, 65
choice of parameters, 64
simple implementation, 63
use in Monte Carlo inference, 180
simulating
beta, 171
beta-binomial, 196
binomial, 171
exponential, 173
gamma
by rejection, 173
multinomial, 279
normal, 171
by rejection, 173
Poisson, 171
simulation, 11
aim of, 169
simulation to approximate likelihood
surface, 179
size-limited model for fertilisation, 40
skewness, 151
smoothed bootstrap, 188
alternative to, 189
performance of, 189
smoothers, 248
smoothing parameter
choice of in kernel density estimation,
283
smoothing spline, 249, 251
Soay sheep, 132
sore throat, 3, 71
spatial statistics, 226
speed of convergence of EM algorithm,
301
splines
cubic, 251
smoothing, 251
use in capture-recapture, 255
stable distribution
of index $\alpha = 1/2$, 120
stable laws, 51
standard errors, 3
convention for display in text, 82

Printed in the United States
by Baker & Taylor Publisher Services